Trigonometric Ratios

Degrees	sin	cos	tan	cot	sec	csc	
0°	.0000	1.000	.0000	—	1.000	—	90°
1°	.0175	.9998	.0175	57.29	1.000	57.30	89°
2°	.0349	.9994	.0349	28.64	1.001	28.65	88°
3°	.0523	.9986	.0524	19.08	1.001	19.11	87°
4°	.0698	.9976	.0699	14.30	1.002	14.34	86°
5°	.0872	.9962	.0875	11.43	1.004	11.47	85°
6°	.1045	.9945	.1051	9.514	1.006	9.567	84°
7°	.1219	.9925	.1228	8.144	1.008	8.206	83°
8°	.1392	.9903	.1405	7.115	1.010	7.185	82°
9°	.1564	.9877	.1584	6.314	1.012	6.392	81°
10°	.1736	.9848	.1763	5.671	1.015	5.759	80°
11°	.1908	.9816	.1944	5.145	1.019	5.241	79°
12°	.2079	.9781	.2126	4.705	1.022	4.810	78°
13°	.2250	.9744	.2309	4.331	1.026	4.445	77°
14°	.2419	.9703	.2493	4.011	1.031	4.134	76°
15°	.2588	.9659	.2679	3.732	1.035	3.864	75°
16°	.2756	.9613	.2867	3.487	1.040	3.628	74°
17°	.2924	.9563	.3057	3.271	1.046	3.420	73°
18°	.3090	.9511	.3249	3.078	1.051	3.236	72°
19°	.3256	.9455	.3443	2.904	1.058	3.072	71°
20°	.3420	.9397	.3640	2.747	1.064	2.924	70°
21°	.3584	.9336	.3839	2.605	1.071	2.790	69°
22°	.3746	.9272	.4040	2.475	1.079	2.669	68°
23°	.3907	.9205	.4245	2.356	1.086	2.559	67°
24°	.4067	.9135	.4452	2.246	1.095	2.459	66°
25°	.4226	.9063	.4663	2.145	1.103	2.366	65°
26°	.4384	.8988	.4877	2.050	1.113	2.281	64°
27°	.4540	.8910	.5095	1.963	1.122	2.203	63°
28°	.4695	.8829	.5317	1.881	1.133	2.130	62°
29°	.4848	.8746	.5543	1.804	1.143	2.063	61°
30°	.5000	.8660	.5774	1.732	1.155	2.000	60°
31°	.5150	.8572	.6009	1.664	1.167	1.942	59°
32°	.5299	.8480	.6249	1.600	1.179	1.887	58°
33°	.5446	.8387	.6494	1.540	1.192	1.836	57°
34°	.5592	.8290	.6745	1.483	1.206	1.788	56°
35°	.5736	.8192	.7002	1.428	1.221	1.743	55°
36°	.5878	.8090	.7265	1.376	1.236	1.701	54°
37°	.6018	.7986	.7536	1.327	1.252	1.662	53°
38°	.6157	.7880	.7813	1.280	1.269	1.624	52°
39°	.6293	.7771	.8098	1.235	1.287	1.589	51°
40°	.6428	.7660	.8391	1.192	1.305	1.556	50°
41°	.6561	.7547	.8693	1.150	1.325	1.524	49°
42°	.6691	.7431	.9004	1.111	1.346	1.494	48°
43°	.6820	.7314	.9325	1.072	1.367	1.466	47°
44°	.6947	.7193	.9657	1.036	1.390	1.440	46°
45°	.7071	.7071	1.000	1.000	1.414	1.414	45°
	cos	sin	cot	tan	csc	sec	Degrees

TRIGONOMETRY FOR TODAY

TRIGONOMETRY FOR TODAY

Robert E. Mosher
CALIFORNIA STATE UNIVERSITY, LONG BEACH

HARPER & ROW, PUBLISHERS
New York Hagerstown San Francisco London

Sponsoring Editor: George J. Telecki
Project Editor: Ralph Cato
Designer: Rita Naughton
Production Supervisor: Will C. Jomarrón
Compositor: Santype International Limited
Printer and Binder: Halliday Lithograph Corporation
Art Studio: Vantage Art, Inc.

Trigonometry for Today

Library of Congress Cataloging in Publication Data

Mosher, Robert E
    Trigonometry for today.

    Includes index.
    1. Trigonometry. I. Title.
QA531.M75   516′.24   75-30689
ISBN 0-06-044630-7

To the greater understanding
of mathematics

Contents

Preface

The experiences of the past few years have revealed a need for balance in the teaching of mathematics. It is important to identify theoretical principles and to describe them in modern nomenclature. However, the understanding of these principles must be based on intuitive concepts and supported with technical skill. This balance must be reflected in the mathematics textbooks for today.

This text is designed for a one-semester course in trigonometry. Intermediate algebra should normally be prerequisite, but there is considerable flexibility in the actual level of the course.

There is a brief review of fundamentals in Chapter 1, including a discussion of functions and relations. Trigonometry is introduced in Chapter 2 from the point of view of the right triangle. The student thus encounters, as early as possible, the complete solution of a legitimate mathematical problem (solution of right triangles) as well as concrete applications.

Chapter 3 turns to the more theoretical development of trigonometric functions. It is important that the graphs of the sine and cosine functions be mentioned early in any trigonometry course, and these graphs are presented here.

After discussion of identities in Chapter 4 and solution of arbitrary triangles in Chapter 5, the graphs of the remaining trigonometric functions are presented in Chapter 6. Here also are more complicated trigonometric graphs and various physical applications.

Chapter 7 is a self-contained treatment of complex numbers and their trigonometric form, including DeMoivre's theorem. Inverse trigonometric functions appear in Chapter 8, with an application to the graph of $y = a \sin x + b \cos x$.

The computational examples and exercises in the text do not assume logarithmic calculation. A detailed explanation of this topic, including exponential functions and logarithmic functions, is provided in Appendix A for use at the instructor's discretion.

Like Appendix A, sections marked with the symbol ◇ are optional; if omitted, no loss of continuity will ensue. The abbreviated table of trigonometric ratios inside the front cover should prove convenient, but a more complete table and a table for logarithms of trigonometric ratios appear in Appendix B. Appendix B also contains an extensive listing of formulas for trigonometry, and for algebra as well.

Throughout the book we have endeavored to unify the theoretical, the geometrical, and the practical aspects of mathematical ideas. For example, our development of the trigonometric functions includes accurate theoretical definitions illustrated geometrically, but preceded by a practical introduction to trigonometry via right triangles and their solution. The student can thus relate the theoretical ideas to familiar computational techniques and geometrical examples. This is what provides the pedagogical balance needed in mathematics textbooks for today.

I would like to take this opportunity to thank Mrs. Gene Chapin, Miss Helen Kuntz, and Mrs. Betty Nosko for their most loyal assistance in preparation of the manuscript and also the Long Beach campus of California State University for its continued cooperation and support.

ROBERT E. MOSHER

TRIGONOMETRY FOR TODAY

1 Numbers and coordinates

Numbers and graphs are the basic tools in the geometric study known as trigonometry. In this first chapter we present the underlying geometric structure of numbers and our number system, together with a brief survey of functions and the coordinate plane.

A picture, or graph, of a mathematical idea is always desirable. In solving a mathematical problem, the first step often is to draw a picture representing the problem.

1-1 Numbers, equality, and order

In trigonometry, as in most branches of mathematics, we usually work with a number system called the *real number system*. The numbers in the system are *real numbers*. They are represented by points on the *number line*.

The number line is a picture, or *graph*, of the real number system, as shown in Figure 1–1. As indicated by the arrows at each end of the line, the number line is a line with no end. It extends in both directions, but never terminates. A point is selected to represent zero. This point is the *origin* of the line. The point representing one is also marked off to show what distance represents one unit. Once the points representing zero and one have been located, it is possible to locate any real number on the scale by going the appropriate distance from the origin.

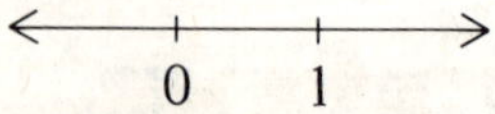

Figure 1–1
The number line

EXAMPLE: We locate the numbers on the number scale by finding the corresponding point. In Figure 1–2 we have located the following numbers on the number scale: -4, $-3\frac{1}{4}$, -2, $-\frac{1}{2}$, $2\frac{1}{2}$, and 4.

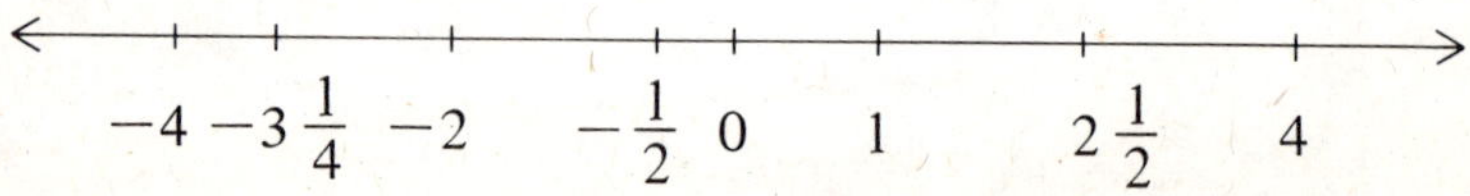

Figure 1–2 ■

The point representing one is usually placed to the right of the origin. Thus all positive numbers lie to the right of the origin, while the negative numbers lie to the left, as shown in Figure 1–3. Zero itself is neither positive nor negative. Any other real number is either positive or negative, but not both.

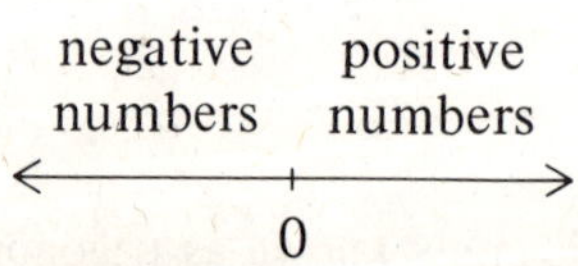

Figure 1–3

Each point on the number line corresponds to exactly one number, and each real number is represented by exactly one point on the number line. Two numbers are *equal* whenever they are represented by the same point on the number line.

Two numbers can be equal even if they are represented by different symbols or numerals.

EXAMPLE: $1 + 1 = 2$ ■

This equation means that $1 + 1$ and 2 are two different ways of specifying the same number; in other words, $1 + 1$ and 2 are different names for the same number. Of course, this number has many other names. A few are $4 \times \frac{1}{2}$, $5 - 3$, and $10 \div 5$.

Given any two real numbers, either they are equal or else one of them is greater than the other.

EXAMPLES: $2 = 1 + 1$
$2 > 1$ (2 is greater than 1)
$1 < 2$ (1 is less than 2) ■

Notice that 1 lies to the left of 2 on the number line.

In general, if a and b are numbers such that a is to the left of b on the number line, then $a < b$ (or $b > a$), as shown in Figure 1–4.

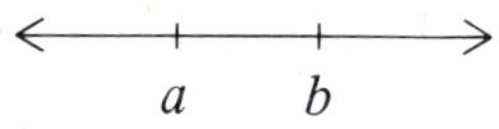

Figure 1–4
$a < b$

The three possibilities are expressed in the following *trichotomy law*:

If a and b are any two real numbers, then exactly one of the following statements is true:

(*i*) $a = b$
(*ii*) $a < b$
(*iii*) $a > b$

If a lies to the left of b and b lies to the left of c, then certainly a lies to the left of c. Thus we have the two *transitive properties of inequalities.*

If $a < b$ and $b < c$, then $a < c$.
If $a > b$ and $b > c$, then $a > c$.

The positive numbers are those which lie to the right of zero. a is positive if $a > 0$; and if $a > 0$, then a is positive. We express this concisely by saying that a is positive *if and only if* $a > 0$. Similarly a is negative if and only if $a < 0$.

EXAMPLES: $2 > 0$, because 2 is positive
$-5 < 0$, because -5 is negative ■

There are several more inequality symbols obtained by combining or modifying ones we have already defined. Their meanings are as follows:

$\leq$ less than or equal to
$\geq$ greater than or equal to
$\not<$ not less than
$\not>$ not greater than
$\not\leq$ not less than or equal to
$\not\geq$ not greater than or equal to

EXAMPLES: $1 \leq 2$ 1 is less than or equal to 2
$2 \not< 1$ 2 is not less than 1
$2 \leq 2$ 2 is less than or equal to 2
$1 \not> 2$ 1 is not greater than 2
$2 \geq 2$ 2 is greater than or equal to 2
$2 \nleq 1$ 2 is not less than or equal to 1
$2 \geq 1$ 2 is greater than or equal to 1
$2 \ngeq 3$ 2 is not greater than or equal to 3 ■

Notice that $<$ and $\ngeq$ mean exactly the same thing. "Less than" is the same as "not greater than or equal to." There are three more pairs of synonyms here. $>$ means the same as $\nleq$, $\leq$ is the same as $\not>$, and $\geq$ is a synonym of $\not<$.

1-1 *Exercises*

Draw a number line and locate these numbers on it.

1. 2
2. 0
3. -4
4. $-3\frac{1}{4}$
5. $\frac{1}{2}$
6. -1

Which of the points v, w, x, y, and z on the number line shown below represent the following?

7. positive numbers
8. negative numbers

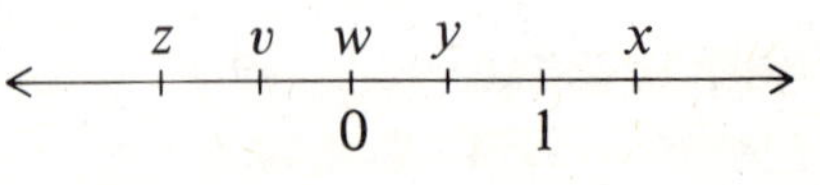

True or false?

9. $-3 > 1\frac{1}{2}$
10. $2 \leq 2$
11. $4 \not> -6$
12. $4 > -7\frac{1}{4}$
13. $0 \leq -3\frac{1}{2}$
14. $4 \geq 0$
15. $1 \nleq 3$
16. $4 \not< 3\frac{1}{2}$
17. $-5 \leq -2$
18. $-4 \geq -7\frac{1}{4}$
19. $-3\frac{1}{4} \geq -3\frac{1}{4}$
20. $-4 > 0$

Write using a synonymous symbol.

21. $1 < 4$
22. $2 \leq 2$
23. $7\frac{1}{2} > -3$
24. $5 \nleq -6\frac{1}{4}$
25. $-3 \not< -5$
26. $x < 3$
27. $2 \geq y$
28. $x + 1 \leq y$
29. $x \not< y + \frac{1}{2}$

True or false?

30. If x is negative and y is positive, then $x < y$
31. If $x < 7$ and $y \geq 7$, then $x \ngeq y$

32. If $x < 6$ and $y > 7$, then $y > x$
33. If $x < 5$, then $x \nleq 6$
34. If $x \leq y$, then $x \ngeq y$
35. If $x < 4$, then $x \neq 4$
36. If $x \geq 3$, then $x \neq 1$
37. Is there a greatest positive number?
38. Is there a least positive number?

1-2 Subsets of the number system

Consider the sets

$A = \{1, 2\}$

and

$B = \{x \mid 1 \leq x \leq 2, x \text{ a real number}\}$

Then $1 \in B$ and $2 \in B$, and therefore each element of A is an element of B. We express this relationship by saying that A is a *subset* of B, or, in symbols,

$A \subseteq B$

Notice in this case that although A is a subset of B, B is not a subset of A, for it is not true that each element of B is an element of A. For example, $1\frac{1}{2} \in B$, but $1\frac{1}{2} \notin A$. In symbols,

$B \nsubseteq A$

For each set A, it is apparent that $A \subseteq A$, because each element of A is an element of A.

EXAMPLE: $\{1, 2\} \subseteq \{1, 2\}$ ■

Another apparent property of subsets is the *transitivity property.*

If $A \subseteq B$ and $B \subseteq C$, then $A \subseteq C$.

For each set A, it is true that $\varnothing \subseteq A$, where $\varnothing$ is the null set, because surely each element of the null set is an element of A (try naming one that isn't). Thus $\varnothing$ is a subset of every set.

Just as the "is less than or equal to" symbol $\leq$ can be reversed to yield the "is greater than or equal to" symbol $\geq$, so can the "subset" symbol $\subseteq$ be reversed to yield a "superset" symbol $\supseteq$. Thus the statement $B \supseteq A$ is read "B is a superset of A", and has exactly the same meaning as $A \subseteq B$.

EXAMPLE: If $A = \{1, 2\}$ and $B = \{x \mid 1 \leq x \leq 2\}$, then $B \supseteq A$. ■

The real number system is a particularly important set, and it has some very important subsets. The real number system itself is usually denoted by R.

Of course, R itself and $\varnothing$ are among the subsets of R, but some other subsets are especially important. The first of these is the set of *natural numbers* or *counting numbers*. These are the numbers 1, 2, 3, ..., starting with 1 and continuing without end. The symbol for the set of natural or counting numbers is N.

EXAMPLE: The graph of N is shown in Figure 1–5. Notice that $0 \notin N$.

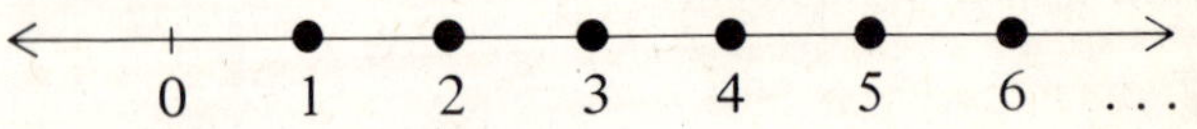

Figure 1–5
$N = \{\text{natural numbers}\}$ ■

Another important set of numbers is the set J of *integers*.[1] The set J consists of the counting numbers 1, 2, 3, ..., the corresponding negative numbers −1, −2, −3, ..., and also 0. Thus N is a subset of J, and J is a superset of N.

EXAMPLE: The graph of J is as shown in Figure 1–6.

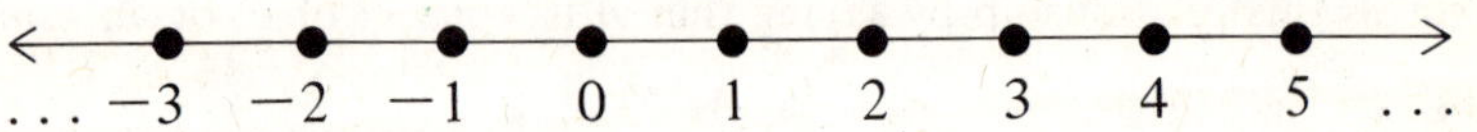

Figure 1–6
$J = \{\text{integers}\}$ ■

The *rational numbers* Q are those numbers which can be written as fractions in which both the numerator and the denominator are integers, and such that the denominator is not zero. Each integer is a rational number, and therefore $J \subseteq Q$

EXAMPLE: $4 \in Q$, because 4 can be written as the fraction 4/1. ■

EXAMPLES: $\frac{1}{2} \in Q$
$\frac{2}{3} \in Q$
$-1\frac{1}{4} \in Q$, because $-1\frac{1}{4} = -\frac{5}{4}$ ■

None of these last examples is an integer.

It is not practical to make a graph of Q, because each segment on the number line contains infinitely many rational numbers. On the other hand, each segment also contains infinitely many real numbers which are not rational. These numbers are called *irrational numbers*.

EXAMPLES: $\pi = 3.14159\ldots$ is an irrational number
$\sqrt{2} = 1.414\ldots$ is an irrational number ■

[1] In many books the set of integers is denoted by Z.

In fact, the square root of any natural number is an irrational number, unless that natural number is itself a square.

EXAMPLES: $\sqrt{3}$ is an irrational number
$\sqrt{4} = 2$ because $4 = 2^2$ ■

Except for the set of irrational numbers, these subsets of the real number system fit together in a chainlike pattern:

$\varnothing \subseteq N \subseteq J \subseteq Q \subseteq R$

In this book, when not otherwise specified, a variable may stand for any real number. However, it is possible to use set notation to restrict the variable to some subset of the number system.

EXAMPLE: $\{x \mid x \in J, \ -3 \leq x \leq 2\}$ means the set of all integers x such that $-3 \leq x \leq 2$. Therefore, $\{x \mid x \in J, \ -3 \leq x \leq 2\} = \{-3, -2, -1, 0, 1, 2\}$. ■

The following examples give a further reminder of the way sets of numbers are graphed on the number line.

EXAMPLE: Let $A = \{1, 2\}$. The graph of A is shown in Figure 1–7.

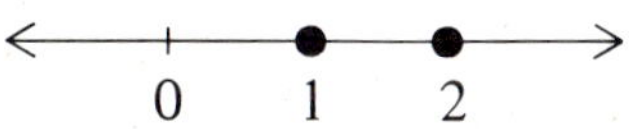

Figure 1–7
$A = \{1, 2\}$ ■

EXAMPLE: Let $B = \{x \mid x < 2\}$. The graph of B is shown in Figure 1–8. The hollow dot at 2 indicates that $2 \notin B$. Compare this with the next example.

Figure 1–8
$B = \{x \mid x < 2\}$ ■

EXAMPLE: Let $C = \{x \mid x \leq 2\}$. The graph of C is as in Figure 1–9. The solid dot shows that $2 \in C$.

Figure 1–9
$C = \{x \mid x \leq 2\}$ ■

EXAMPLES: Graph $D = \{x \mid x > -1\}$. The graph is shown as Figure 1–10.
The graph of $E = \{x \mid 1 \leq x < 2\}$ is given in Figure 1–11.

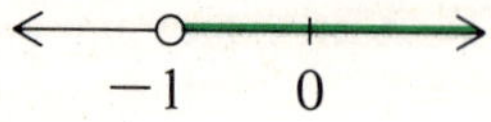

Figure 1–10
$D = \{x \mid x > -1\}$

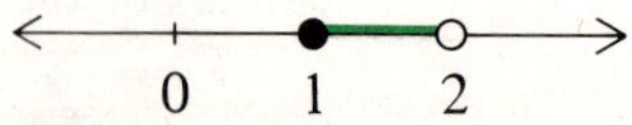

Figure 1–11
$E = \{x \mid 1 \le x < 2\}$ ■

1-2 *Exercises*

Is it true or false that $A \subseteq B$?

1. $A = \{x \mid 1 \le x \le 4\}$
   $B = \{2, 3, 4\}$
2. $A = \{x \mid 1 < x < 4\}$
   $B = \{2, 3, 4\}$
3. $A = \{x \mid x \in N\}$
   $B = \{x \mid x \in J\}$
4. $A = \{x \mid x \ge 6\}$
   $B = \{x \mid x \in Q, x \ge 6\}$
5. $A = \{x \mid x \in J, 1 \le x \le 5\}$
   $B = \{x \mid x \in N, 1 \le x \le 5\}$

6. For each of Exercises 1 through 5, is it true or false that $A \supseteq B$?

True or false?

7. $1 \in \varnothing$
8. $\pi \notin R$
9. $0 \in Q$
10. $-4 \notin J$
11. $\sqrt{2} \in Q$
12. $0 \notin N$
13. $3\frac{1}{2} \in N$
14. $-2\frac{1}{4} \in Q$
15. $-2 \in \{x \mid x \in Q, -3 < x < 2\}$
16. $-2 \notin \{x \mid x \in N, -3 < x < 2\}$

Graph the following.

17. $\{x \mid x \in N, 2 \le x \le 4\}$
18. $\{x \mid x \in J, 2 \le x \le 4\}$
19. $\{x \mid x \in N, -4 < x < -2\}$
20. $\{x \mid x \in J, -4 < x < -2\}$

Graph the indicated sets of numbers.

21. $\{x \mid x > 4\}$
22. $\{-\frac{1}{2}, 2; 3\}$
23. $\{x \mid x = 2\}$
24. $\{x \mid x \neq 0\}$
25. $\{x \mid x \le 3\frac{1}{2}\}$
26. $\{x \mid -3 < x < -1\frac{1}{4}\}$
27. $\{x \mid -1 < x \le 4\}$
28. $\{x \mid x \ngeq 6\}$

29. $\{x \mid 0 < x < -2\}$

30. $\{x \mid x \neq 2\frac{1}{4}\}$

31. $\{x \mid x \nless -2\}$

Use set notation to describe the set pictured.

32.

33.

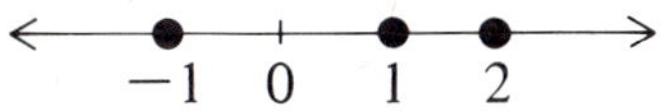

34.

35.

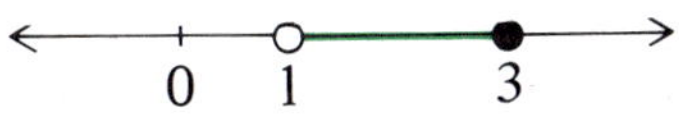

36.

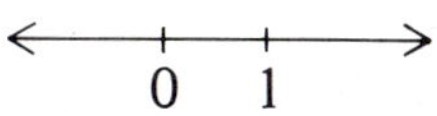

37.

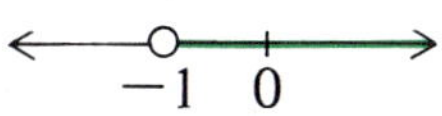

38.

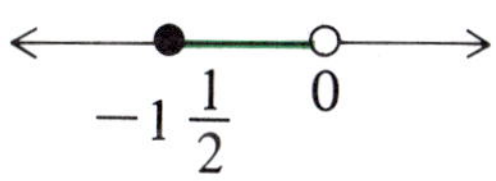

39.

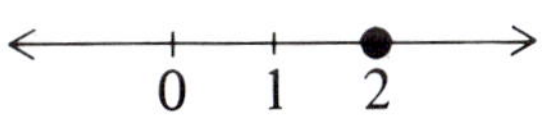

40.

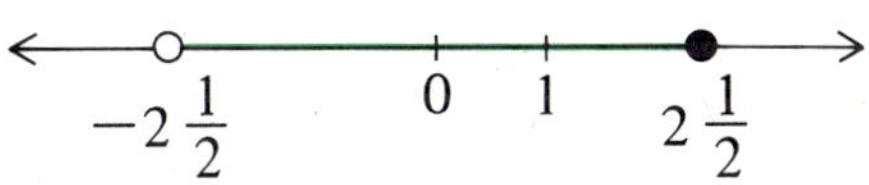

41. Is there a greatest positive rational number?
42. Is there a least positive rational number?
43. Is there a greatest positive integer?
44. Is there a least positive integer?
45. Is there a greatest negative integer?

1-3 Absolute value

The *absolute value* of a number is the distance between that number and the origin on the number line.

EXAMPLE: The absolute value of 2 is 2, because 2 is 2 units away from zero, as shown in Figure 1–12.

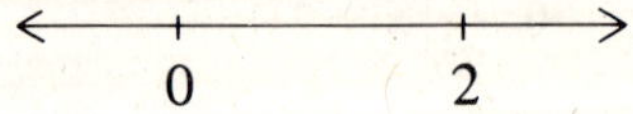

Figure 1–12
The distance from zero is 2 ■

EXAMPLE: The absolute value of -2 is 2, as we see in Figure 1–13. Note that -2 is still 2 units away from zero.

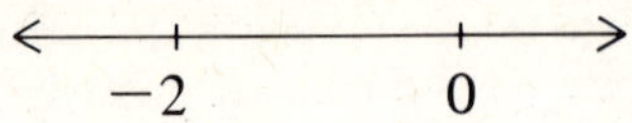

Figure 1–13
The distance from zero is 2 ■

Absolute value is indicated by enclosing the number in vertical bars $|\ |$.

EXAMPLES: $|2| = 2$
$|-2| = 2$ ■

Because absolute value is a distance, $|a|$ is never negative. $|0| = 0$, but if $a \neq 0$, then $|a| > 0$.

If a is a positive number, then $|a| = a$, because a will be a units away from the origin, as in Figure 1–14. But if a is a negative number, then $|a| = -a$, because $-a$ will be positive and will be the distance from a to the origin (Figure 1–15).

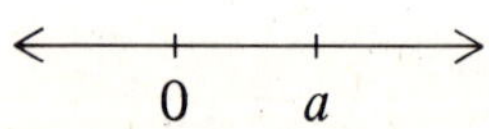

Figure 1–14
The distance from zero is a

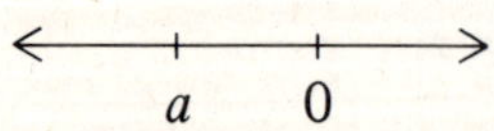

Figure 1–15
The distance from zero is $-a$

EXAMPLE: $|-4| = -(-4) = 4$ ■

The formal definition of absolute value is as follows:

If a is a real number, then
$|a| = a$ if $a \geq 0$
$|a| = -a$ if $a < 0$

We have already seen the following:

If a is a real number, then $|a| \geq 0$, and $|a| = 0$ if and only if $a = 0$.

Because $-a$ and a are the same distance from the origin, we have the following property:

For each real number a, $|a| = |-a|$.

Absolute value notation can often be used to describe sets of numbers.

EXAMPLE: Graph $\{x \mid |x| \leq 2\}$, and describe this set without using absolute value notation.

From the graph in Figure 1–16, we see that $\{x \mid |x| \leq 2\} = \{x \mid -2 \leq x \leq 2\}$.

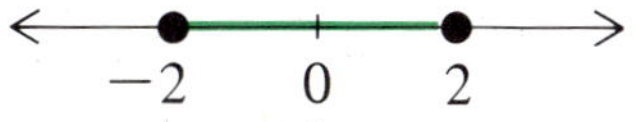

Figure 1–16
$\{x \mid |x| \leq 2\}$ ■

EXAMPLE: Graph $\{x \mid |x| > 2\}$, and describe the set without using absolute value notation.

From the graph in Figure 1–17, we see that

$\{x \mid |x| > 2\} = \{x \mid x > 2 \text{ or } x < -2\}$.

The distance between any two real numbers can be expressed in terms of absolute value, according to the following:

If a and b are any two real numbers, then $|a - b|$ is equal to the distance between a and b on the number line.

Note that for $b = 0$, we have $|a - 0| = |a|$, the distance between a and 0 on the number line.

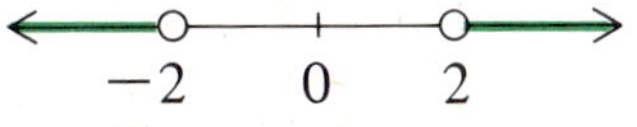

Figure 1–17
$\{x \mid |x| > 2\}$ ■

EXAMPLE: The distance between -1 and 2 is $|-1 - 2| = 3$, as in Figure 1–18. This distance can also be computed as $|2 - (-1)| = 3$.

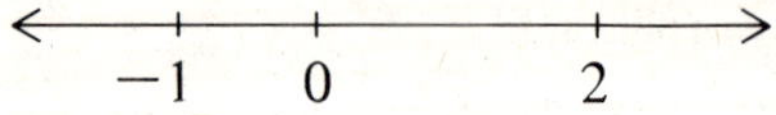

Figure 1–18
The distance between −1 and 2 is 3 ■

1-3 *Exercises*

Write without using absolute value notation.

1. $|3|$
2. $|-5|$
3. $|-2\frac{1}{4}|$
4. $|0|$
5. $|5-4|$
6. $|4-7\frac{1}{4}|$
7. $|\frac{1}{2}-5+2|$

Graph and express without absolute value notation.

8. $\{x \mid |x| = 2\}$
9. $\{x \mid |x| = 0\}$
10. $\{x \mid |x| = -2\}$
11. $\{x \mid |x| < 1\}$
12. $\{x \mid |x| > 4\}$
13. $\{x \mid |x| \leq 0\}$
14. $\{x \mid |x| \geq 2\}$

Express as a distance.

15. $|8-3|$
16. $|6+4|$
17. $|4-x|$
18. $|2+y|$
19. $|z-0|$

Express, using absolute value notation, the distance between the given numbers.

20. 4 and 7
21. -2 and -3
22. -5 and x
23. 3 and $-y$

What is the relationship between $|ab|$ and $|a|\,|b|$

24. when $a = 2$ and $b = 3$?
25. when $a = -2$ and $b = 3$?
26. when $a = 2$ and $b = -3$?
27. when $a = -2$ and $b = -3$?
28. in general?
29. Prove your answer to Exercise 28.

Express using absolute value notation.

30. $\{x \mid -a \leq x \leq a\}$ where $a > 0$

31. $\{x \mid x \geq a \text{ or } x \leq -a\}$, where $a > 0$

1-4 The coordinate plane

Graphs of subsets of real numbers, plotted on the number line, help us to understand statements involving one variable. In a very similar way, sets of ordered pairs can be plotted in the *coordinate plane*. These graphs help us to understand statements involving two variables.

To draw the coordinate plane, we draw two number lines, one horizontal and one vertical. The point where they intersect is the *origin*. The origin represents zero on each line.

The horizontal line represents the first variable. As usual, the positive direction is to the right. The vertical line represents the second variable. Its positive direction is up. These two number lines are called *axes*. In the coordinate plane shown in Figure 1–19, the horizontal line is the x-axis, and the vertical line is the y-axis. When x and y are the variables, this is the usual arrangement.

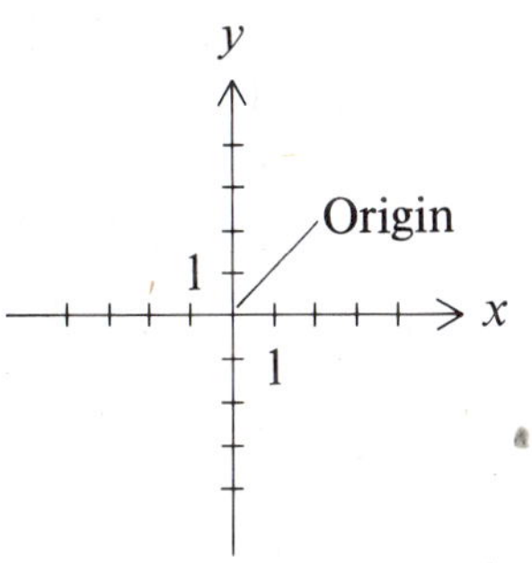

Figure 1–19
The coordinate plane

Notice that the number 1 is marked on each axis to show what length represents one unit. This length may or may not be the same for the two axes.

To plot the point (x, y) in the coordinate plane, we first locate the x-coordinate on the x-axis and mark it. Similarly, we mark off the y-coordinate on the y-axis. Finally, the point (x, y) lies directly above (or below) the x-coordinate on the x-axis, but at a height corresponding to the y-coordinate that was marked on the y-axis.

EXAMPLE: The point $(-4, -1)$ lies one unit below the point on the x-axis with x-coordinate of -4, as shown in Figure 1–20.

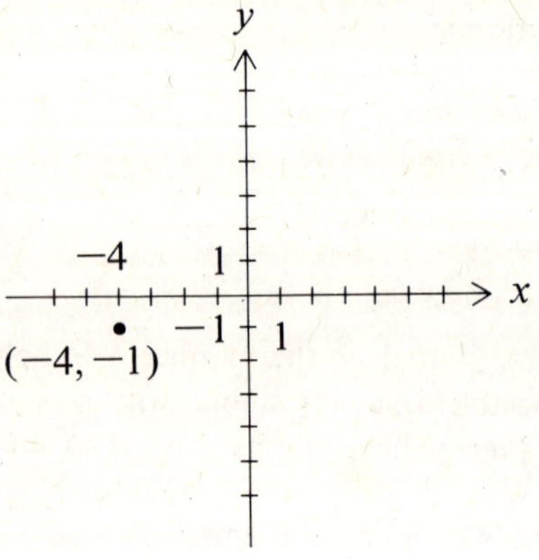

Figure 1–20

EXAMPLE: The points (1, 2), $(4, -\frac{1}{2})$, and $(-3, 0)$ are shown plotted in the plane in Figure 1–21. Note that the point $(-3, 0)$ lies on the x-axis.

We know that each ordered pair (x, y) represents a point in the xy-plane. Therefore, a set of ordered pairs can be graphed as a set of points in the plane.

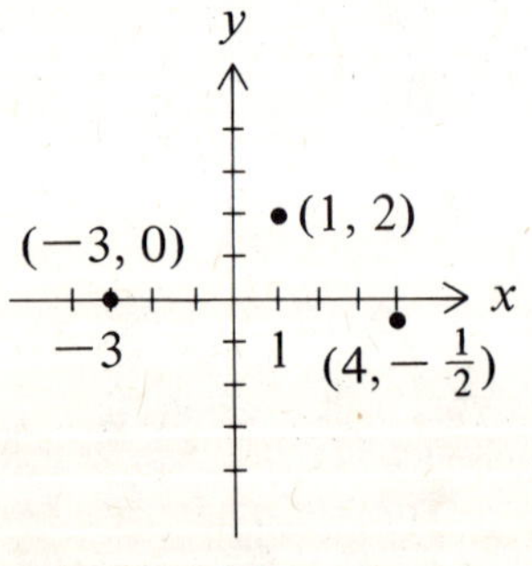

Figure 1–21

We may determine a subset of the plane by specifying algebraic properties of the corresponding ordered pairs (x, y). The general problem of graphing is to translate algebraic conditions on ordered pairs (x, y) into geometric properties of the graph.

Some very important subsets of the plane are determined by simple conditions on the ordered pairs (x, y).

EXAMPLE: Graph the set of ordered pairs (x, y) such that $x > 0$.

On the number line, the positive numbers are those to the right of zero. Thus this set of points in the plane is the set of points lying to the right of the y-axis, as shown in Figure 1–22.

The graph in the preceding example is called the *right half-plane*. In set-theoretic notation, it could be described as the graph of $\{(x, y) | x > 0\}$.

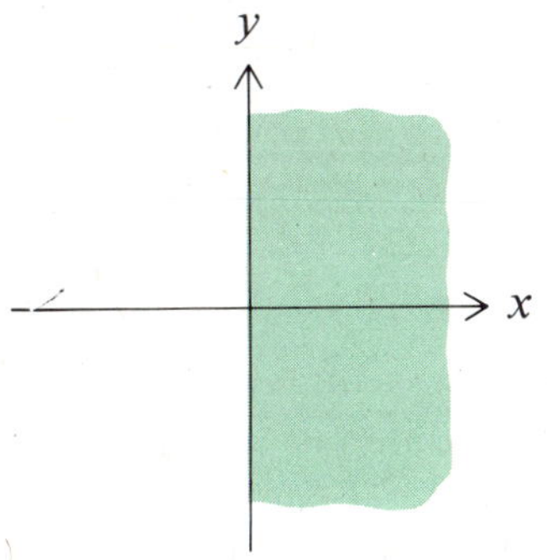

Figure 1–22
$\{(x, y) | x > 0\}$ ■

EXAMPLE: Graph $\{(x, y) | y < 0\}$.

Because the negative direction on the y-axis is downward, this set is the set of points lying below the x-axis, as shown in Figure 1–23. This graph is called the *lower half-plane*.

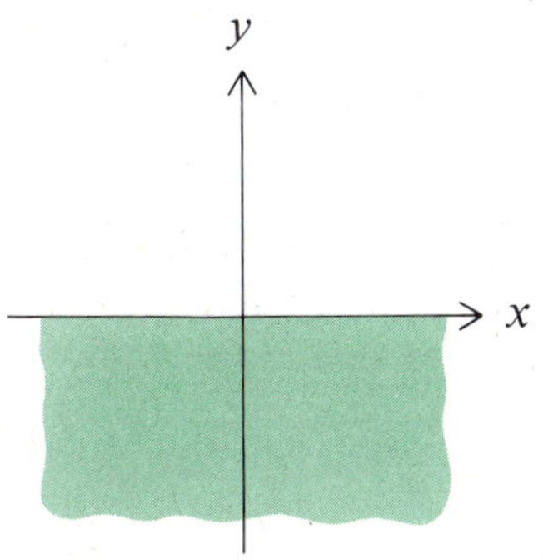

Figure 1–23
$\{(x, y) | y < 0\}$ ■

The four *quadrants* are very important subsets of the plane. They are named as shown in Figure 1–24.

Let us determine the conditions for a point (x, y) to lie in the first quadrant. Each point (x, y) in the first quadrant lies in the right half-plane and so satisfies $x > 0$. Similarly, each point in the first quadrant satisfies $y > 0$, because the first quadrant is also a subset of the upper half-plane. In fact, the first quadrant is the intersection of the right half-plane and the upper half-plane, as shown in Figure 1–25. Thus (x, y) lies in the first quadrant if and only if both $x > 0$ and $y > 0$.

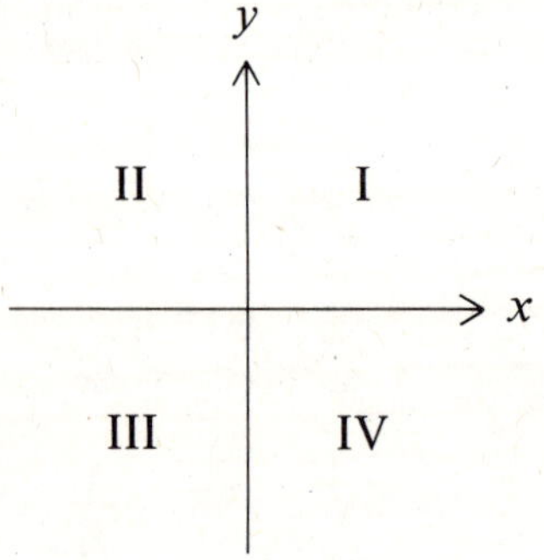

Figure 1–24
The four quadrants

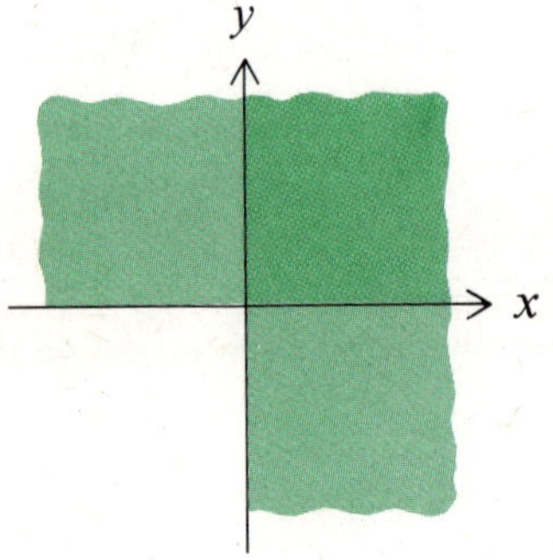

Figure 1–25

Similarly, each of the quadrants may be thought of as the intersection of two half-planes. Here are the inequalities which define each quadrant.

First quadrant: $\{(x, y) | x > 0 \text{ and } y > 0\}$
Second quadrant: $\{(x, y) | x < 0 \text{ and } y > 0\}$
Third quadrant: $\{(x, y) | x < 0 \text{ and } y < 0\}$
Fourth quadrant: $\{(x, y) | x > 0 \text{ and } y < 0\}$

Note that the axes are not considered as part of any quadrants. However, at this point we can describe the axes algebraically as follows:

Consider a point (x, y). If $y > 0$, this point lies above the x-axis, whereas if $y < 0$, the point is below the x-axis. The remaining possibility, $y = 0$, is the condition which must be satisfied for the point (x, y) to lie on the x-axis. Thus the x-axis may be described as $\{(x, y) | y = 0\}$. More briefly, we say that $y = 0$ is the equation of the x-axis. Similarly, the point (x, y) lies on the y-axis if and only if $x = 0$. Therefore, $x = 0$ is the equation of the y-axis.

EXAMPLE: The point $(0, -4)$ lies on the y-axis, as shown in Figure 1–26, because the x-coordinate of this point is 0.

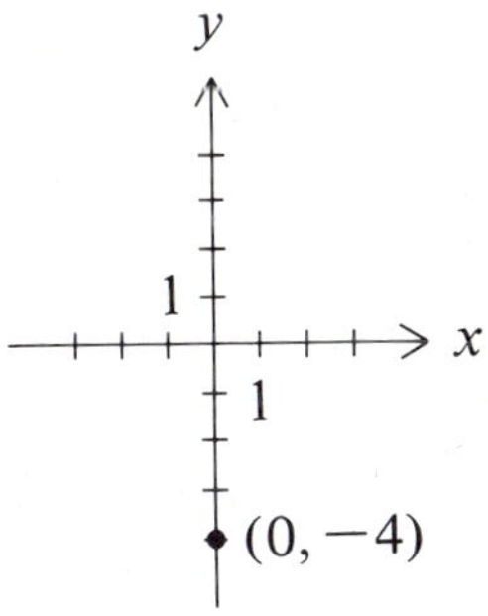

Figure 1–26 ■

We can easily find the graphs of other subsets of the plane specified by simple equations or inequalities. Here are some examples:

EXAMPLE: Find the graph in the xy-plane described by the equation $x = 2$.

This graph consists of those points (x, y) such that the x-coordinate is 2. The graph is the vertical line shown in Figure 1–27.

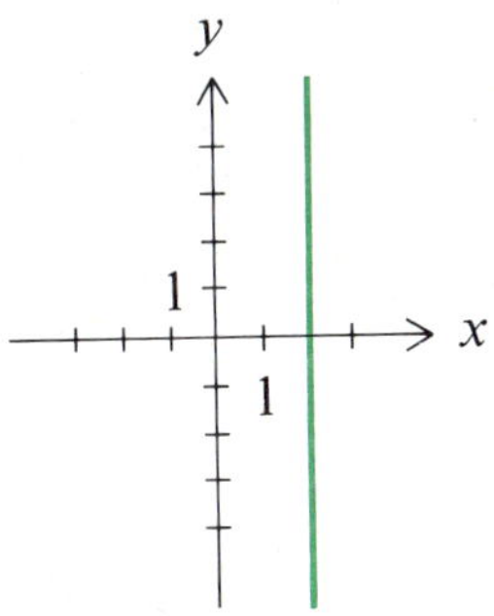

Figure 1–27
$\{(x, y) | x = 2\}$ ■

EXAMPLE: Find the graph of $x < 2$ in the plane.

This graph is the half-plane lying to the left of the line $x = 2$, as shown in Figure 1–28. The line $x = 2$ is dashed to indicate that this line is not part of the graph.

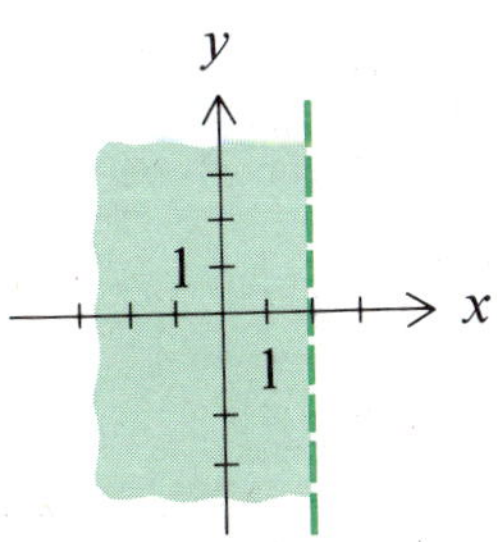

Figure 1–28
$\{(x, y) | x < 2\}$ ■

EXAMPLE: Graph $y = -1$ in the plane.

This graph is a horizontal line, as shown in Figure 1–29.

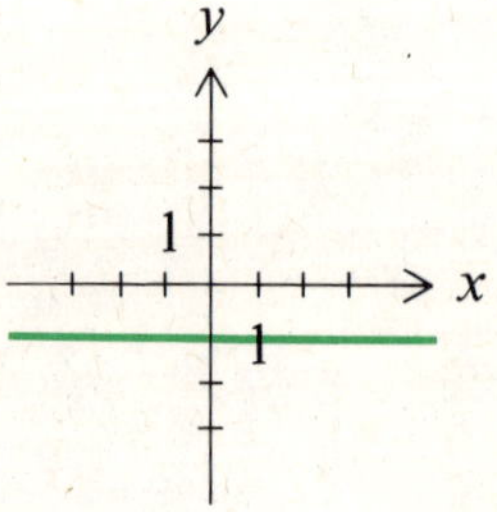

Figure 1–29
$\{(x, y) \mid y = -1\}$ ■

EXAMPLE: Graph $-1 < y < 1$ in the plane.

This graph consists of points (x, y) that satisfy both of the conditions $-1 < y$ and $y < 1$. Therefore, the points on this graph lie above the line $y = -1$, but below the line $y = 1$. The graph is a horizontal strip, as shown in Figure 1–30.

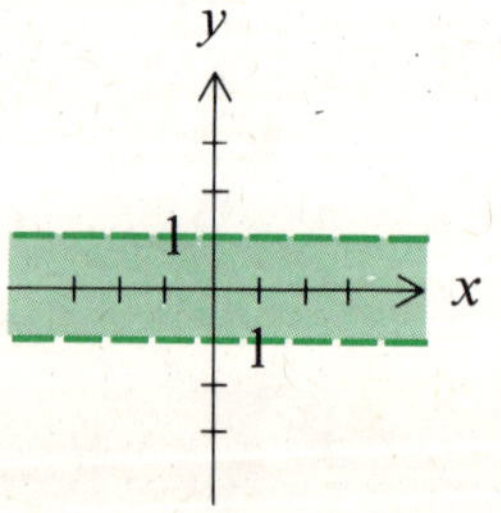

Figure 1–30
$\{(x, y) \mid -1 < y < 1\}$ ■

1-4 *Exercises*

Plot each of the following ordered pairs as points in the plane.

1. $(\frac{1}{2}, -\frac{1}{2})$
2. $(4, 0)$
3. $(0, 0)$
4. $(1.5, 2.5)$
5. $(-3, -1\frac{1}{2})$
6. $(0.7, 2.3)$

7. Find the coordinates of each point illustrated.

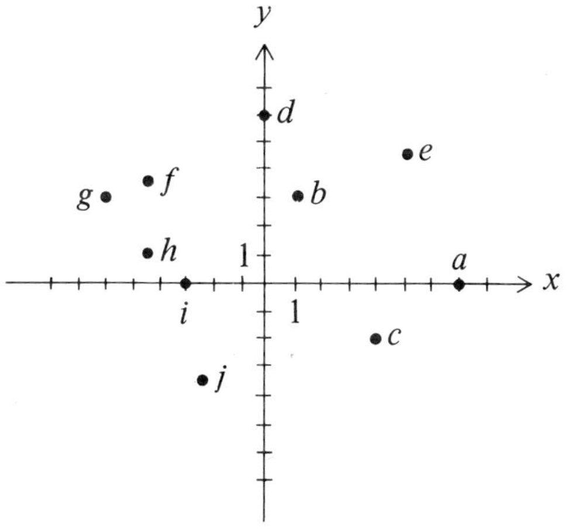

A point (x, y) is called a *lattice point* if both x and y are integers. Plot all lattice points (x, y) with the property that

8. $x^2 + y^2 \leq 0$
9. $x^2 + y^2 \leq 1$
10. $x^2 + y^2 \leq 2$
11. $x^2 + y^2 \leq 3$
12. $x^2 + y^2 \leq 4$

For each of the following subsets of the plane, find the equation or inequality for which the given subset is the graph.

13. the x-axis
14. the y-axis
15. the right half-plane
16. the lower half-plane
17. a vertical line two units to the right of the y-axis
18. a horizontal line five units below the x-axis
19. the second quadrant
20. the fourth quadrant

For each figure, find an equation or inequality for which the indicated set is the graph.

21.

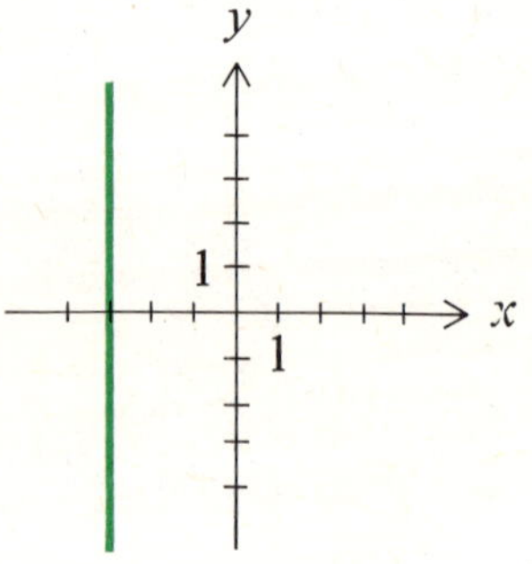

22.

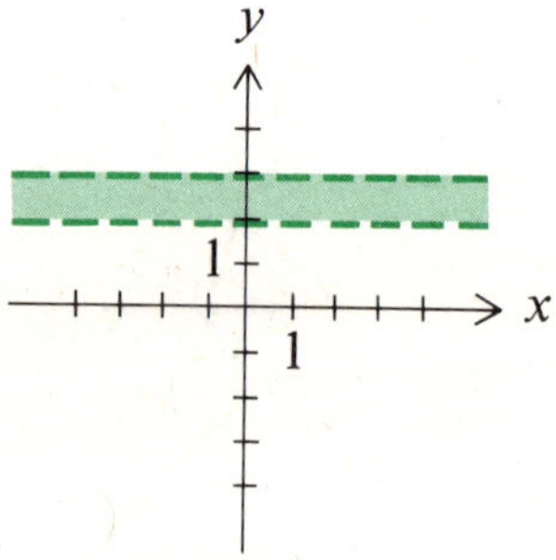

23.

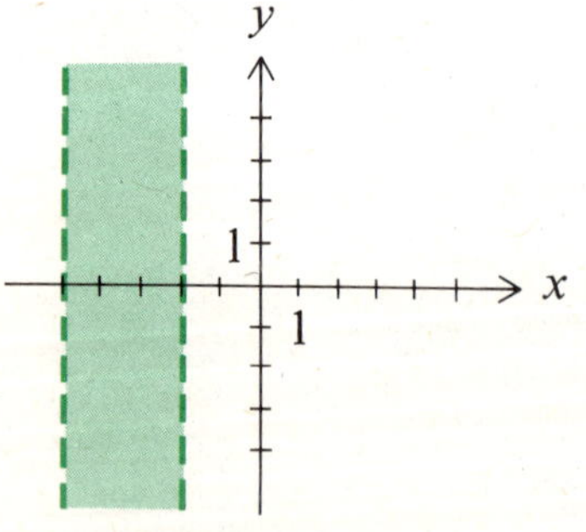

24.

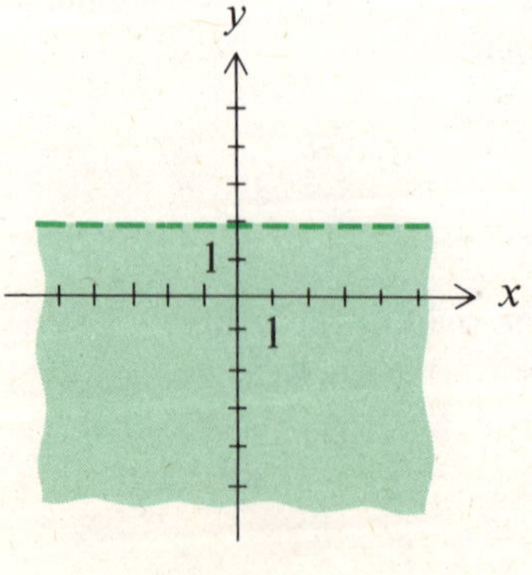

Sketch the graph of the given equation or inequality.

25. $x < 2\frac{1}{2}$

26. $y > -3\frac{1}{2}$

27. $-2 < x < 5$

28. $3\frac{1}{2} < y < 4\frac{1}{2}$

29. $y = -\frac{3}{4}$

30. $x = 2.2$

Describe geometrically the following sets.

31. $\{(x, y) \mid y > 0\}$

32. $\{(x, y) \mid -4 < x < 0\}$

33. $\{(x, y) \mid -1\frac{1}{2} < y < 3\frac{1}{2}\}$

34. $\{(x, y) \mid x = 7\}$

35. $\{(x, y) \mid y = 0\}$

36. $\{(x, y) \mid xy > 0\}$

37. $\{(x, y) \mid xy \geq 0\}$

38. $\{(x, y) \mid |x + y| \geq 0\}$

39. $\{(x, y) \mid |x + y| \geq 1\}$

1-5 Functions and relations

Many mathematical statements express relationships between variables. Perhaps the single most important kind of mathematical relationship is a *function.* Roughly speaking, we have a function whenever the value of the second variable is completely determined by the value of the first variable. We say, then, that the second variable is a function of the first variable.

As an example, let us consider the relationship between x and y existing for points (x, y) on a straight line. We shall assume that this line is not vertical, such as in Figure 1–31.

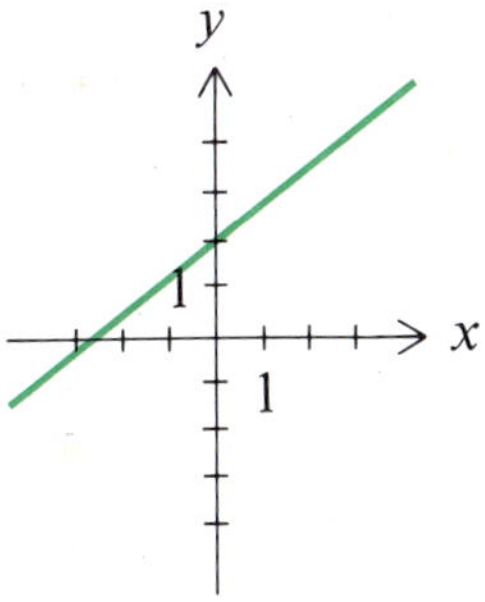

Suppose we choose a specific value, say x_1, for x. Then, as we examine the graph, we see that there is exactly one value for y, say y_1, such that (x_1, y_1) lies on the graph. This value y_1 is found as follows: We first locate x_1 on the x-axis. We then find exactly one point on the line having x_1 as its first coordinate; this point lies on the vertical line $x = x_1$. Now y_1 is the y-coordinate of this point, as shown in Figure 1–32. Thus for each value of x there is a unique value of y such that (x, y) lies on the graph. We express this relationship by saying that the graph determines *y as a function of x.*

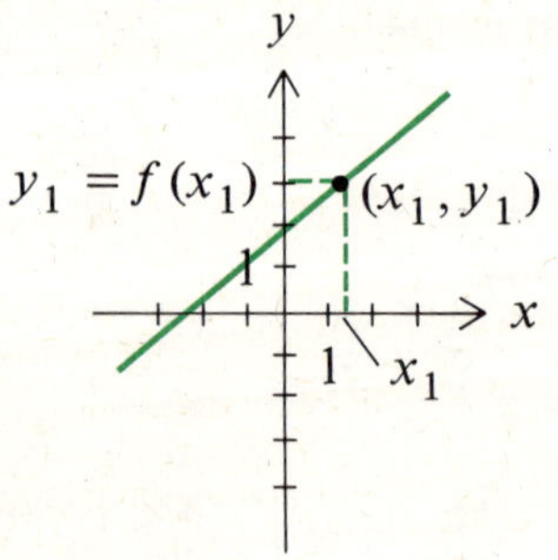

Figure 1–32

Often the functional relationship is expressed by writing

$y = f(x)$ (read " y equals f of x")

Here f is the name of the function. The notation emphasizes that for y to be a function of x, the value of y must be uniquely determined by the value of x.

EXAMPLE: Express y as a function of x for points on the line for which the equation is

$3x + y = 6$

We must express y in terms of x. Therefore, the equation is solved for y, yielding

$y = -3x + 6$

Therefore,

$y = f(x)$

where f is the function described by the formula

$f(x) = -3x + 6$ ■

In this example, we should think of the function f as the *correspondence* which associates to each value of x the corresponding value

$f(x) = -3x + 6$

Whenever a function f is described by a formula for $f(x)$, we have the following:

The value of a function f at a specific value of x is found by substituting that specific value for x into the formula for $f(x)$.

EXAMPLE: Find $f(1)$ if

$f(x) = -3x + 6$

The function f associates to x the number $-3x + 6$. Therefore, f associates to 1 the number $f(1) = (-3)(1) + 6 = 3$. ■

EXAMPLE: Find $f(x+1)$ if

$f(x) = -3x + 6$

Substituting $x + 1$ for x, we see that

$f(x+1) = -3(x+1) + 6 = -3x + 3$ ■

We can now formulate a more precise definition of function:

A function f is a correspondence which associates to each number x a unique number $f(x)$.

EXAMPLES: The formula $y = x + 2$ expresses y as a function of x. Here $y = f(x)$ where $f(x) = x + 2$.

The formula $y = \frac{1}{2}x$ expresses y as a function of x. Here $y = f(x)$ where $f(x) = \frac{1}{2}x$.

The formula $y = x^2$ expresses y as a function of x. Here $y = f(x)$ where $f(x) = x^2$.

The formula $y = 2x^2 + 0.984x - 4.15\pi$ expresses y as a function of x. Here $y = f(x)$ where $f(x) = 2x^2 + 0.984x - 4.15\pi$.

The correspondence described by

$$y = \begin{cases} 0 \text{ if } x \text{ is a rational number} \\ 1 \text{ if } x \text{ is an irrational number} \end{cases}$$

is a function, because each value of x determines a unique value of y. ■

We emphasize that in each example the y-value is uniquely determined by the x-value. The formula or correspondence tells us how to find $f(x)$ if we know x. Therefore, we have a correspondence which associates to each value of x a unique value of y.

We can recognize a geometric property which is shared by each graph which defines y as a function of x. A given value of x must uniquely determine the value of y such that (x, y) lies on the graph. Therefore, if for any value of x more than one point with this x-value lies on the graph, the graph does not determine y as a function of x. Geometrically, for each fixed value of x, the corresponding vertical line must not meet the graph in more than one point.

EXAMPLE: The graphs in Figures 1–33 and 1–34 each define y as a function of x. Each vertical line meets the graph in one point.

EXAMPLE: The graphs in Figures 1–35 and 1–36 do not define y as a function of x. There exists, for each graph, at least one vertical line which meets the graph in more than one point.

A function is a special case of a more general concept, known as a *relation*. We have seen that certain subsets of the plane are graphs of functions, but any subset of the plane may be thought of as the graph of a relation. Each point in the plane

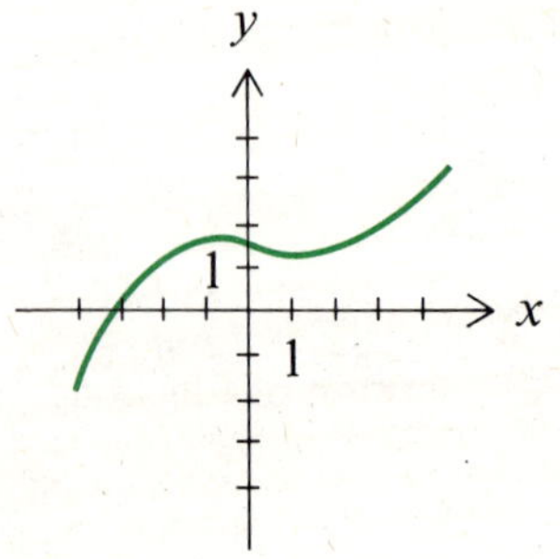

Figure 1–33
The graph of a function $y = f(x)$

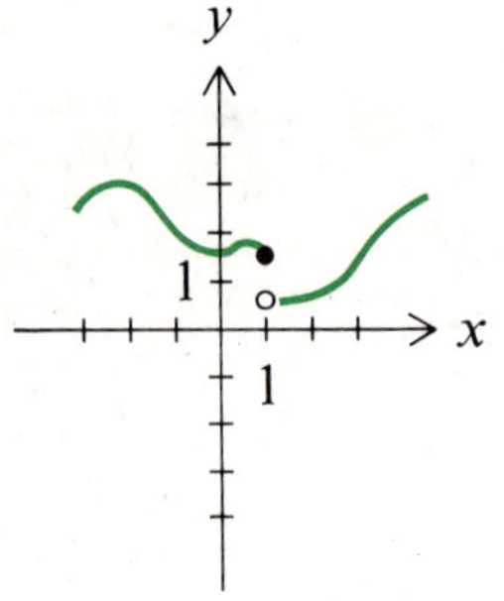

Figure 1–34
Another function ■

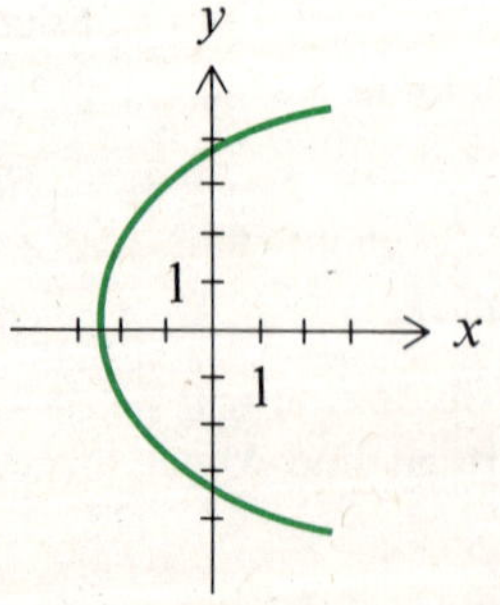

Figure 1–35
The graph does not determine y as a function of x

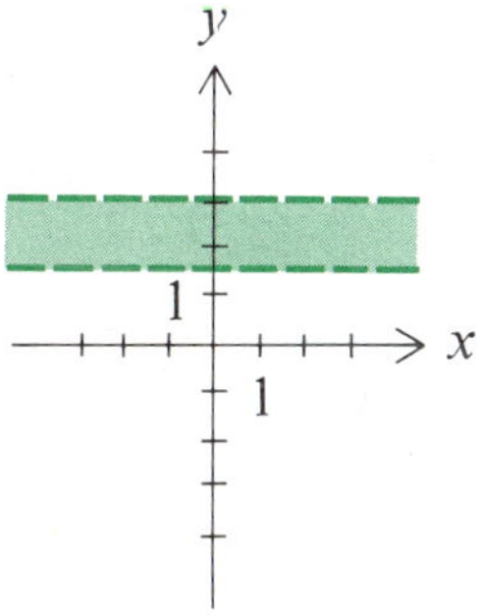

Figure 1–36
Not the graph of a function ■

corresponds to an ordered pair of real numbers. Therefore, a relation in the plane consists of any set of ordered pairs of real numbers, according to the following definition:

Any set of ordered pairs of real numbers is a relation.

The symbol $\mathscr{R}$ is often used to denote a relation.

EXAMPLE: The set of all ordered pairs (x, y) where x and y are real numbers is a relation. The graph of this relation is the entire plane. ■

EXAMPLE: Let $\mathscr{R}$ be the relation $\{(1, 3), (1, 4), (2, 3)\}$. The graph of $\mathscr{R}$ is shown in Figure 1–37.

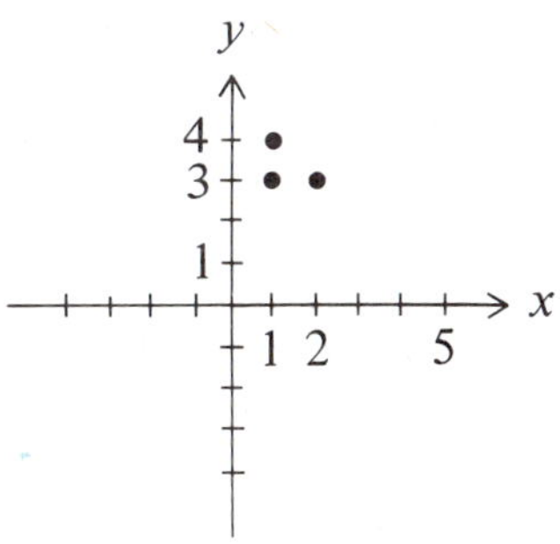

Figure 1–37
$\{(1, 3), (1, 4), (2, 3)\}$ ■

In this example, 1 and 2 are the only x-values which occur in ordered pairs belonging to the relation $\mathscr{R}$. We express this by saying that $\{1, 2\}$ is the *domain* of the relation $\mathscr{R}$, according to the following definition:

The domain of a relation $\mathscr{R}$ is the subset of the set of real numbers obtained as follows: The real number x is a member of the domain provided there exists a real number y such that $(x, y) \in \mathscr{R}$.

Thus the domain of $\mathscr{R}$ is the set of all first entries of ordered pairs of $\mathscr{R}$.

In the same example (Figure 1–37), the only y-values which appear in any ordered pair are 3 and 4. We say that $\{3, 4\}$ is the *range* of the relation $\mathscr{R}$, according to the following definition:

The range of a relation $\mathscr{R}$ is the subset of the set of real numbers obtained as follows. The real number y is an element of the range provided there exists a real number x such that $(x, y) \in \mathscr{R}$.

Thus the range of $\mathscr{R}$ is the set of all second entries of ordered pairs of $\mathscr{R}$.

EXAMPLE: Let $\mathscr{R}$ be the relation illustrated in Figure 1–38. The domain of $\mathscr{R}$ is $\{2, 3\}$. The range of $\mathscr{R}$ is $\{y \mid 0 \leq y \leq 2\}$.

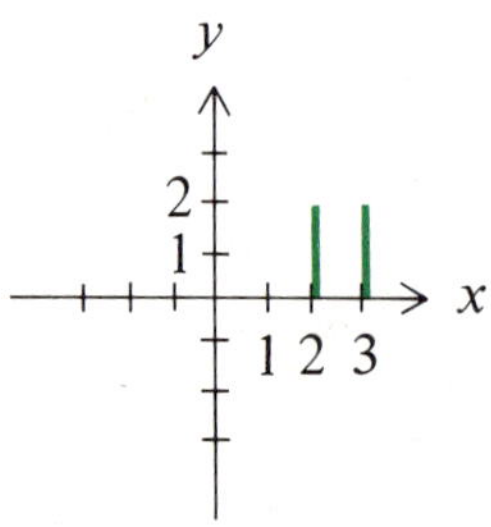

Figure 1–38

EXAMPLE: Consider the relation $\mathscr{R}$ illustrated in Figure 1–39. The domain of $\mathscr{R}$ is $\{x \mid 0 \leq x \leq 1\}$, and the range is $\{y \mid 1 \leq y \leq 2\}$.

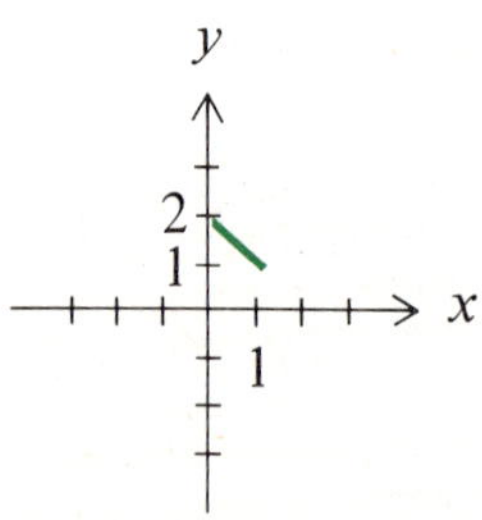

Figure 1–39

Let us now determine the connection between relations and functions. A relation is any set of ordered pairs (x, y), while a function f must assign to each x a unique y-value $f(x)$. We may think of the function f as being the set of all ordered pairs (x, y) such that $y = f(x)$. In symbols, we view f as $f = \{(x, y) \mid y = f(x)\}$ or even $f = \{(x, f(x))\}$. With this interpretation, *every function is a relation.*

However, not every relation is a function. To see this, we return to the relation $\mathscr{R}$ of Figure 1–37, with $\mathscr{R} = \{(1, 3), (1, 4), (2, 3)\}$. Because (1, 3) and (1, 4) are elements of $\mathscr{R}$, there are two different y-values attached to $x = 1$, and so we do not know whether 3 or 4 should be the y-value corresponding to $x = 1$. No *unique* y-value has been assigned to the x-value 1. The trouble is that there are two different ordered pairs of $\mathscr{R}$ having the same first entry.

If, however, no two different ordered pairs of a relation $\mathscr{R}$ have the same first entry, then the relation does define a function f. There will then be, for each x in the domain of $\mathscr{R}$, exactly one y such that $(x, y) \in \mathscr{R}$. The function f associates to x this single y-value.

Geometrically, we recognize a function as follows. If the graph of a relation defines y as a function of x, then for each value of x in the domain, the vertical line with that x-value meets the graph in just one point. If any vertical line meets the graph in more than one point, the relation is not a function.

In the discussion above, we required that a function f assign a unique y-value to *each* real number x. In other words, thinking of a function as a relation, we required its domain to be the whole real number system. We now, though, will find it more convenient to allow the domain of a function to be *any subset* of the set of real numbers. With this agreement, we see that any relation in which no two different ordered pairs have the same first entry defines a function. The function assigns to each x-value in its domain a unique y-value. The set of all such y-values constitutes the range of the function.

EXAMPLE: The relation $\mathscr{R}$ illustrated in Figure 1–39 defines a function f. The domain of f is $\{x \mid 0 \leq x \leq 1\}$, and the range of f is $\{y \mid 1 \leq y \leq 2\}$. The function f (like every function) associates to each element of its domain a unique element of its range. ■

-5 *Exercises*

Does the graph define y as a function of x?

1.

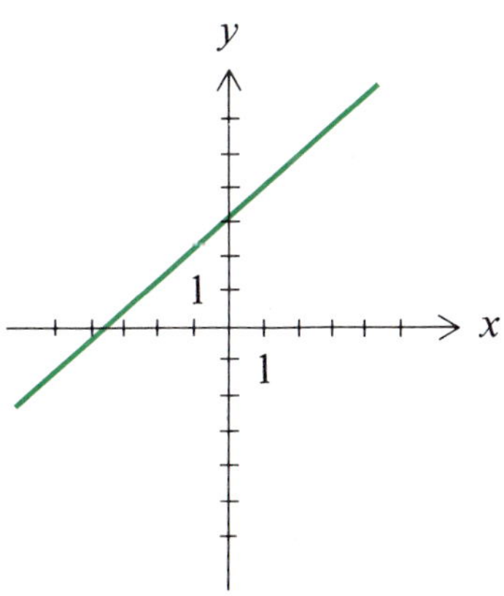

2.

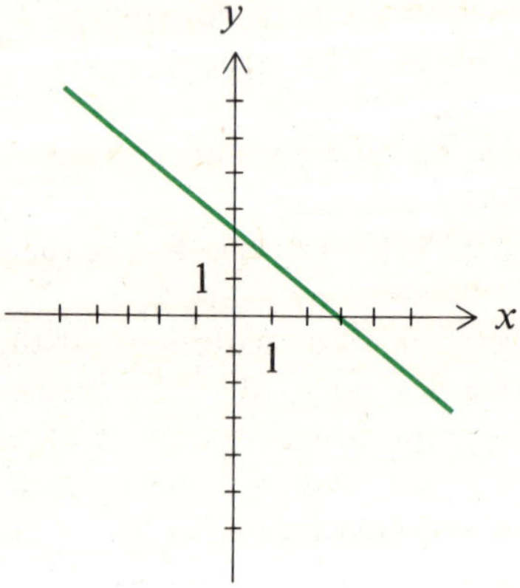

3.

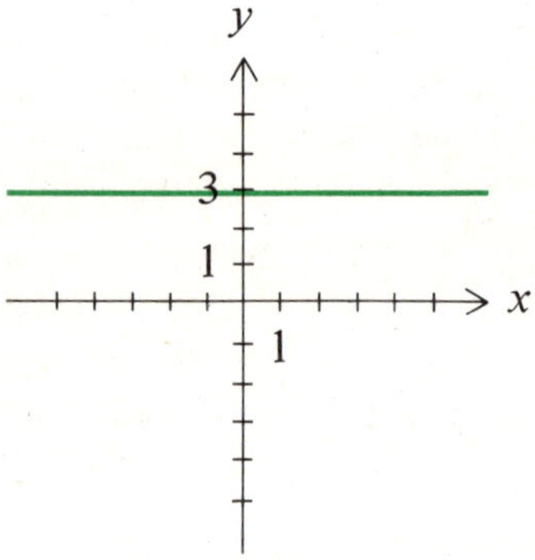

4.

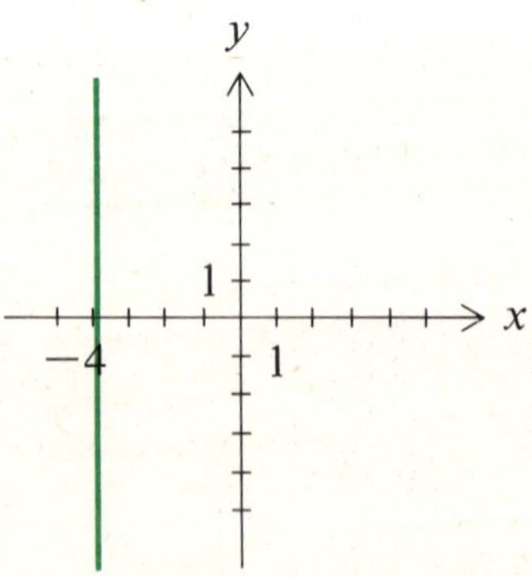

5. What geometrical property of a graph assures that it defines y as a functio of x?

Does the equation define y as a function of x?

6. $x = |y|$

7. $x = 4y - 5$
8. $y = 2$
9. $x = y^2$

If f is the function with formula $f(x) = x + 2$, find each of the following:

10. $f(0)$
11. $f(2)$
12. $f(-1)$
13. $f(t)$
14. $f(x - 1)$
15. $f(4x)$
16. $f(u + v)$
17. Work Exercises 10 through 16 for $f(x) = 3x - 4$.
18. Work Exercises 10 through 16 for $f(x) = x^2 + 2$.
19. Work Exercises 10 through 16 for $f(x) = |x| + 3$.

Given the graph, find the domain and range of the relation.

20.

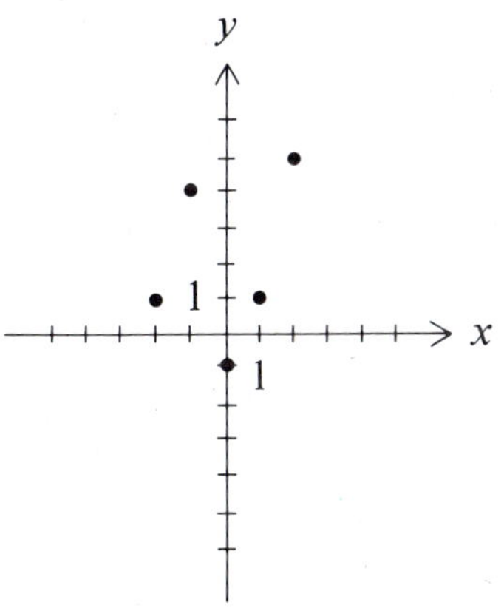

21.

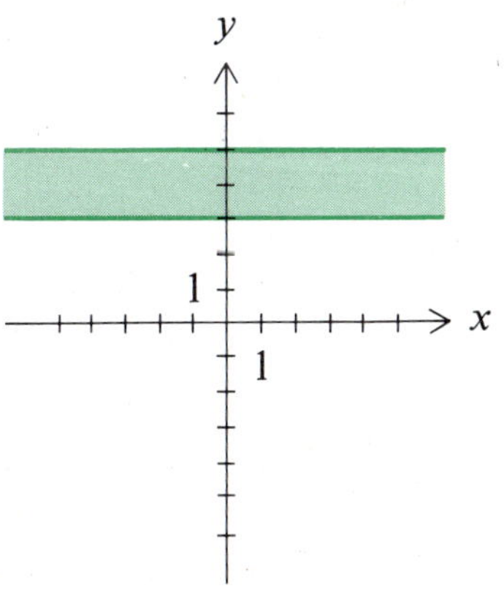

22.

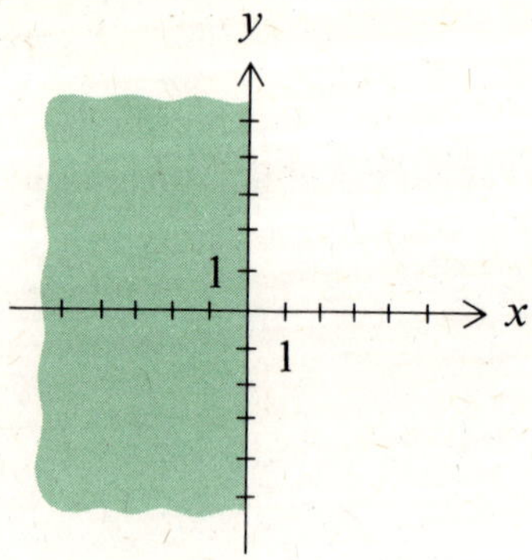

23.

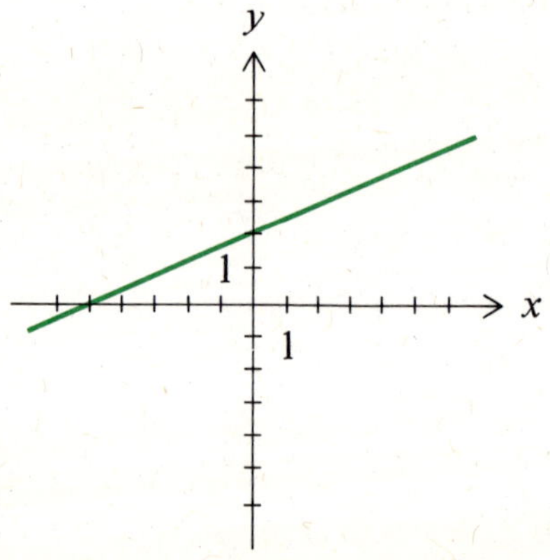

24.

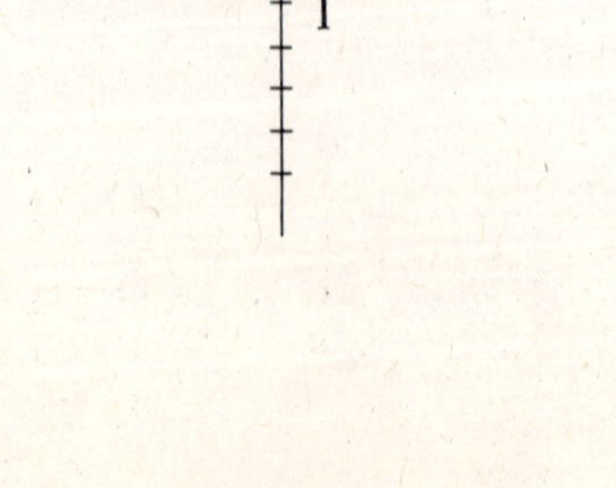

25.

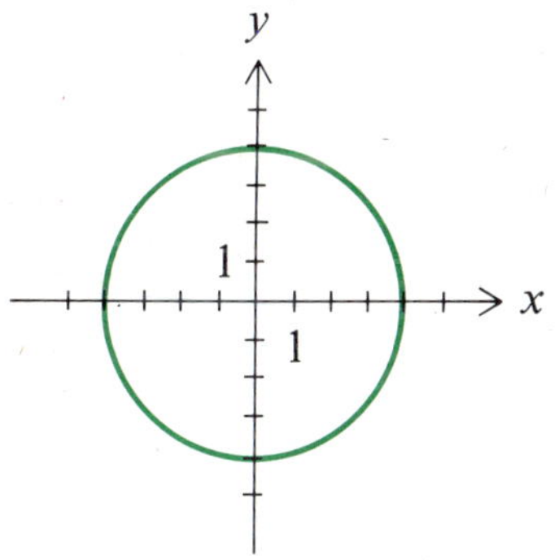

26. For each graph in Exercises 20 through 25, tell whether the given relation is a function.

Graph the indicated relation.

27. $\{(4, 1), (3, 2), (4, 3), (3, 4)\}$
28. $\{(-1, 1), (0, 1), (1, 1), (2, 1)\}$
29. $\{(1, 1), (2, 3), (3, 5), (4, 7)\}$
30. $\{(2, 0), (4, 0), (6, 0), (8, 3)\}$
31. $\{(x, -x + 5)\}$
32. $\{(x, 4x + 1)\}$
33. $\{(x, f(x)) \mid f(x) = 2x - 3\}$
34. $\{(x, f(x)) \mid f(x) = x - 2\}$
35. $\{(x, 3)\}$
36. $\{(-2, y)\}$
37. $\{(x, y) \mid y < 0\}$
38. $\{(x, y) \mid x < y\}$
39. For the relation given in Exercises 27 through 38, find the domain and range.
40. For the relation given in Exercises 27 through 38, tell whether the relation is a function.

Review exercises (Chapter 1)

Explain, define, or illustrate with an example.

1. $\geq$
2. $\nleq$
3. $\{x \mid x \in J, -1 \leq x \leq 2\}$
4. rational number
5. subset
6. absolute value
7. axes
8. origin
9. quadrant
10. function
11. relation
12. domain
13. range

Graph on the number line

14. $\{-\frac{1}{2}\}$

15. $\{2\frac{1}{4}\}$
16. $\{-2\frac{1}{4}, -\frac{1}{2}\}$
17. $\{x \mid -3 \leq x \leq 2\frac{1}{2}\}$

Sketch the graph in the plane
18. $\{(-3, -8), (2, -4)\}$
19. $\{(x, y) \mid -6 < x < 2\}$
20. $\{(x, y) \mid y = 0\}$

True or false? What is your reasoning?
21. If $x \nleq 4$, then $x < 4$
22. $Q \subseteq N$
23. $1\frac{1}{4} \in J$
24. $\pi \notin Q$
25. $\sqrt{7} \in Q$
26. $|-6| = |6|$
27. Is there a greatest natural number?
28. Is there a least natural number?

Let $f(x) = |5x - 2|$
29. Is f a function? Explain.
30. What is the domain of f?
31. What is the range of f?
32. Find $f(0)$
33. Find $f(1)$
34. Find $f(4)$
35. Find $f(-1)$
36. Find $f(x + 1)$
37. Find $f(x + y)$

2

Introduction to trigonometry

Today the properties of trigonometric functions have a wide variety of applications, ranging from calculus and other advanced mathematics through disciplines such as physics and economics and on to practical use in music, engineering, and mechanics.

Historically, though, trigonometry arose as the computational branch of geometry. The word trigonometry itself refers to triangle measurement; the subject was concerned with the ratios that occur naturally in any right triangle.

This chapter is about the trigonometry of a right triangle. Later we will move on to a more theoretical approach to trigonometry.

2-1 Trigonometric ratios in a right triangle

We begin with a right triangle, triangle ABC, as shown in Figure 2–1. Usually, a right triangle ABC is labeled so that the right angle is at vertex C. The length of a side is denoted by the lower-case letter corresponding to the opposite angle. Thus, in Figure 2–1, a is the length of the side opposite angle A, and c is the length of the hypotenuse. The symbol for triangle is $\triangle$, and the symbol for angle is $\angle$.

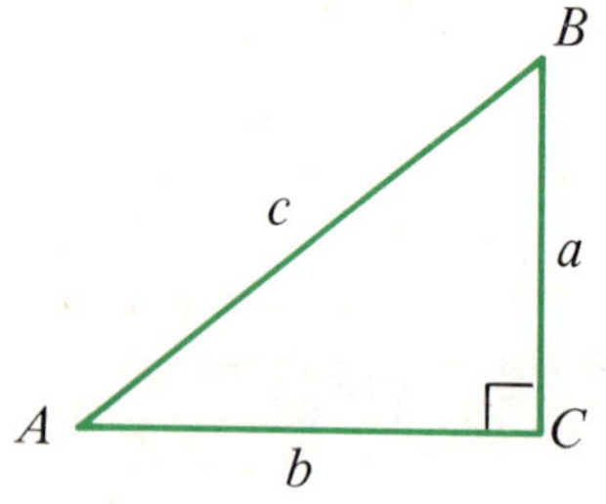

Figure 2–1
Right $\triangle ABC$

In $\triangle ABC$ (Figure 2–1), we define the trigonometric ratios of $\angle A$ as follows:

$$\text{sine of } \angle A = \frac{\text{side opposite } \angle A}{\text{hypotenuse}} = \frac{a}{c}$$

$$\text{cosine of } \angle A = \frac{\text{side adjacent to } \angle A}{\text{hypotenuse}} = \frac{b}{c}$$

$$\text{tangent of } \angle A = \frac{\text{side opposite } \angle A}{\text{side adjacent to } \angle A} = \frac{a}{b}$$

$$\text{cotangent of } \angle A = \frac{\text{side adjacent to } \angle A}{\text{side opposite } \angle A} = \frac{b}{a}$$

$$\text{secant of } \angle A = \frac{\text{hypotenuse}}{\text{side adjacent to } \angle A} = \frac{c}{b}$$

$$\text{cosecant of } \angle A = \frac{\text{hypotenuse}}{\text{side opposite } \angle A} = \frac{c}{a}$$

These ratios are usually abbreviated as follows:

sine of $\angle A = \sin A$
cosine of $\angle A = \cos A$
tangent of $\angle A = \tan A$
cotangent of $\angle A = \cot A$ (or, in some books, ctn A)
secant of $\angle A = \sec A$
cosecant of $\angle A = \csc A$

The six trigonometric ratios that are assigned to $\angle A$ are each real numbers.

Although we used $\triangle ABC$ in the definition of these ratios, their values depend only on the measure of $\angle A$. In fact, let $\triangle A'B'C'$ be another right triangle, with $\angle A' = \angle A$. From geometry, we know that $\triangle A'B'C'$ is similar to $\triangle ABC$. These two similar triangles are shown in Figure 2–2.

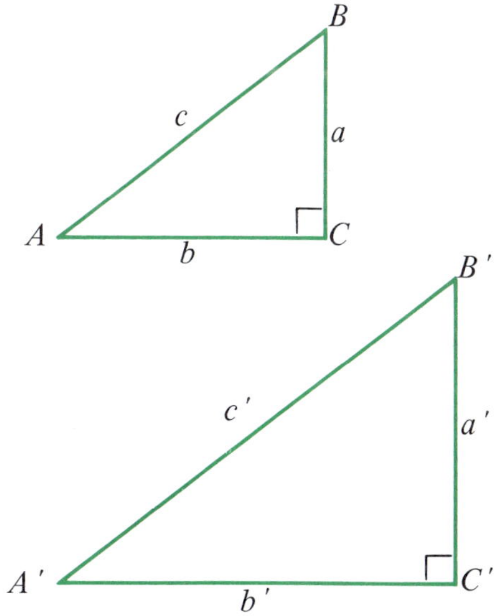

Figure 2–2
$\triangle A'B'C'$ is similar to $\triangle ABC$

Also from geometry, we know that the ratios of the lengths of the corresponding sides of similar triangles are equal. Thus, for example, in Figure 2–2 we have

$$\frac{a'}{c'} = \frac{a}{c}$$

and therefore

$$\sin A' = \sin A$$

Similarly, $b/c = b'/c'$, and so $\cos A = \cos A'$, and so on.

EXAMPLE: Let $\triangle ABC$ be a right triangle with $a = 3$, $b = 4$, and $c = 5$, as illustrated in Figure 2–3. We compute the trigonometric ratios of $\angle A$ as follows:

$$\sin A = \frac{\text{side opposite}}{\text{hypotenuse}} = \frac{3}{5}$$

$$\cos A = \frac{\text{side adjacent}}{\text{hypotenuse}} = \frac{4}{5}$$

$$\tan A = \frac{\text{side opposite}}{\text{side adjacent}} = \frac{3}{4}$$

$$\cot A = \frac{\text{side adjacent}}{\text{side opposite}} = \frac{4}{3}$$

$$\sec A = \frac{\text{hypotenuse}}{\text{side adjacent}} = \frac{5}{4}$$

$$\csc A = \frac{\text{hypotenuse}}{\text{side opposite}} = \frac{5}{3} \blacksquare$$

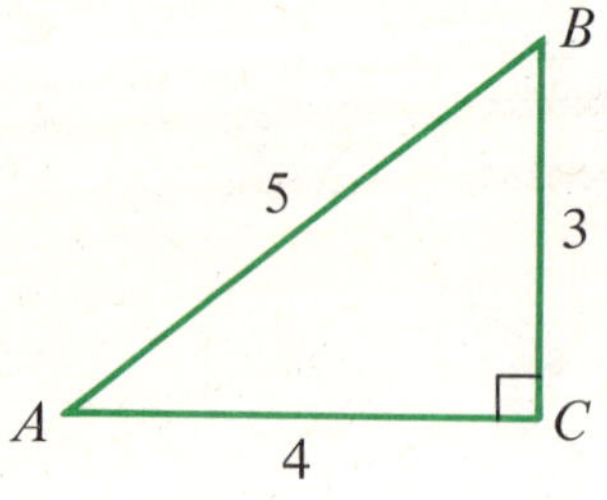

Figure 2–3

EXAMPLE: For the same $\triangle ABC$ (Figure 2–3), the trigonometric ratios of $\angle B$ are as follows:

$\sin B = \frac{4}{5}$
$\cos B = \frac{3}{5}$
$\tan B = \frac{4}{3}$
$\cot B = \frac{3}{4}$
$\sec B = \frac{5}{3}$
$\csc B = \frac{5}{4}$ ■

From geometry, let us again recall the theorem of Pythagoras:

If $\triangle ABC$ is a right triangle, with the right angle at C, then

$a^2 + b^2 = c^2$

Further, if a, b, and c are positive numbers such that $a^2 + b^2 = c^2$, then $\triangle ABC$ is a right triangle with c as the hypotenuse.

EXAMPLE: We verify that the triangle of Figure 2–3 is a right triangle as follows:

We have $a = 3$, $b = 4$, and $c = 5$. Thus, $a^2 + b^2 = 3^2 + 4^2 = 9 + 16 = 25 = 5^2$, and so $a^2 + b^2 = c^2$, as required. ■

EXAMPLE: $\triangle ABC$, shown in Figure 2–4, with $a = 12$, $b = 5$, and $c = 13$ is a right triangle, since $12^2 + 5^2 = 144 + 25 = 169 = 13^2$. The trigonometric ratios of $\angle A$ are as follows:

$\sin A = \frac{12}{13}$
$\cos A = \frac{5}{13}$
$\tan A = \frac{12}{5}$
$\cot A = \frac{5}{12}$
$\sec A = \frac{13}{5}$
$\csc A = \frac{13}{12}$ ■

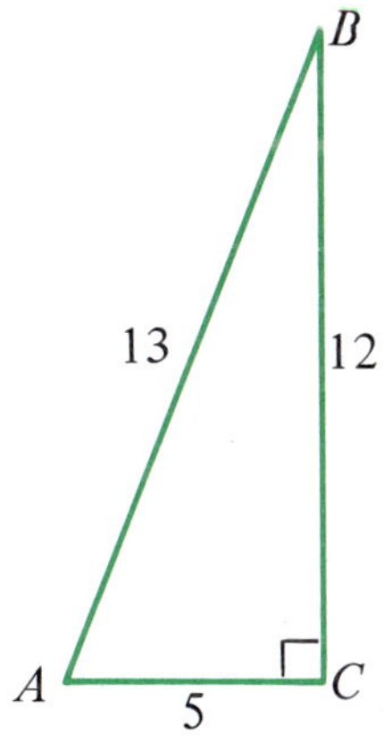

Figure 2–4

We would like to be able to calculate trigonometric ratios. For most angles, this is a difficult job and it requires advanced mathematical techniques. However, in a few special cases there are simple methods using plane geometry. We consider first an angle of measure 45°.

If $\angle A = 45°$ in right triangle $\triangle ABC$, we also have $\angle B = 45°$, and $\triangle ABC$ is an isoceles right triangle.

We draw side a of length 1, and so $b = 1$, also. By the theorem of Pythagoras, $c^2 = 1^2 + 1^2 = 2$, and so $c = \sqrt{2}$. The triangle is shown in Figure 2–5.

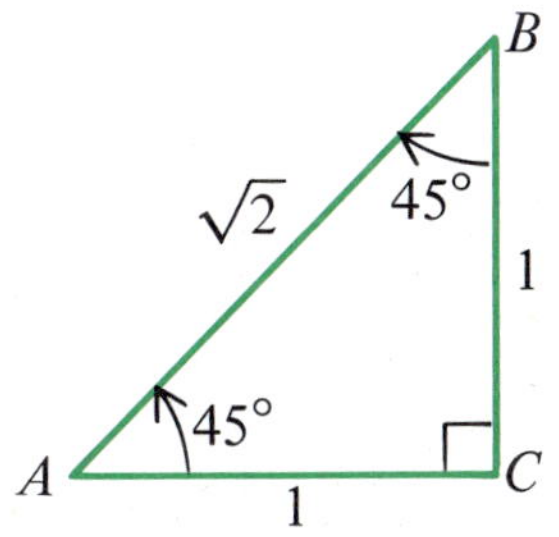

Figure 2–5

We now can write down the trigonometric ratios of $\angle A$. Since $\angle A$ has measure 45° we express these ratios simply as sin 45°, cos 45°, and so on as follows:

$$\sin 45° = \frac{1}{\sqrt{2}} = \frac{\sqrt{2}}{2} \qquad \text{(approximately equals 0.707)}$$

$$\cos 45° = \frac{1}{\sqrt{2}} = \frac{\sqrt{2}}{2}$$

$$\tan 45° = \frac{1}{1} = 1$$

$$\cot 45° = \frac{1}{1} = 1$$

$$\sec 45° = \frac{\sqrt{2}}{1} = \sqrt{2} \qquad \text{(approximately equals 1.414)}$$

$$\csc 45° = \frac{\sqrt{2}}{1} = \sqrt{2}$$

Next, we find the trigonometric ratios for an angle of measure 60°. We start with the equilateral triangle $\triangle ABD$ shown in Figure 2–6. Each angle of $\triangle ABD$ has measure 60°. We draw $\triangle ABD$ so that the length of each side is 2.

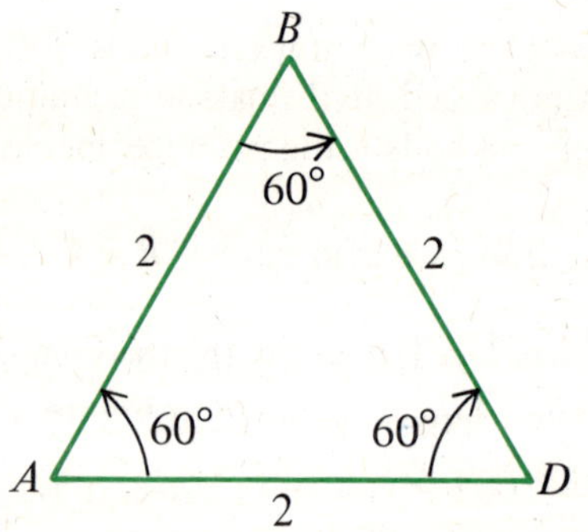

Figure 2–6

Let C be the mid-point of AD, and draw BC as in Figure 2–7. Then $\angle ACB$ is a right angle and the length of AC is 1. We thus have the right triangle $\triangle ABC$ shown in Figure 2–8. In right $\triangle ABC$, $a^2 = c^2 - b^2 = 2^2 - 1^2 = 4 - 1 = 3$, and so $a = \sqrt{3}$.

From Figure 2–8, we read off the trigonometric ratios as follows:

$$\sin 60° = \frac{\sqrt{3}}{2} \qquad \text{(or about 0.866)}$$

$$\cos 60° = \frac{1}{2}$$

$$\tan 60° = \sqrt{3} \qquad \text{(approximately equals 1.732)}$$

$$\cot 60° = \frac{1}{\sqrt{3}} = \frac{\sqrt{3}}{3} \qquad \text{(about 0.577)}$$

$\sec 60° = 2$

$\csc 60° = \frac{2}{\sqrt{3}} = \frac{2\sqrt{3}}{3}$ (approximately equals 1.155)

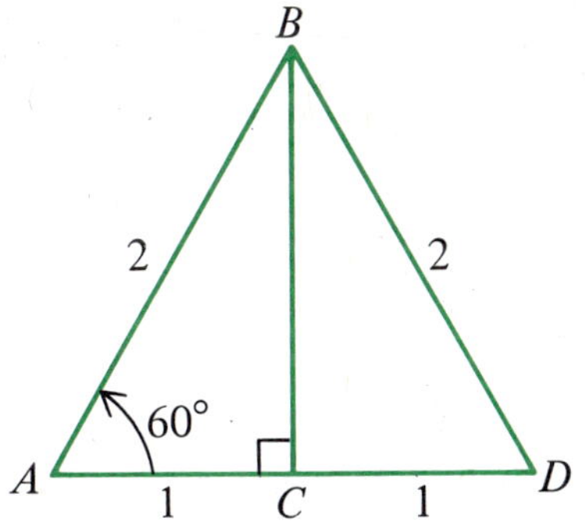

Figure 2–7
C is the midpoint of *AD*

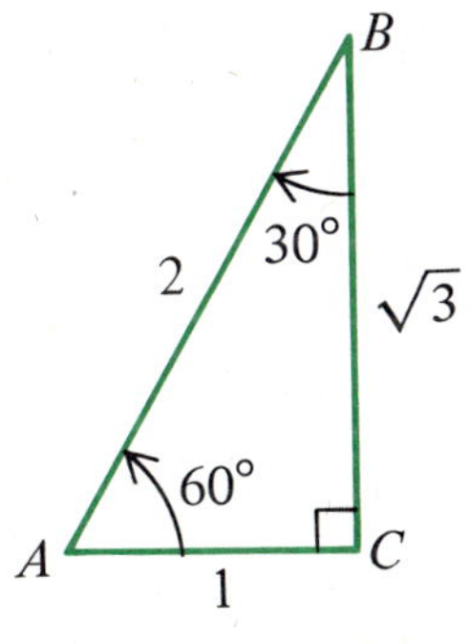

Figure 2–8

From Figure 2–8, it is also possible to find the trigonometric ratios of an angle of measure 30°. You are asked to do this in Exercises 1 through 6.

2–1 *Exercises*

Use Figure 2–8 to determine the following.

1. sin 30°
2. cos 30°
3. tan 30°
4. cot 30°
5. sec 30°
6. csc 30°

The lengths of the sides of $\triangle ABC$ are as given. Is $\triangle ABC$ a right triangle? If so, which is the right angle?

7. $a = 2$, $b = 3$, $c = 4$
8. $a = 7$, $b = 25$, $c = 24$
9. $a = 2$, $b = 1$, $c = \sqrt{5}$
10. $a = 10$, $b = 6$, $c = 8$
11. $a = 2$, $b = \sqrt{21}$, $c = 5$
12. $a = \sqrt{7}$, $b = 5$, $c = 4$

For the right triangle shown, find
13. $\sin A$
14. $\cos A$
15. $\tan A$
16. $\cot A$
17. $\sec A$
18. $\csc A$
19. $\sin B$
20. $\cot B$
21. $\sec B$
22. $(\sin A)^2 + (\cos A)^2$

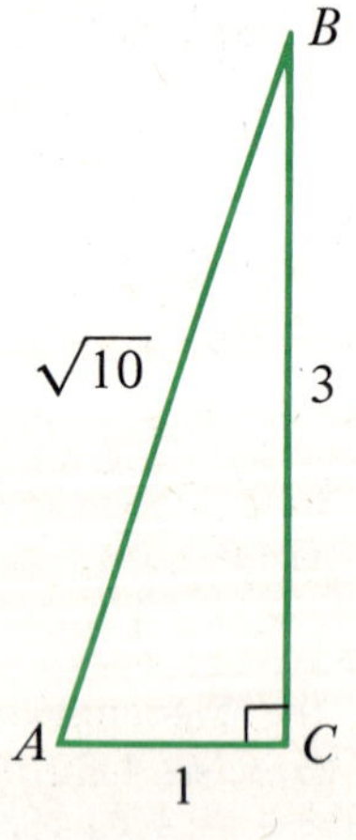

23. Work Exercises 13 through 22 for a right triangle $\triangle ABC$ with $a = \sqrt{15}$, $b = 1$, and $c = 4$.

In the right triangle $\triangle ABC$ shown on page 41, find
24. $\sin A$
25. b
26. $\cos A$
27. $\sin (90° - A)$
28. $\tan A$
29. $(\sin A)^2 + (\cos A)^2$

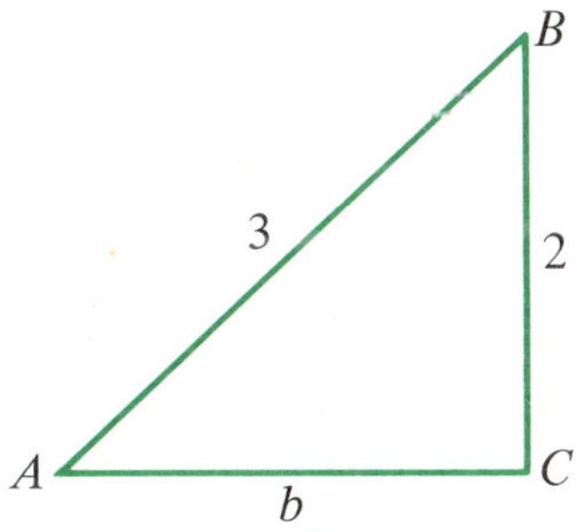

Suppose, in right $\triangle ABC$, that $\tan A = 1/5$. Draw the triangle with $a = 1$ and $b = 5$ and use it to find

30. $\sin A$
31. $\cos A$
32. $\cot B$
33. $(\sin A)^2 + (\cos A)^2$

Suppose, in right $\triangle ABC$, that $\cos A = \frac{2}{3}$. Find

34. $\sin A$
35. $\tan A$
36. $\sin B$
37. $(\sin A)^2 + (\cos A)^2$

2-2 Identities among trigonometric ratios

Our objective now is to develop identities among the various trigonometric ratios. These identities are used to simplify trigonometric expressions and to evaluate all the trigonometric ratios for an acute angle, once the value for just one ratio is known.

The first set of identities expresses the tangent, cotangent, secant, and cosecant of an angle in terms of the sine and cosine of the same angle. We call these the *simplification identities.*

For any acute[1] angle A, we have

$$\tan A = \frac{\sin A}{\cos A}$$

$$\cot A = \frac{\cos A}{\sin A}$$

$$\sec A = \frac{1}{\cos A}$$

[1] An *acute* angle is an angle with measure between 0° and 90°

$$\csc A = \frac{1}{\sin A}$$

EXAMPLE: Find tan A if $\sin A = \frac{4}{5}$, and $\cos A = \frac{3}{5}$.

We have

$$\tan A = \frac{\sin A}{\cos A} = \frac{4/5}{3/5} = \frac{4}{3} \blacksquare$$

EXAMPLE: Find cot A if $\tan A = 2$.

We have

$$\cot A = \frac{1}{\tan A} = \frac{1}{2} \blacksquare$$

EXAMPLE: Find sin A if $\csc A = 4$.

Because $\csc A = 1/\sin A$, we have $\sin A = 1/\csc A$. Therefore, $\sin A = \frac{1}{4}$. $\blacksquare$

To justify the simplification identities, we need only refer to the definitions of the trigonometric ratios in right triangle $\triangle ABC$ (see Figure 2–9).

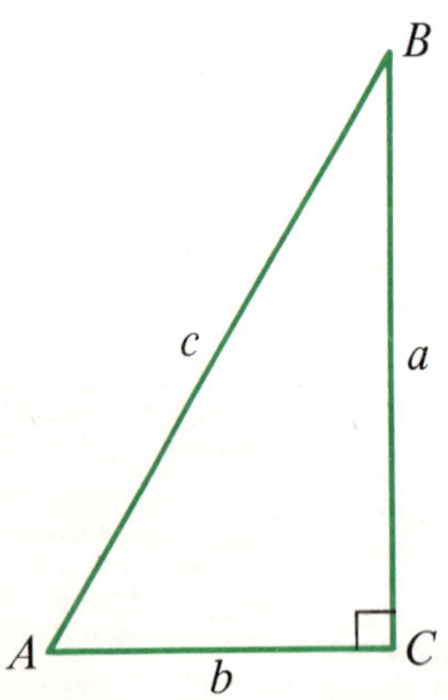

Figure 2–9
Right $\triangle ABC$

Thus, for example, $\tan A = a/b$, and so

$$\tan A = \frac{a}{b} = \frac{a/c}{b/c} = \frac{\sin A}{\cos A}$$

and the first identity is valid. The other justifications are similar and are left to you in Exercises 1, 2, and 3.

The next set of identities connect the trigonometric ratios of an acute angle with the ratios of its complement.[2] These are the *cofunction*[3] *identities.*

For any acute angle $\angle A$,

$$\sin(90° - A) = \cos A$$
$$\cos(90° - A) = \sin A$$
$$\tan(90° - A) = \cot A$$
$$\cot(90° - A) = \tan A$$
$$\sec(90° - A) = \csc A$$
$$\csc(90° - A) = \sec A$$

EXAMPLE: $\sin 40° = \cos(90° - 40°) = \cos 50°$. ■

EXAMPLE: $\tan 45° = \cot(90° - 45°) = \cot 45°$. ■

EXAMPLE: $\csc 73° = \sec(90° - 73°) = \sec 17°$. ■

There is a convenient way to remember the cofunction identities. We think of sine and *co*sine as being *cofunctions* of each other. Similarly tangent and *co*tangent are a pair of cofunctions, and also secant and *co*secant. We may then summarize the cofunction identities as follows.

A trigonometric ratio of an acute angle has the same value as the cofunction of the complement of the angle.

By the use of the cofunction identities, trigonometric ratios for acute angles with measure greater than 45° can always be rewritten in terms of acute angles with measure less than 45°. The following examples illustrate the application of both the simplification and the cofunction identities.

EXAMPLE: Express csc 81° in terms of sine and/or cosine of an angle with measure less than 45°.

We have $\csc 81° = \sec(90° - 81°) = \sec 9° = 1/\cos 9°$. ■

EXAMPLE: Express tan 64° in terms of sine and/or cosine of an angle with measure less than 45°.

We have $\tan 64° = \cot(90° - 64°) = \cot 26° = (\cos 26°)/\sin 26°$. ■

Justification of the cofunction identities is not difficult. We refer to right triangle $\triangle ABC$ (Figure 2–9) and observe that angles $\angle A$ and $\angle B$ are complementary, so that $B = (90° - A)$. Thus, for example,

$$\sin(90° - A) = \sin B = \frac{\text{side opposite } \angle B}{\text{hypotenuse}} = \frac{b}{c} = \cos A$$

[2] Two acute angles are *complementary* if the sum of their measures is 90°. Thus, the two acute angles in a right triangle are complementary, and each is the complement of the other.

[3] Strictly speaking, we should call these identities *coratio*, not *cofunction*. Later we will justify the use of "function" terminology for the trigonometric ratios.

and so the first cofunction identity is valid. The proofs of the others are similar and are left as Exercises 4 through 8.

Still another important set of identities is available to us. These are referred to as the *Pythagorean identities.* (We follow the usual convention and write $(\sin A)^2$ as $\sin^2 A$, etc.)

For any acute angle $\angle A$,

$$\sin^2 A + \cos^2 A = 1$$
$$1 + \tan^2 A = \sec^2 A$$
$$1 + \cot^2 A = \csc^2 A$$

EXAMPLE: We verify the first Pythagorean identity for an angle of measure 60°. We know (see Section 2–1) that $\sin 60° = \frac{1}{2}\sqrt{3}$ and $\cos 60° = \frac{1}{2}$. Thus,

$$\sin^2 60° + \cos^2 60° = \left(\frac{\sqrt{3}}{2}\right)^2 + \left(\frac{1}{2}\right)^2 = \frac{3}{4} + \frac{1}{4} = 1 \blacksquare$$

EXAMPLE: Find $\sec A$ if A is an acute angle such that $\tan A = 2$.

By the second Pythagorean identity,

$$\sec^2 A = 1 + \tan^2 A = 1 + 2^2 = 1 + 4 = 5$$

and so $\sec A = \sqrt{5}$. ■

EXAMPLE: Find $\sin A$ if A is an acute angle with $\cos A = \frac{1}{3}$.

By the first Pythagorean identity,

$$\sin^2 A = 1 - \cos^2 A = 1 - (\tfrac{1}{3})^2 = 1 - \tfrac{1}{9} = \tfrac{8}{9}$$

Thus,

$$\sin A = \sqrt{\frac{8}{9}} = \frac{2\sqrt{2}}{3}. \blacksquare$$

EXAMPLE: Find $\sin A$, if A is an acute angle with $\sec A = 4$.

Since $\cos A = 1/\sec A$, we have $\cos A = \frac{1}{4}$. Hence,

$$\sin^2 A = 1 - \cos^2 A = 1 - (\tfrac{1}{4})^2 = 1 - \tfrac{1}{16} = \tfrac{15}{16}$$

and so

$$\sin A = \sqrt{\frac{15}{16}} = \frac{\sqrt{15}}{4}. \blacksquare$$

To justify the Pythagorean identities, we begin with the formula

$$a^2 + b^2 = c^2$$

valid in right triangle $\triangle ABC$ (Figure 2–9).

Dividing each side by c^2, we obtain

$$a^2/c^2 + b^2/c^2 = c^2/c^2$$

or

$$(a/c)^2 + (b/c)^2 = 1$$

But $a/c = \sin A$ and $b/c = \cos A$. Hence, we have

$$\sin^2 A + \cos^2 A = 1$$

and the first Pythagorean identity is proved. The other two are checked similarly, and the proof is left to you (see Exercises 9 and 10).

2–2 Exercises

Use the definitions of the trigonometric ratios to justify the identities.

1. $\cot A = \dfrac{\cos A}{\sin A}$

2. $\sec A = \dfrac{1}{\cos A}$

3. $\csc A = \dfrac{1}{\sin A}$

4. $\cos(90° - A) = \sin A$

5. $\tan(90° - A) = \cot A$

6. $\cot(90° - A) = \tan A$

7. $\sec(90° - A) = \csc A$

8. $\csc(90° - A) = \sec A$

Use the Pythagorean theorem and the definitions of the trigonometric ratios to justify the identities.

9. $1 + \tan^2 A = \sec^2 A$
10. $1 + \cot^2 A = \csc^2 A$

Express in terms of sin A and/or cos A.

11. $\sec A$
12. $\tan A$
13. $\cot A + \csc A$
14. $(\csc A)/(\cot A)$
15. $1 + \tan^2(90° - A)$
16. $\sin(90° - A)\sec(90° - A)$

Express in terms of the sine and/or cosine of an angle with measure of at most 45°.

17. $\cot 32°$
18. $\sin 11° \sec 79°$
19. $\tan 48° + \cot 48°$
20. $\cos 71° \tan 71°$

Express in terms of the sine of an angle with measure of at most 45°.

21. $\cos 11°$
22. $\tan 55°$

23. sec 65°
24. sec 25°

Given that A is an acute angle with sin $A = 3/7$, use identities to find
25. cos A
26. tan A
27. cot A
28. sec A
29. csc A

Given that A is an acute angle with $\csc^2(90° - A) = \frac{49}{33}$, use identities to find
30. sin A
31. cos A
32. tan A
33. cot A
34. sec A
35. csc A

Express in terms of tan A, where A is an acute angle.
36. sin A
37. cos A
38. cot A
39. sec A
40. csc A

2-3 Values of trigonometric ratios

For any acute angle, we have defined the trigonometric ratios. The next questio is: How do we find the numerical value of a trigonometric ratio for a specific angle

For a few special angles, the problem can be solved by geometry. In fact, w did this in Section 2–1 for angles of measure 30°, 45°, and 60°. Unfortunately for most angles, this would not be possible.

Just as in the case of logarithmic functions, tables of values of the trigonometri ratios have been computed by methods of advanced mathematics. You will fin this table inside the front cover. For angles between 0° and 45° the measure of th angle is given in the left column. The value of the trigonometric ratio is the found in the column under the name of the ratio.

EXAMPLE: Find sin 35° from the table inside the front cover of the book.

We find 35° in the left column. Directly across from 35°, under sin, we fin 0.5736. Thus an approximate value for sin 35° is 0.5736. ■

Note that the values in the table are four-place approximations, not exac values. Still, it is customary to write the preceding result as sin 35° = 0.5736 eve though this statement is not exactly accurate.

EXAMPLE: Find cot 21°.

In the table, in the column headed by cot, directly to the right of 21°, we find that cot 21° = 2.6051. ■

We use the cofunction identities to find trigonometric ratios for angles between 45° and 90°.

EXAMPLE: Find sec 73°.

From the cofunction identities, we have

$$\sec 73° = \csc (90° - 73°) = \csc 17°$$

From the table,

$$\csc 17° = 3.4203$$

Therefore,

$$\sec 73° = 3.4203$$ ■

Angle measures from 45° to 90° are listed in the *right* column of the table of values. These entries are to be associated with the function names at the *bottom* of the columns. This is an automatic procedure for taking advantage of the cofunction identities.

EXAMPLE: We again evaluate sec 73°.

We find 73° in the right column of the table. Looking to the left of 73°, in the column labeled sec at the bottom, we again find sec 73° = 3.4203. ■

EXAMPLE: Find sin 52°.

From the table we find directly that sin 52° = 0.7880. ■

In the traditional system of angle measurement, each degree is divided into sixty minutes, and each minute is divided into sixty seconds. The table inside the front cover gives four-place values of the trigonometric ratios at intervals of one degree. A more complete table, giving four-place values at intervals of ten minutes, and called Trigonometric Ratios, is included in Appendix B. For angles expressed to the nearest minute, linear interpolation (as explained in Section A-5) can be used to find approximate function values.

Note: The entries for angles of 0° and 90°, as well as the columns headed "Radians" in the table in Appendix B, will become self-explanatory later on.

2–3 *Exercises*

Find values of the following using the table inside the front cover.

1. sin 8°
2. tan 85°
3. csc 39°
4. cot 14°

5. cos 63°
6. sec 76°

In the triangle shown, the measure of angle $\angle A$ is 41°. Measure sides a, b, and c and find approximate values for
7. sin 41°
8. sin 49°
9. tan 41°
10. sec 49°

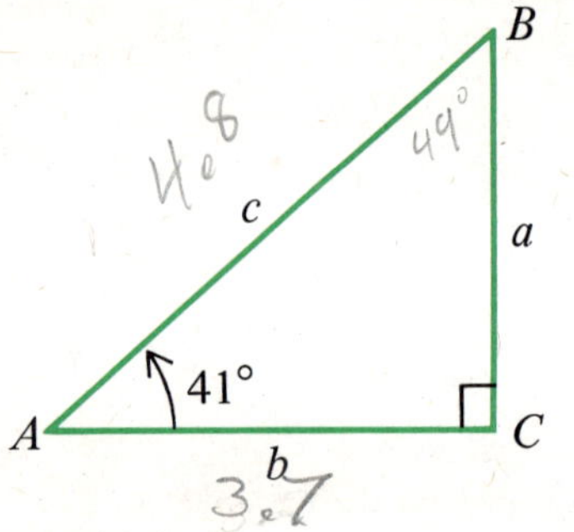

Find the values from the table and compare with your approximations obtained in Exercises 7 through 10.
11. sin 41°
12. sin 49°
13. tan 41°
14. sec 49°

Use a protractor to draw a right triangle $\triangle ABC$ with $\angle A = 20°$. Measure sides a, b, and c and find approximate values for
15. sin 20°
16. cos 20°
17. tan 20°
18. sec 20°

Find the values from the table and compare with your approximations obtained in Exercises 15 through 18.
19. sin 20°
20. cos 20°
21. tan 20°
22. sec 20°

23. From the table inside the front cover, sin 11° = 0.1908 and cos 11° = 0.9816. We compute $\sin^2 11° + \cos^2 11°$ as

$$\begin{aligned}\sin^2 11° + \cos^2 11° = (0.1908)^2 + (0.9816)^2 &= 0.03640464 + 0.96353856 \\ &= 0.99994320\end{aligned}$$

From the Pythagorean identities, we should have

$\sin^2 11° + \cos^2 11° = 1$

Explain the discrepancy.

2-4 Solution of right triangles

Practical applications of trigonometry often require the *solution of a triangle*—that is, finding the measure of the unknown sides or angles.

For the present, we focus our attention on right triangles. (Later, we will discuss the solution of any triangle.) As usual, we work with a right triangle $\triangle ABC$, with the right angle at C and the sides labeled as in Figure 2–10. Thus, we know

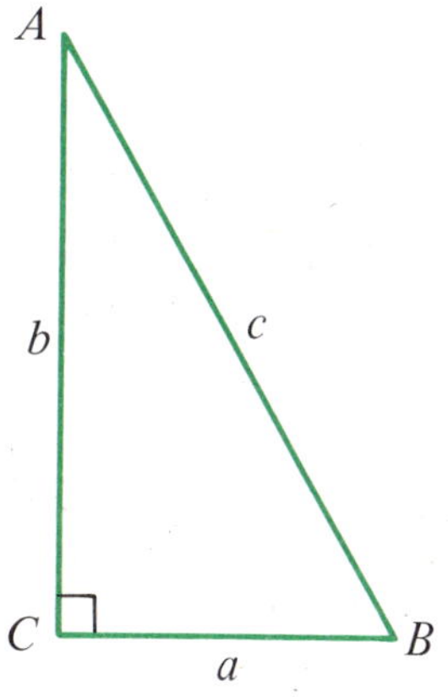

Figure 2–10
Right $\triangle ABC$

$\angle C = 90°$. The remaining parts of the triangle are the acute angles $\angle A$ and $\angle B$, the legs a and b, and the hypotenuse c. In a typical application, the measures of some of these are known; the others are to be found. Trigonometric ratios are used to express the unknown measures in terms of the known data, as in the following illustrations.

EXAMPLE: Let $\angle A = 28°$ and $c = 8$ units, as in Figure 2–11. Find the measures of a, b, and $\angle B$.

We first note that $\angle A + \angle B + \angle C = 180°$. Hence $\angle A + \angle B = 90°$, and so $\angle B = 90° - 28° = 62°$.

In right triangle $\triangle ABC$, $\sin A = a/c$. Therefore, we have $\sin 28° = a/8$, or $a = 8 \sin 28°$. From the table inside the front cover, $\sin 28° = 0.4695$. Therefore, $a = (8) \cdot (0.4695) = 3.756$.

Similarly, $\cos A = b/c$, or $b = c \cos A$. Hence, in this triangle $b = 8 \cos 28° = (8) \cdot (0.8829) = 7.063$. ■

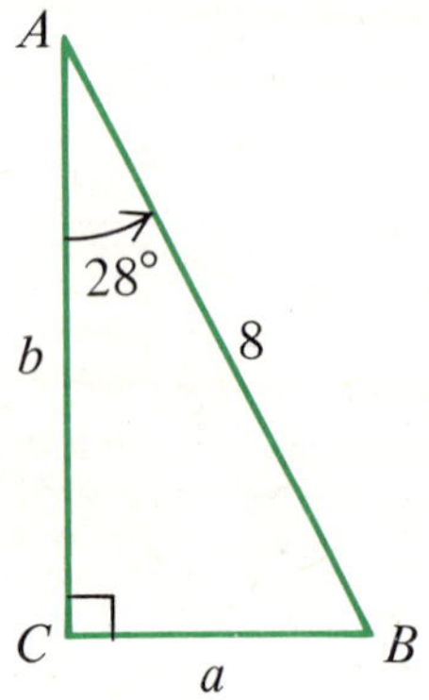

Figure 2–11

In the preceding example, we should note that the results $a = 8 \sin 28°$ anc $b = 8 \cos 28°$ are exact, but the numerical values $a = 3.756$ and $b = 7.063$ ar approximate. This is because the values for sin 28° and cos 28° taken from th table are themselves approximations. The values given in the table have fou significant digits. Numerical answers based on values from the table should als be rounded to four significant digits.

The arithmetic required in these computations can be done directly, or b logarithms (see Appendix A), or with a calculator or slide rule. For some purpose the exact answer (such as $a = 8 \sin 28°$) is the most desirable. It should then b simple, at least in principle, to obtain a numerical result.

EXAMPLE: Solve right triangle $\triangle ABC$, where $a = 14.2$ feet and $\angle B = 11°$.

It is helpful to draw the picture first, as in Figure 2–12. We immediately hav $\angle A = 90° - 11° = 79°$. From $b/a = \tan B$, we obtain $b = a \tan B = 14.2 \tan 11° =$ $(14.2) \cdot (0.1944) = 2.760$ feet.

Since $c/a = \sec B$, we have $c = a \sec B = 14.2 \sec 11° = (14.2) \cdot (1.019) =$ 14.47 feet. ■

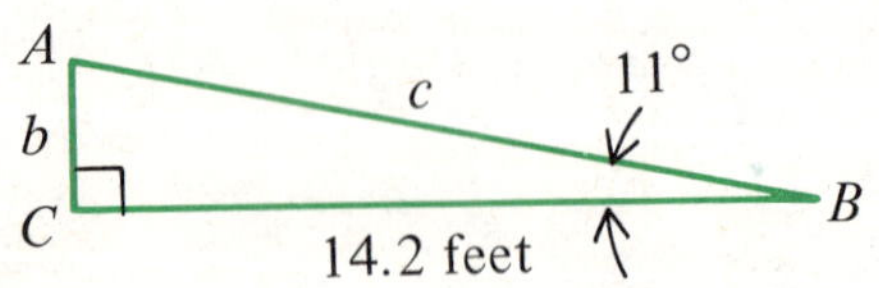

Figure 2–12

EXAMPLE: Solve right $\triangle ABC$, where $a = 3$ and $b = 7$.

Because $\tan A = a/b$, we have $\tan A = 3/7$. Rounding to four significant digits, we obtain

$$\tan A = 0.4286$$

We look for 0.4286 in the tangent columns of the table inside the front cover. The closest entry is $0.4245 = \tan 23°$, and so $\angle A = 23°$, correct to the nearest degree. We then have $\angle B = 67°$.

To obtain c, we might use the relation $\csc A = c/a$, and so $c = a \csc A$. Another possibility, more convenient in this case, is to use the Pythagorean formula $a^2 + b^2 = c^2$. Thus

$$c^2 = 3^2 + 7^2 = 9 + 49 = 58$$

and so

$$c = \sqrt{58} = 7.616$$ ■

The results of the preceding example, correct to the nearest degree, might not be accurate enough for some applications. In the example, we had $\tan A = 0.4286$ and we referred to the table inside the front cover to find $\angle A$. Let us instead use the more complete table of Trigonometric Ratios of Appendix B. The closest value of tan is $0.4279 = \tan 23°\ 10'$, and so $\angle A = 23°\ 10'$, correct to the nearest ten minutes. The corresponding value for $\angle B$ is $90° - 23°\ 10' = 66°\ 50'$.

Still more accuracy can be obtained by interpolation (see Section A–5). Using interpolation and the table Trigonometric Ratios of Appendix B, our results are $\angle A = 23°\ 12'$ and $\angle B = 66°\ 48'$.

The degree of accuracy which should be sought depends on several factors.

1. *The given data.* Our answer cannot reasonably be expected to have more significant digits than the input data (which may, for example, be physical measurements). For this purpose, an angle measurement expressed to the nearest degree is considered to have two significant digits. Angles expressed to the nearest ten minutes have three significant digits, and angles given to the nearest minute have four.

2. *The accuracy of the tables.* Obviously the accuracy of our results is limited by the number of significant digits for values found in a table.

3. *The computational method.* Rapid hand calculation will not yield four-place accuracy. Nor is a slide rule dependable to more than two significant figures. Working with logarithms, using four-place tables, one can easily obtain three significant digits, and a fourth is available by interpolation. With a calculator, you can readily get four significant digits, and you may get eight if your calculator is sophisticated enough.

4. *The usage desired.* It is pointless to struggle for the fourth significant digit when you are finding the length, in feet, of a ladder.

Our policy in this text will be to give the maximum accuracy, consistent with the given data and obtained from the tables inside the front cover, with no interpolation.[4] Your teacher will advise you what accuracy is expected in your solutions to the assigned numerical problems.

Another application of trigonometry is to find the area of a right triangle. For any triangle, the area is given by

$$\text{area} = \tfrac{1}{2} \cdot (\text{base}) \cdot (\text{altitude})$$

The symbol for area of a triangle is K. Therefore, for right triangle $\triangle ABC$, we have the formula

$$K = \tfrac{1}{2}ab$$

We also know that $a = c \sin A$. Thus we have the alternate formula

$$K = \tfrac{1}{2}(c \sin A)b$$

or

$$K = \tfrac{1}{2}bc \sin A$$

Hence the area is one-half the product of two sides and the sine of the angle included between them. Other expressions for the area are given in the following formulas, which you are asked to check in the exercises.

$$K = \tfrac{1}{2}ac \sin B$$
$$K = \tfrac{1}{2}c^2 \sin A \cos A$$
$$K = \tfrac{1}{2}c^2 \sin A \sin B$$

EXAMPLE: The legs of a right triangle are 10 feet and 15 feet, respectively. What is the area of the triangle?

From the formula $K = \tfrac{1}{2}ab$, we find that the area is $\tfrac{1}{2} \cdot 10 \cdot 15 = 75$ square feet. ■

EXAMPLE: In right $\triangle ABC$, we have that $\angle A = 23°$ and the hypotenuse is 14 inches. What is the area?

From the formula $K = \tfrac{1}{2}c^2 \sin A \cos A$, we see that

$$\begin{aligned} K &= (\tfrac{1}{2}) \cdot (14)^2 \cdot \sin 23° \cos 23° \\ &= (\tfrac{1}{2}) \cdot (196) \cdot (0.3907) \cdot (0.9205) = 35.24 \text{ square inches} \end{aligned}$$ ■

2–4 *Exercises*

Given in right triangle $\triangle ABC$ that $\angle A = 49°$ and $c = 23$ units

1. Find $\angle B$
2. Find a
3. Find b

[4] An exception to this occurs in Appendix A. There, when four-place data are given, four-place results are expected.

Given that $a = 37.1$ and $\angle B = 33°$

4. Find $\angle A$
5. Find b
6. Find K

Given that $a = 11$ and $b = 14$

7. Find $\angle A$
8. Find c
9. Find K

Given that $\angle A = 38°$ and $a = 2380$ yards

10. Find b
11. Find c
12. Find K

Given that $\angle A = 59°$ and $b = 62.7$ meters

13. Find $\angle B$
14. Find a
15. Find c

Given that $a = 98.2$ feet and $c = 142$ feet

16. Find $\angle A$
17. Find b
18. Find K

19. Suppose you are given the measures of $\angle B$ and side b in right triangle $\triangle ABC$. Find formulas which express the unknown measures in this triangle in terms of b and $\angle B$.
20. Assume the measures of side b and the hypotenuse c are known in right triangle $\triangle ABC$. Find a formula for the area of the triangle in terms of b and c.

Derive the following alternate formulas for the area of right triangle $\triangle ABC$,

21. $K = \frac{1}{2}ac \sin B$
22. $K = \frac{1}{2}c^2 \sin A \cos A$
23. $K = \frac{1}{2}c^2 \sin A \sin B$

2-5 Applications

Many practical problems can be solved by the trigonometry of a right triangle. In these applications, as with all word problems, it is helpful to draw a picture whenever possible.

EXAMPLE: The top of a ladder 16 feet long is leaning against the top of a vertical wall. If the ladder makes an angle of 40° with the ground, how high is the wall?

We first make a sketch of the situation, as in Figure 2–13. Because the wall is vertical, we recognize that a right triangle exists in Figure 2–13 and label it as in Figure 2–14. The height of the wall is a.

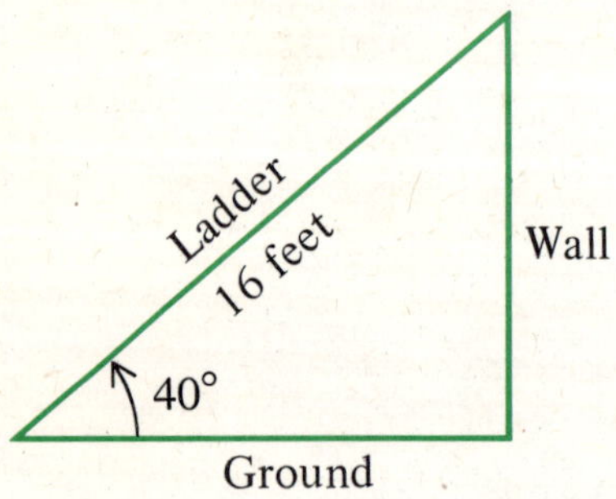

Figure 2–13

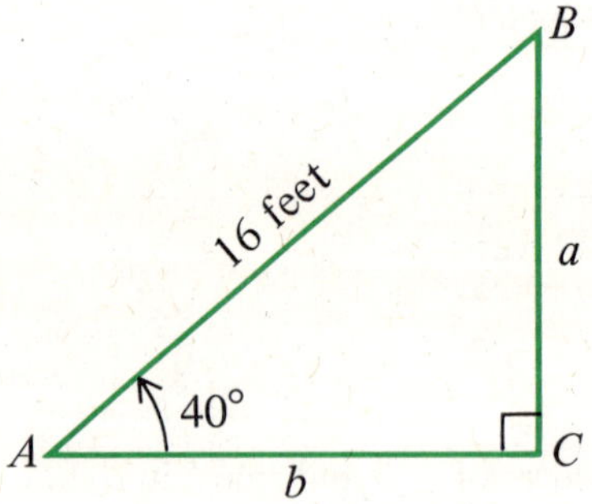

Figure 2–14

From $\triangle ABC$, we have $a/c = \sin A$, and so $a = c \sin A = 16 \sin 40° = (16) \cdot (0.6428) = 10.3$. Thus the wall is 10.3 feet high. ■

In the next example, reference is made to an *angle of elevation*. For a point located higher than the observer, this is the angle between the line of sight to that point and the horizontal (see Figure 2–15). When the point is lower than the observer, this angle is known as the *angle of depression*.

EXAMPLE: A flagpole casts a shadow 15 meters long at a time when the angle of elevation of the sun is 34°. How tall is the flagpole?

A picture is sketched in Figure 2–16. From right $\triangle ABC$, we have $a/b = \tan A$, and so $a = b \tan A = 15 \tan 34° = (15) \cdot (0.6745) = 10.1$. Thus the flagpole is 10.1 meters high. ■

In navigation, the direction of a course is given in degrees, measured *clockwise* from the north. The direction of a course that points toward a specific location is known as the *bearing* of that location (from the observer).

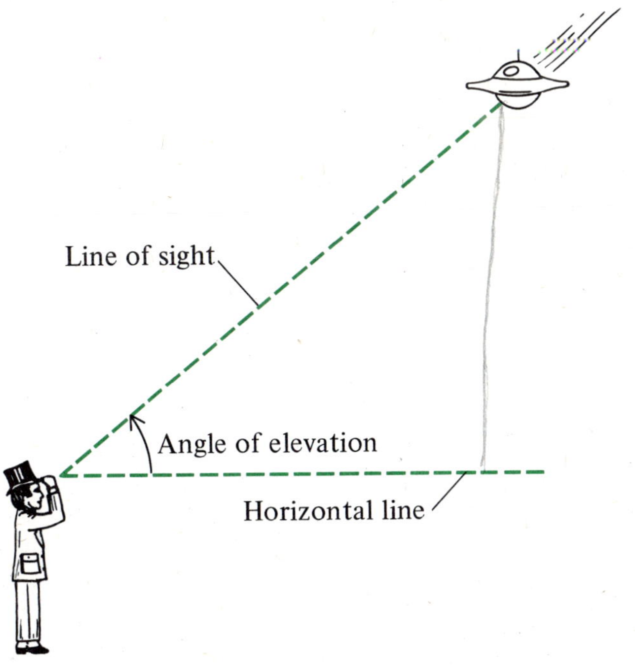

Figure 2–15
Angle of elevation

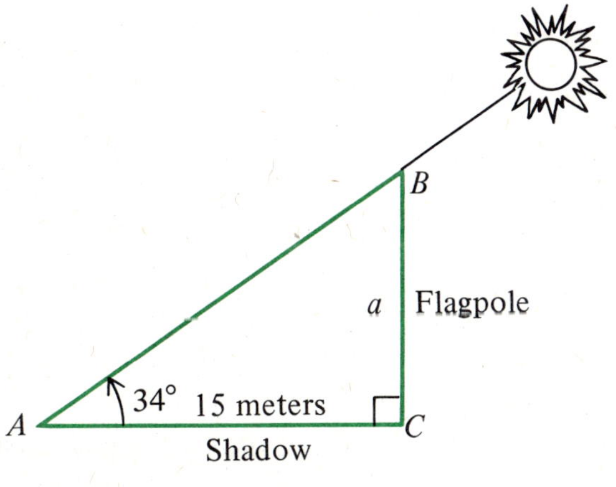

Figure 2–16

EXAMPLE: A marker is located 1000 yards due north of the dock. A sailboat leaves the dock on a course of 340°. How far has the boat gone when the bearing of the marker is 90°?

The situation is illustrated in Figure 2–17. The boat starts at the dock A, with the marker C to the north. When the boat reaches point B, the marker is due east of the boat, and so its bearing is 90°. Thus, in right $\triangle ABC$, the distance traveled by the boat is c.

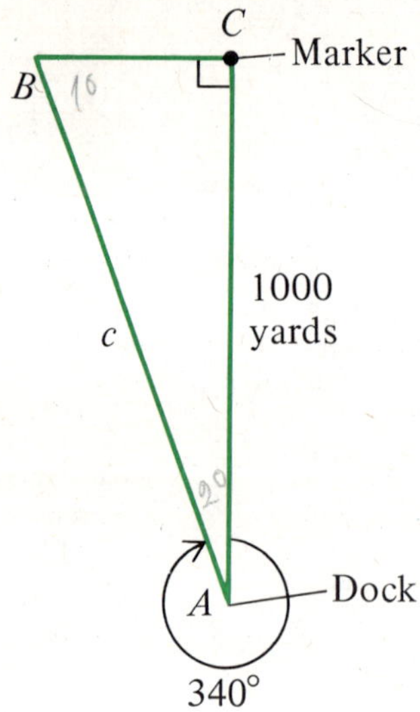

Figure 2–17

From $\triangle ABC$, we have

$$\frac{c}{b} = \sec A$$

and, in the triangle, we have $\angle A = 360° - 340° = 20°$. Therefore

$$\begin{aligned} c &= b \sec A \\ &= 1000 \sec 20° \\ &= (1000) \cdot (1.064) \\ &= 1064 \end{aligned}$$

Thus the boat has traveled 1064 yards from the dock. ■

In physics, a quantity which must be described by its direction as well as by its size or magnitude is known as a *vector*. For example, wind velocity is a vector because to describe it, we need to know both the speed and the wind direction.

Geometrically, a vector is represented by an arrow. The sum, or *resultant*, of two vectors also has a geometric interpretation.

The sum of two vectors α and β is illustrated in Figure 2–18.[5] Here are the steps for addition of vectors.

1. Draw α.
2. Draw β, beginning β where α leaves off.
3. Complete a triangle with an arrow going from the beginning of α to the end of β. This arrow, which is shown dashed in Figure 2–18, represents the resultant vector $\alpha + \beta$.

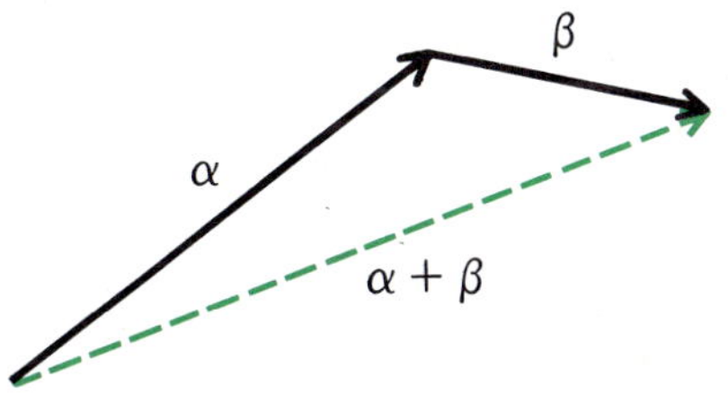

Figure 2–18
Addition of vectors

EXAMPLE: An airplane flies with an air speed of 120 miles per hour due North, but in a 40 mile per hour crosswind blowing from West to East. In what direction, and how fast, is the plane flying with respect to the ground?

The ground speed is represented by the resultant of the air speed vector and the wind velocity vector, as shown in Figure 2–19.

In right triangle $\triangle ABC$, we have

$$c = \sqrt{40^2 + 120^2} = \sqrt{16000} = 126.5$$

The angle $\angle A$ satisfies

$$\tan A = \tfrac{40}{120} = 0.3333$$

and so $\angle A = 18°$. Hence, with respect to the ground, the plane is traveling on a course of 18° at a speed of 126.5 miles per hour. ■

2–5 *Exercises*

1. A ladder is leaning against the side of a building. The ladder makes an angle of 30° with the ground, and the base of the ladder is 15 feet from the building. How long is the ladder?

[5] The symbols α and β are the Greek letters *alpha* and *beta*; respectively.

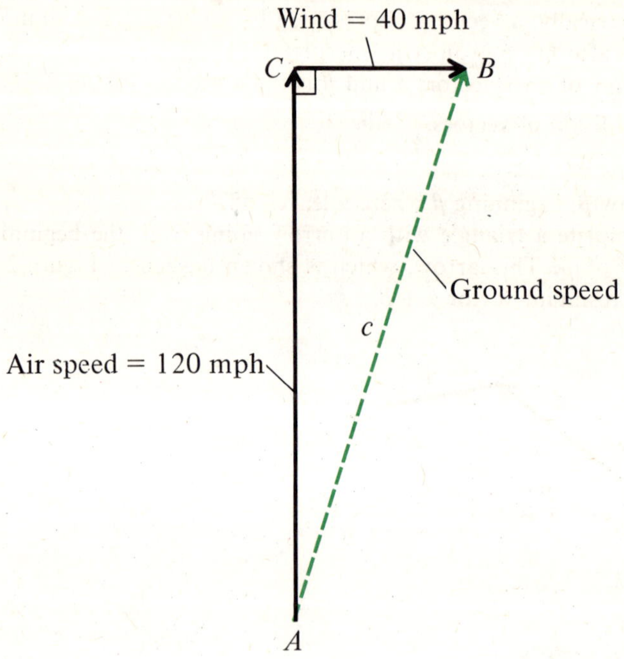

Figure 2–19

2. The angle of elevation of the top of a tower is 47°. If the tower is known to be 325 feet tall, how far away from it is the observer?

3. Frank stands at the top of the Washington Monument, 555 feet high. The angle of depression to his friend Thomas on the ground is 73°. How far is Thomas from the center of the base of the monument?

4. To determine the distance between two points P and Q on opposite sides of the river, the surveyor follows these steps. He starts by standing with his transit at P, sights across to point Q, and turns the transit through 90° so that he is sighting another point R. He measures the distance from P to R to be 312 feet. Then he takes his transit to R and measures $\angle PRQ$ as 43°. Find the distance from P to Q.

5. From a helicopter hovering 400 feet over a floating space capsule, the angle of depression of the water line at the bow of the recovery ship is 77°. How far is the capsule from the bow?

6. For safe use, the manufacturer recommends that the maximum angle made by a ladder with the ground be 75°. If the ladder is 24 feet long, how high above the ground can its top reach safely?

7. A sailboat on a course of 270° passes a buoy 4000 yards due south of the lighthouse and continues on course. When the bearing of the lighthouse is 54°, how far is it from the boat?

8. A pilot is planning his flight from Springfield to Lincolnville, 83 miles away at a bearing of 338°, but on the other side of the mountain. To avoid the mountain, he is going to start on a course of 18° until a 90° turn will head him directly to Lincolnville. How far should he go before he turns?

9. A television transmitting tower is located on top of a 1000-foot tall building. From a point a block away, Geoffrey measures the angle of elevation of the top of the building to be 72°, and that of the top of the tower is 75°. How high is the top of the transmitter?

10. Chuck, from the window of his fourth-floor apartment, measures the angle of depression of the base of the office building across the street, 60 feet away, to be 34°, and the angle of elevation of the top as 49°. How tall is the office building?

11. A microwave tower 80 feet high stands on the edge of a cliff. From the valley below, Steve measures the angle of elevation of the base of the tower to be 64°, and the angle of elevation of the top is 70°. How far is Steve from the base of the cliff?

12. A cruise ship doing 8 knots (i.e., moving at a rate of 8 nautical miles per hour) on a course of 122° measures the bearing of a lighthouse as 80°. An hour later, the bearing is 69°. If the ship continues on course, what will be the bearing of the lighthouse when the ship is closest to it? How far away will it be?

13. Angelo's remote-controlled toy motorboat is capable of going 8 miles per hour in still water. He points the boat to go directly across the river, but Angelo neglects to take account of the current, which is flowing at the rate of 5 miles per hour. What angle will the boat's actual course make with the riverbank? How fast will the boat travel?

14. Linda plans to fly her twin-engine plane at an air speed of 160 miles per hour. She wants to go due West over the ground, but the wind is blowing from North to South at a speed of 20 miles per hour. What will be her ground speed?

Review exercises (Chapter 2)

In $\triangle ABC$, given that $a = 2$, $b = 7$, and $\angle C = 90°$, find

1. $\sin A$
2. $\tan A$
3. $\sec A$
4. $\sin B$
5. $\tan B$
6. $\sec B$

Evaluate without reference to the table of Trigonometric Ratios:

7. $\sin 30°$
8. $\sin 60°$
9. $\sin 45°$

Complete the identities:

10. $\tan A = (\sin A)/?$

11. $\csc (90° - A) = ?$

12. $\sin^2 A + \cos^2 A = ?$

Use the table inside the front cover to evaluate

13. $\cos 57°$

14. $\tan 40°$

Given in $\triangle ABC$ that $\angle A = 33°$, $b = 15.3$, and $\angle C = 90°$,

15. find a

16. find the area of $\triangle ABC$

17. A ramp 25 feet long leads from ground level to the entrance to the library 10 feet above. What angle does the ramp make with the horizontal?

18. Alice, looking out a window of the humanities tower, measures the angle of depression of the base of the statue on the lawn outside to be 60°. From the floor below, ten feet lower, Kathleen finds that the angle of depression is 55°. How far is the statue from the building?

3

Trigonometric functions

The modern approach to trigonometry is an approach that is more sophisticated than that of studying the ratios in a right triangle. For applications to calculus and other branches of advanced mathematics, as well as physics and engineering, trigonometric *functions* are required. The development of these functions is the main goal of this chapter.

As so often is the case, the geometry of the situation is especially important. In fact, modern trigonometry can be interpreted as a means of working with geometric concepts like angle, rotation, and arc length in the framework of algebra.

3-1 The distance formula and the equation of a circle

Let (x_1, y_1) and (x_2, y_2) be two points in the plane. Our object is to find a formula for the distance d between them. See Figure 3–1.

We form a right triangle by connecting each of these points to the point (x_2, y_1), as shown in Figure 3–2.

The length of the horizontal leg of this right triangle is $|x_2 - x_1|$, and the

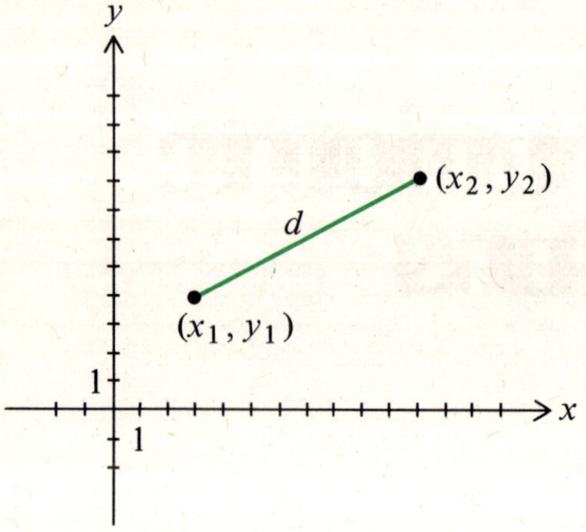

Figure 3–1

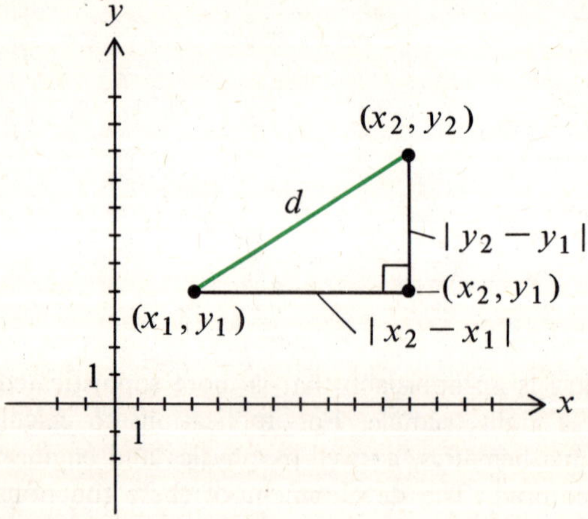

Figure 3–2

length of the vertical leg is $|y_2 - y_1|$. The length of the hypotenuse of this triangle is the distance d.

According to the Pythagorean theorem,

$$d^2 = |x_2 - x_1|^2 + |y_2 - y_1|^2$$

Because either $|x_2 - x_1| = x_2 - x_1$ or $|x_2 - x_1| = -(x_2 - x_1)$, we have

$$|x_2 - x_1|^2 = (x_2 - x_1)^2$$

Similarly,

$$|y_2 - y_1|^2 = (y_2 - y_1)^2$$

Thus,

$$d^2 = (x_2 - x_1)^2 + (y_2 - y_1)^2$$

Because the distance d cannot be a negative number, we conclude that

$$d = \sqrt{(x_2 - x_1)^2 + (y_2 - y_1)^2}$$

To summarize, we have the following *distance formula*:

The distance d between the two points (x_1, y_1) and (x_2, y_2) is given by

$$d = \sqrt{(x_2 - x_1)^2 + (y_2 - y_1)^2}$$

EXAMPLE: To find the distance d between (2, 3) and (6, 4), we use the formula. Therefore,

$$d = \sqrt{(6 - 2)^2 + (4 - 3)^2} = \sqrt{16 + 1} = \sqrt{17} \blacksquare$$

EXAMPLE: We find the distance d between $(-6, -1)$ and $(-3, 5)$ to be

$$d = \sqrt{[-3 - (-6)]^2 + [5 - (-1)]^2} = \sqrt{9 + 36} = \sqrt{45} = 3\sqrt{5} \blacksquare$$

In Exercise 7 you are asked to verify that the distance d between (x_1, y_1) and (x_2, y_2) can also be expressed as

$$d = \sqrt{(x_1 - x_2)^2 + (y_1 - y_2)^2}$$

Therefore, in using the distance formula, it makes no difference which point is labelled the first and which the second point.

EXAMPLE: We again find the distance d between $(-6, -1)$ and $(-3, 5)$.

$$d = \sqrt{[-6 - (-3)]^2 + (-1 - 5)^2} = \sqrt{9 + 36} = \sqrt{45} = 3\sqrt{5}$$

This is the same as the distance between $(-6, -1)$ and $(-3, 5)$ found above. ■

A useful special case occurs when one of the two points is the origin. In Exercise 8 you are asked to check that:

The distance d from the point (x, y) to the origin is given by $d = \sqrt{x^2 + y^2}$.

EXAMPLE: The distance d from the point $(-3, 4)$ to the origin is given by $d = \sqrt{(-3)^2 + 4^2} = \sqrt{9 + 16} = \sqrt{25} = 5$. ■

EXAMPLE: The distance d from $(1, -1)$ to the origin is found to be $d = \sqrt{1^2 + (-1)^2} = \sqrt{1 + 1} = \sqrt{2}$. ■

We now are ready to find the equation of a circle (with center at the origin).

Geometrically, a circle is determined by a point (the center) and a positive number r (the radius). The circle consists of those points in the plane which are at distance r from the center (see Figure 3–3). The center itself is not a point of the circle.

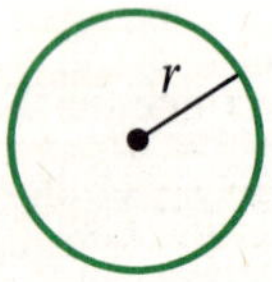

Figure 3–3
A circle of radius r

We want to translate the geometric description of the circle into an algebraic equation satisfied by the points on the circle. We assume that the center of the circle is at the origin. Then the point (x, y) lies on the circle if and only if its distance from the origin is r. (See Figure 3–4.)

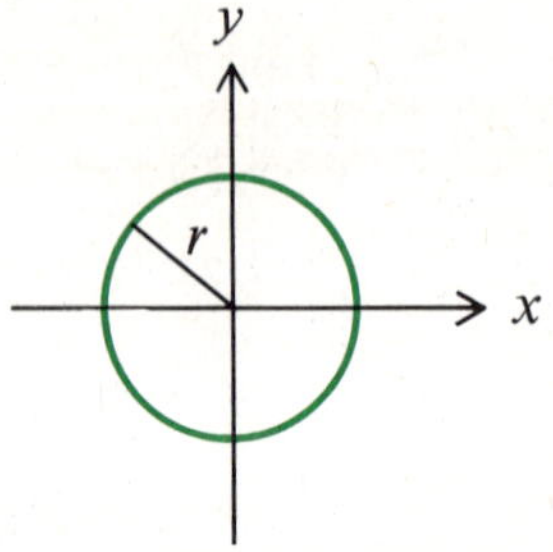

Figure 3–4

Using the distance formula, we see that the distance from (x, y) to the origin is r if and only if

$$r = \sqrt{x^2 + y^2}$$

It follows that $r^2 = x^2 + y^2$. On the other hand, if we are given an equation of form $r^2 = x^2 + y^2$, where r is positive, we have

$$r = \sqrt{x^2 + y^2}$$

Therefore, we may state the following results:

Let r be a positive number. Then the graph of the equation $x^2 + y^2 = r^2$ is a circle of radius r with its center at the origin.

EXAMPLE: To find the graph of the equation $x^2 + y^2 = 9$ we observe that this equation is of the form $x^2 + y^2 = r^2$ with $r = 3$. Therefore, the graph is a circle with center at the origin and radius 3. The graph is shown in Figure 3–5. ■

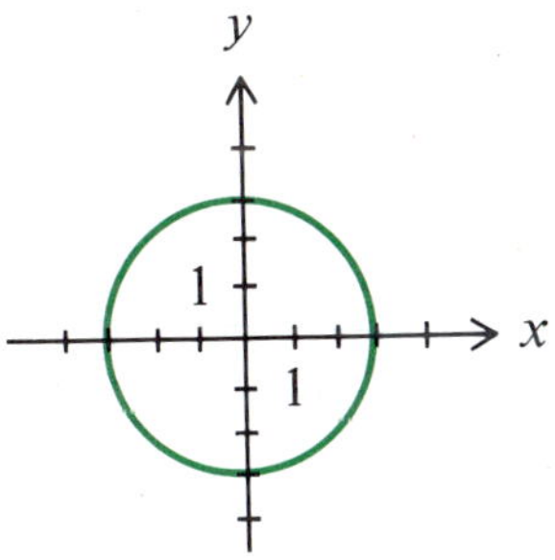

Figure 3–5
$\{(x, y) | x^2 + y^2 = 9\}$

EXAMPLE: We find the equation of the circle with center at the origin and radius $\sqrt{7}$ to be $x^2 + y^2 = (\sqrt{7})^2$ or $x^2 + y^2 = 7$. ■

EXAMPLE: Let us graph the equation

$$x^2 + y^2 = -1$$

This equation cannot be put in the form $x^2 + y^2 = r^2$ for any positive number r, because -1 is not the square of any positive number. In fact, for any pair (x, y) of real numbers, we have $x^2 + y^2 \geq 0$. It follows that the solution set to the equation $x^2 + y^2 = -1$ is the null set or $\varnothing$. ■

EXAMPLE: The graph of the equation

$$x^2 + y^2 = 1$$

is a circle of radius 1 with center the origin. See Figure 3–6. This circle will have special importance in our study of trigonometry; it is known as the *unit circle*. ■

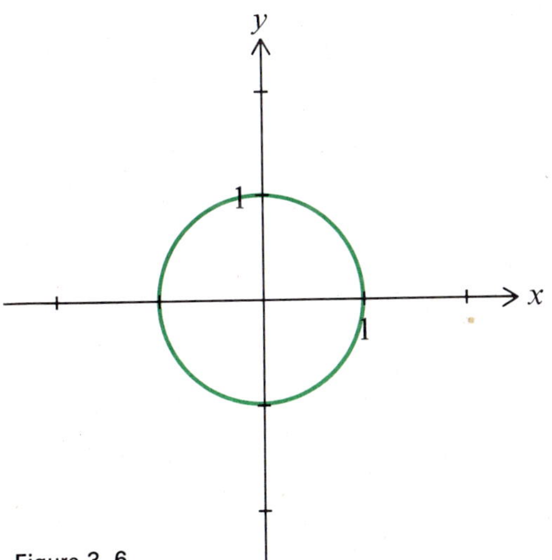

Figure 3–6
The unit circle

It should be observed that the equation of a circle does not define y as a function of x. As an example, consider the circle having the equation $x^2 + y^2 = 25$, as shown in Figure 3–7. It is not true that each value of x determines a unique value of y. For each x such that $-5 < x < 5$, there are two points on the graph, one above the x-axis and one below. Thus, as we see in Figure 3–8, there are two points on the graph which lie above the point $x = 3$ on the x-axis. These points are (3, 4) and (3, −4). Thus, the graph does not determine y as a function of x.

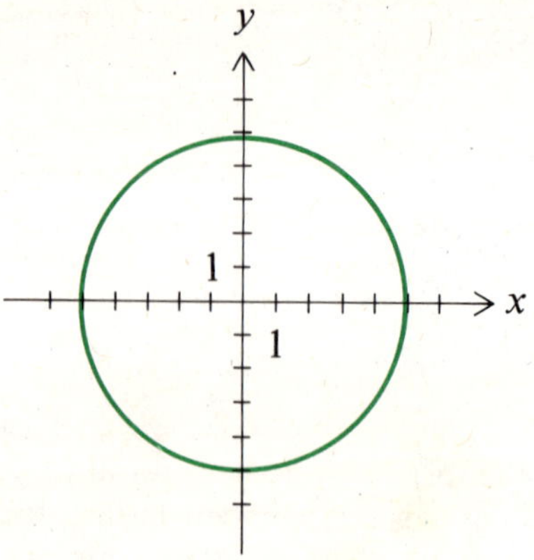

Figure 3–7
$\{(x, y) | x^2 + y^2 = 25\}$

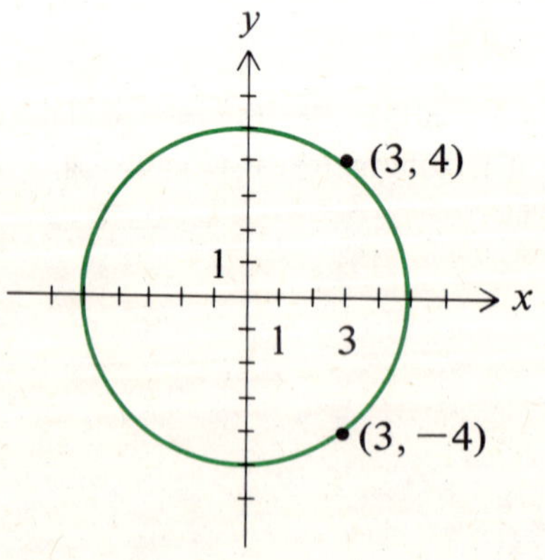

Figure 3–8

The same observation could be made algebraically without using the graph. To obtain y as a function of x, we would try to solve the equation $x^2 + y^2 = 25$ for y. The steps would be

$$x^2 + y^2 = 25$$
$$y^2 = 25 - x^2$$
$$y = \pm\sqrt{25 - x^2}$$

Thus we obtain two values for y corresponding to each x such that $25 - x^2 > 0$. As a result, the equation

$$x^2 + y^2 = 25$$

does not define y as a function of x.

3–1 *Exercises*

Find the distance between the given points.

1. (5, 6) and (1, 3)
2. (3, 4) and (7, 9)
3. (1, −2) and (−2, 3)
4. (0, −6) and (3, 0)
5. (−1, −3) and (−4, −5)
6. (3, −2) and (2, −5)

7. Verify that the distance d between (x_1, y_1) and (x_2, y_2) can also be expressed as

$$d = \sqrt{(x_1 - x_2)^2 + (y_1 - y_2)^2}$$

8. Use the distance formula with $(x_1, y_1) = (x, y)$ and $(x_2, y_2) = (0, 0)$ to verify the formula for the distance between (x, y) and the origin.

Find the distance from the origin to the given point.

9. (−3, 4)
10. (7, −24)
11. (3, 0)
12. (0, −6)
13. (1, −2)
14. (−2, −2)
15. (−12, −5)
16. (6, 2)

Find the equation of the circle with center at the origin and with the radius given.

17. 2
18. 11
19. 8
20. $\sqrt{3}$
21. $\sqrt{15}$
22. $\sqrt{7}$

Sketch the graph.

23. $x^2 + y^2 = 36$
24. $x^2 + y^2 = 0$
25. $x^2 + y^2 = 16$
26. $x^2 + y^2 = -4$
27. $x^2 + y^2 = 17$
28. $x^2 + y^2 = 50$

29. Which of the equations given in Exercises 23 through 28 define y as a function of x?

30. **(a)** Use the distance formula to express the distance from (x, y) to $(3, 5)$.
(b) What equation must (x, y) satisfy if its distance from $(3, 5)$ is 6?
(c) What is the equation of the circle of radius 6 with center at $(3, 5)$?

31. **(a)** Use the distance formula to express the distance from (x, y) to (x_0, y_0).
(b) What equation must (x, y) satisfy if its distance from (x_0, y_0) is r?
(c) What is the equation of the circle of radius r with center at (x_0, y_0)?

3-2 Angles and rotation

Our goal is to define the trigonometric ratios for all angles, not just for the angles that occur as the acute angles in a right triangle. To do this, we need a good understanding of angle measurement. The key to that is *rotation.*

Let us, then, rotate the plane, with the origin kept fixed as the center of rotation. The positive half of the x-axis will be moved by the rotation to a new position, as shown in Figure 3–9.

We think of the positive x-axis as the *initial ray* of the rotation. In its final position, this axis becomes the *terminal ray* of the rotation. The angle formed by these rays is called the angle produced by the rotation.

We can measure the rotation in degrees, according to this basic principle:

one complete revolution = 360°

The measure of the rotation is positive if the direction is counterclockwise. For a clockwise rotation, the measure is negative.

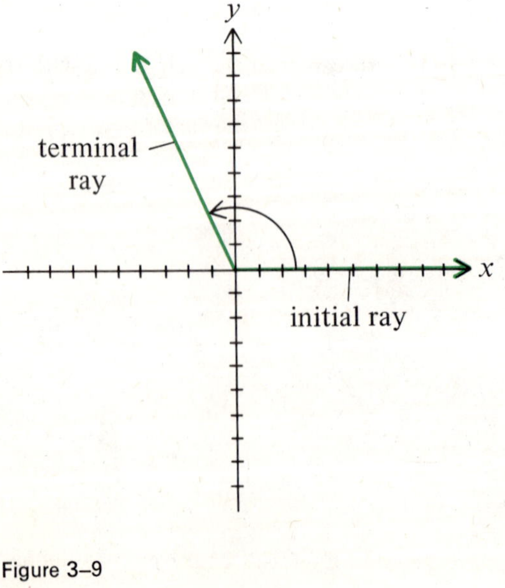

Figure 3–9

EXAMPLE: The rotation illustrated in Figure 3–10 has measure 90°. ■

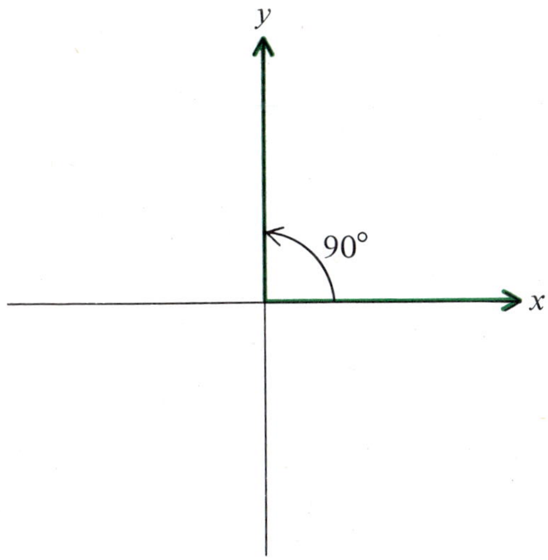

Figure 3–10
A rotation of measure 90°

EXAMPLE: The rotation shown in Figure 3–11 has measure −135°. ■

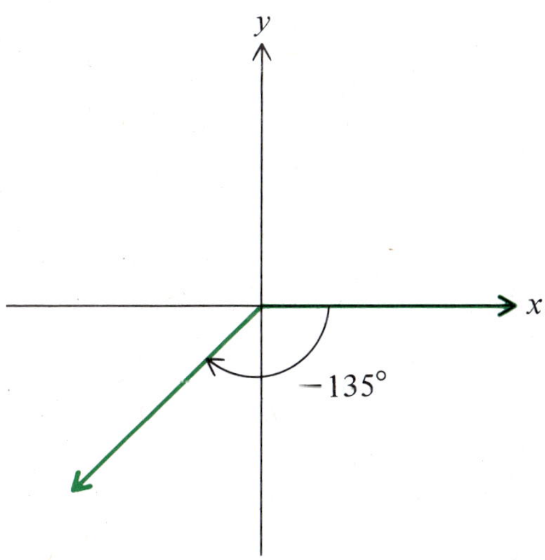

Figure 3–11
A rotation of measure −135°

EXAMPLE: In Figure 3–12 we see a rotation of 540°. ■

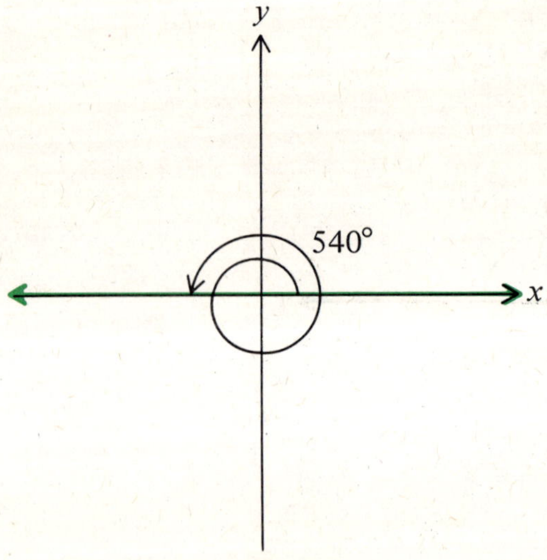

Figure 3–12
A rotation of measure 540°

EXAMPLE: A rotation of 60° is shown in Figure 3–13. ■

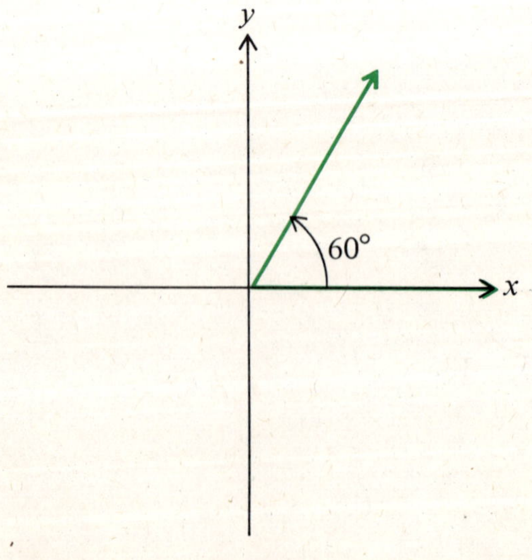

Figure 3–13
A rotation of measure 60°

Besides using degree measure, we can measure rotation in a different unit, the *radian*. For theoretical purposes, radian measure is almost always preferable. Here is the way it is defined.

Let Q be the point (1, 0) on the positive x-axis. During the rotation the point Q always remains at distance one from the origin. Therefore Q traces out an arc of the unit circle, as shown in Figure 3–14. We then use the length of this arc to make the following definition.

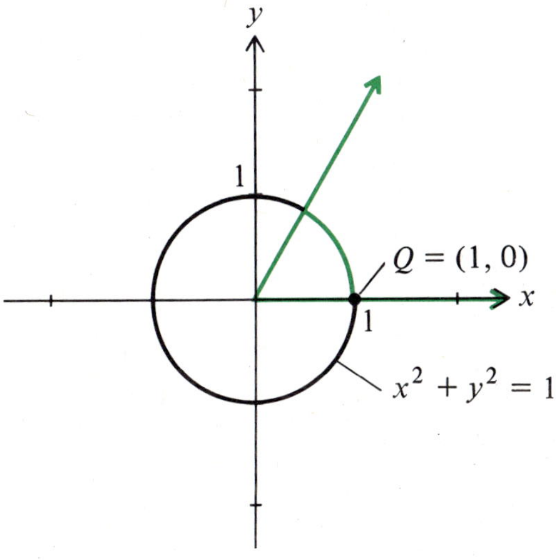

Figure 3–14

For a counterclockwise rotation the length of the arc traced out by Q is the *radian measure* of the rotation.

The radian measure of a clockwise rotation is the negative of the length of the arc traced out by Q.

We know from geometry that the formula for the circumference C of a circle of radius r is

$$C = 2\pi r$$

Because the point Q is on the unit circle (radius 1), it follows that in one revolution Q travels through 2π units of arc length. Thus we have the fundamental relationship:

$$2\pi \text{ radians} = 1 \text{ revolution}$$

Because 1 revolution = 360°, we have the basic conversion formula

$$2\pi \text{ radians} = 360°$$

To convert from degree measure to radian measure, it is usually most convenient to use the following equation, derived from the conversion formula by division by 360.

$$1° = \frac{\pi}{180} \text{ radians}$$

EXAMPLE: We convert 45° to radians as follows.

$$45° = 45 \cdot 1° = 45 \cdot \left(\frac{\pi}{180}\right) \text{ radians} = \frac{45\pi}{180} \text{ radians}$$

$$= \frac{\pi}{4} \text{ radians. } \blacksquare$$

EXAMPLE: $-60° = -60 \cdot \left(\frac{\pi}{180}\right) \text{ radians} = -\frac{\pi}{3} \text{ radians } \blacksquare$

To convert from radians to degrees, the process is similar, but we use the following equation, obtained from the basic conversion formula by division by 2π.

$$1 \text{ radian} = \left(\frac{180}{\pi}\right)^{\circ}$$

Substituting the approximate value $\pi = 3.142$, we find that one radian is approximately fifty-seven degrees.

EXAMPLE: We express $-4\pi/3$ radians in degrees.

$$-\frac{4\pi}{3} \text{ radians} = -\frac{4\pi}{3} \cdot 1 \text{ radian} = -\frac{4\pi}{3} \cdot \left(\frac{180}{\pi}\right)^{\circ}$$

$$= -240° \blacksquare$$

Earlier in this section the measures of several rotations were given in degrees. The corresponding radian measure is found in each of the following examples.

EXAMPLE: The rotation shown in Figure 3–10 has measure

$$90° = 90 \cdot \left(\frac{\pi}{180}\right) \text{ radians} = \frac{\pi}{2} \text{ radians } \blacksquare$$

EXAMPLE: The rotation shown in Figure 3–11 has measure

$$-135° = -135 \cdot (\pi/180) \text{ radians} = -3\pi/4 \text{ radians } \blacksquare$$

EXAMPLE: In Figure 3–12, we saw a rotation of measure

$$540° = 540 \cdot (\pi/180) \text{ radians} = 3\pi \text{ radians } \blacksquare$$

EXAMPLE: The rotation of Figure 3–13 has measure

$$60° = 60 \cdot (\pi/180) \text{ radians} = \pi/3 \text{ radians } \blacksquare$$

When degree measure is used, the symbol for degrees is needed to so indicate. *When no unit of measure is specified, radian measure is understood.*

EXAMPLE: The measure of the rotation shown in Figure 3–13 is $\pi/3$. ■

Now we are ready to define the measure of any angle, as follows.

The measure of an angle produced by rotation is defined to be the same as the measure of the rotation itself.

This measurement may be in degrees or in radians. It may be positive or negative, depending on the direction of the rotation.

EXAMPLE: The angle produced by rotation shown in Figure 3–15 has measure 405°, or $9\pi/4$. ■

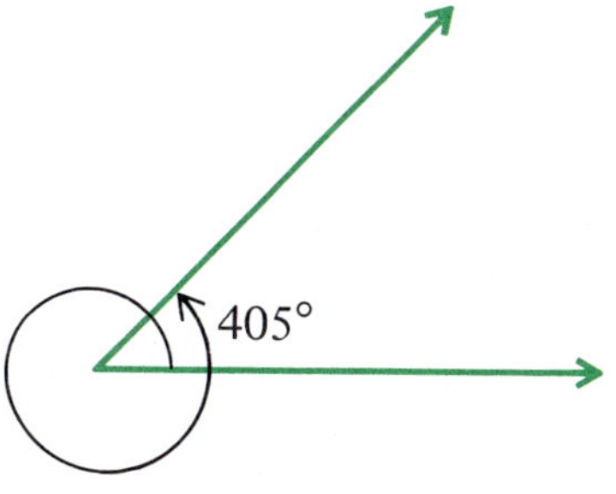

Figure 3–15
An angle of measure 405°

EXAMPLE: The angle illustrated in Figure 3–16 is produced by rotation of $-30°$. Therefore the measure of this angle is $-30°$, or $-\pi/6$. ■

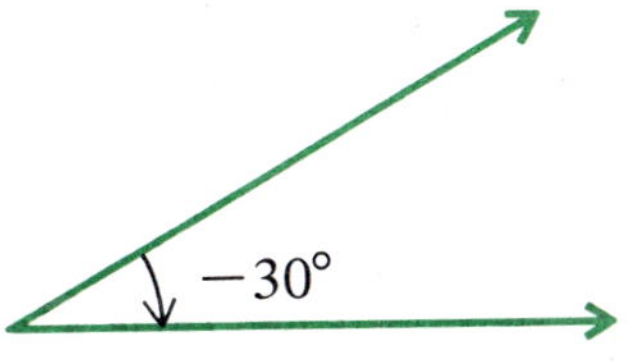

Figure 3–16
An angle of measure $-30°$

For an angle formed by two rays without use of rotation, we assume that the angle was produced by the smallest possible positive rotation.

EXAMPLE: The angle shown in Figure 3–17 has measure 30°, or $\pi/6$. Since no rotation is specified, we assume that this angle was produced by the smallest possible positive rotation. ■

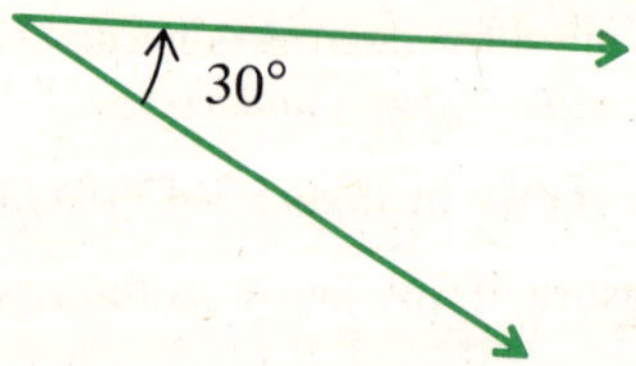

Figure 3–17
An angle of measure 30°

3–2 *Exercises*

Sketch the initial ray and terminal ray for the following rotations.

1. 45° **2.** $-45°$
3. 180° **4.** 0°
5. $-120°$ **6.** 150°
7. π **8.** $\pi/3$
9. 2π **10.** $-\pi$
11. $-2\pi/3$ **12.** $-3\pi/4$

Express in degrees
13. $\frac{1}{2}$ revolution
14. 2/3 revolution
15. $5\pi/4$
16. $2\pi/5$
17. $\pi/10$
18. $-3\pi/10$

Express in radians
19. 3/4 revolution
20. 4 revolutions
21. 5°
22. $-15°$
23. 210°
24. $-270°$

Sketch an angle with the given measure.
25. 390°
26. $-330°$
27. 5π
28. $-8\pi/3$
29. 1°
30. 1

Let Q be the point (1, 0) on the initial ray of a rotation. Through how many units of arc length does Q travel on the unit circle in the given rotation?
31. 2π **32.** 2

33. -5π **34.** -5

35. $-72°$ **36.** $3/5$ revolution

37. $-1/12$ revolution **38.** $-54°$

3-3 Trigonometric ratios for angles

Now we are ready to define trigonometric ratios for any angle θ.[1]

The angle θ is first placed in *standard position*, with the positive x-axis as the initial ray of θ (see Figure 3–18). We then select $P = (x, y)$ to be any point, except for the origin, on the terminal ray of θ. The distance from the origin to P is denoted by r.

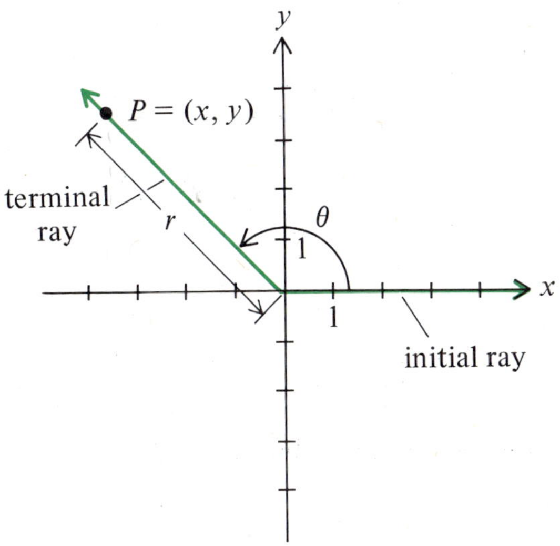

Figure 3–18
P is selected on the terminal ray

The trigonometric ratios of θ are then defined by:

$$\sin\theta = \frac{y}{r}$$

$$\cos\theta = \frac{x}{r}$$

$$\tan\theta = \frac{y}{x}, \quad x \neq 0$$

[1] The symbol θ is the Greek letter *theta*.

$$\cot \theta = \frac{x}{y}, \quad y \neq 0$$

$$\sec \theta = \frac{r}{x}, \quad x \neq 0$$

$$\csc \theta = \frac{r}{y}, \quad y \neq 0$$

EXAMPLE: **We find the trigonometric ratios for an angle of measure 135° ($=3\pi/4$ radians).**

An angle of measure 135° is shown in standard position in Figure 3–19, with the point $P = (-1, 1)$ selected on the terminal ray. By the distance formula (Section 3–1) we find that

$$r = \sqrt{x^2 + y^2} = \sqrt{(-1)^2 + 1^2} = \sqrt{1+1} = \sqrt{2}$$

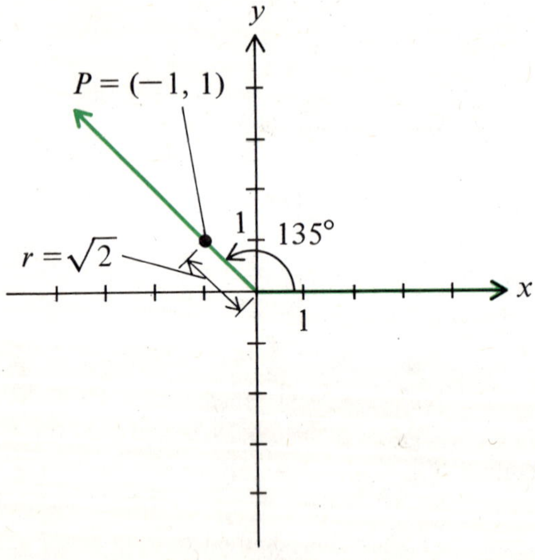

Figure 3–19

Thus

$$\sin 135° = \frac{y}{r} = \frac{1}{\sqrt{2}} = \frac{\sqrt{2}}{2}$$

$$\cos 135° = \frac{x}{r} = \frac{-1}{\sqrt{2}} = \frac{-\sqrt{2}}{2}$$

$$\tan 135° = \frac{y}{x} = \frac{1}{-1} = -1$$

$$\cot 135° = \frac{x}{y} = \frac{-1}{1} = -1$$

$$\sec 135° = \frac{r}{x} = \frac{\sqrt{2}}{-1} = -\sqrt{2}$$

$$\csc 135° = \frac{r}{y} = \frac{\sqrt{2}}{1} = \sqrt{2} \blacksquare$$

When we find the trigonometric ratios of θ, we might select a different point $P' = (x', y')$ on the terminal ray of θ, as in Figure 3–20. The same ratios will be obtained. In fact, from the similar triangles in Figure 3–20, we see that

$$\frac{y'}{r'} = \frac{y}{r}, \frac{x'}{r'} = \frac{x}{r}, \frac{y'}{x'} = \frac{y}{x}$$

and so on.

If the point P is such that $x = 0$, then $\tan \theta$ and $\sec \theta$ are undefined. If $y = 0$, then $\cot \theta$ and $\csc \theta$ are undefined.

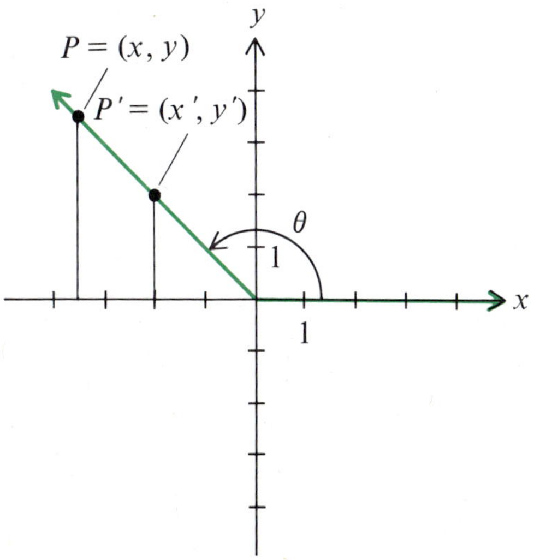

Figure 3–20
P and *P′* are on the terminal ray

EXAMPLE: We calculate the trigonometric ratios for an angle of measure 0 (= 0 radians = 0°).

The positive x-axis is both the initial and terminal ray when this angle is placed in standard position. We select the point $P = (1, 0)$, as in Figure 3–21 We then have $x = 1$, $y = 0$, and $r = 1$. Hence,

$$\sin 0 = \frac{y}{r} = 0$$

$$\cos 0 = \frac{x}{r} = 1$$

$$\tan 0 = \frac{y}{x} = 0$$

$$\cot 0 = \frac{x}{y} \text{ is undefined}$$

$$\sec 0 = \frac{r}{x} = 1$$

$$\csc 0 = \frac{r}{y} \text{ is undefined} \blacksquare$$

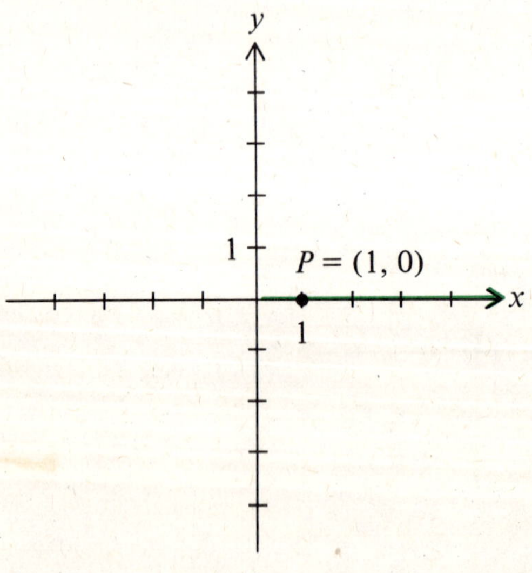

Figure 3–21

EXAMPLE: If the point $P = (-3, -4)$ lies on the terminal ray of θ, find the trigonometric ratios for θ. (We assume θ is in standard position.)

From the distance formula, we have

$$r = \sqrt{x^2 + y^2} = \sqrt{(-3)^2 + (-4)^2} = \sqrt{9 + 16} = \sqrt{25} = 5$$

Hence

$\sin\theta = -\frac{4}{5}$
$\cos\theta = -\frac{3}{5}$
$\tan\theta = \frac{4}{3}$
$\cot\theta = \frac{3}{4}$
$\sec\theta = -\frac{5}{3}$
$\csc\theta = -\frac{5}{4}$ ■

For an angle θ in standard position, the quadrant containing the terminal ray is also called the quadrant of θ. The signs of x, y, and r in the various quadrants are given in Figure 3–22. The signs of the trigonometric ratios of θ are as shown in Figure 3–23.

Quadrant	I	II	III	IV
x	+	−	−	+
y	+	+	−	−
r	+	+	+	+

Figure 3–22

Quadrant of θ	I	II	III	IV
$\sin\theta = \frac{y}{r}$	+	+	−	−
$\cos\theta = \frac{x}{r}$	+	−	−	+
$\tan\theta = \frac{y}{x}$	+	−	+	−
$\cot\theta = \frac{x}{y}$	+	−	+	−
$\sec\theta = \frac{r}{x}$	+	−	−	+
$\csc\theta = \frac{r}{y}$	+	+	−	−

Figure 3–23
Signs of trigonometric ratios

EXAMPLE: Find the sign of cos 230°.

Because 230° represents a third-quadrant angle, cos 230° is negative. ■

If the terminal ray of θ is on one of the axes, then θ is called a *quadrantal angle*. The quadrantal angles do not lie in any quadrant. Their trigonometric ratios are as tabulated in Figure 3–24. We have already calculated these ratios for an angle of measure 0°. In the exercises, you are asked to verify some of the remaining entries in Figure 3–24.

θ, in degrees	0°	90°	180°	270°
θ, in radians	0	$\frac{\pi}{2}$	π	$\frac{3\pi}{2}$
$\sin\theta$	0	1	0	-1
$\cos\theta$	1	0	-1	0
$\tan\theta$	0	undefined	0	undefined
$\cot\theta$	undefined	0	undefined	0
$\sec\theta$	1	undefined	-1	undefined
$\csc\theta$	undefined	1	undefined	-1

Figure 3–24
Trigonometric ratios for quadrantal angles

Finally, note that the terminal ray of θ is unchanged if any number of complete revolutions is added to θ. Thus the trigonometric ratios of θ also remain unchanged. Because

1 revolution = 360° = 2π radians, we have:

$$\sin(\theta + 2\pi) = \sin(\theta + 360°) = \sin\theta$$
$$\cos(\theta + 2\pi) = \cos(\theta + 360°) = \cos\theta$$

and so on for the other trigonometric ratios.

It follows that, for any integer k,

$$\sin(\theta + 2\pi k) = \sin(\theta + k360°) = \sin\theta$$
$$\cos(\theta + 2\pi k) = \cos(\theta + k360°) = \cos\theta$$

and so on.

EXAMPLE: $\sin 9\pi = \sin (9\pi - 4 \cdot 2\pi) = \sin \pi$ ■

EXAMPLE: $\tan -75° = \tan (-75° + 360°) = \tan 285°$ ■

EXAMPLE: $\csc -15\pi/4 = \csc (-15\pi/4 + 2 \cdot 2\pi) = \csc \pi/4$ ■

Angles in standard position with the same terminal ray are called *coterminal.* Thus:

Coterminal angles have the same trigonometric ratios.

3–3 *Exercises*

Determine the sign of the trigonometric ratio.

1. $\cos 28°$
2. $\csc 53°$
3. $\cos 118°$
4. $\tan 140°$
5. $\sec -29°$
6. $\sin -158°$
7. $\cot 6\pi/5$
8. $\sin 5\pi/3$
9. $\sec -\pi/8$
10. $\tan -5\pi/4$

Find all trigonometric ratios for an angle θ in standard position such that P lies on the terminal ray.

11. $P = (3, -4)$
12. $P = (-5, 12)$
13. $P = (1, 2)$
14. $P = (-1, -2)$

15. The point (0, 1) lies on the terminal ray of a 90° angle. Evaluate the trigonometric ratios for this angle.
16. Find a point on the terminal ray of an angle of measure π and use it to evaluate $\cos \pi$.
17. Find a point on the terminal ray of an angle of measure 270° and use it to evaluate $\csc 270°$.

Express using a trigonometric ratio of an angle with measure between 0° and 360° or between 0 and 2π.

18. $\sin 701°$
19. $\tan 1000°$
20. $\cot 11\pi/2$
21. $\cos -21\pi/5$
22. $\sec -93°$
23. $\csc -7\pi/9$

Evaluate

24. $\sin 7\pi/2$
25. $\cos -45°$
26. $\tan -135°$
27. $\sec 5\pi/4$

28. csc 3π
29. cot -3π

3-4 Trigonometric functions

We have defined trigonometric ratios for any angle or any rotation. For mos geometrical applications of trigonometry, no further mathematical sophisticatio would be necessary.

However, for more theoretical work, we should observe that each trigonc metric ratio defines a function with domain a set of real numbers. Each real numbe is viewed as the radian measure of an angle. The trigonometric ratio for that angl is taken as the value of the corresponding trigonometric function.

We therefore have six trigonometric functions. Expressed in set languag these are:

$$\sin = \{(\theta, \sin\theta)\}$$
$$\cos = \{(\theta, \cos\theta)\}$$
$$\tan = \{(\theta, \tan\theta)\}$$
$$\cot = \{(\theta, \cot\theta)\}$$
$$\sec = \{(\theta, \sec\theta)\}$$
$$\csc = \{(\theta, \csc\theta)\}$$

EXAMPLE: If f denotes the sine function, then $f(3)$ is the sine of an angle wit radian measure 3. Thus $f(3) = \sin 3$. ■

EXAMPLE: If f denotes the cosine function, then $f(-\pi) = \cos(-\pi) = -1$ (Se Section 12–3). ■

EXAMPLE: If f denotes the secant function, then $f(8) = \sec 8$, the secant of a angle with radian measure 8. ■

Strictly speaking, we are using each of the symbols sin, cos, and so on, fc two different things. For example, "sin" denotes a trigonometric ratio, so th sin A is the sine of $\angle A$. But also, "sin" is the sine function, which associates t each real number θ the sine of the angle which has θ as its radian measure.

This dual usage may not be exactly proper, but it will not cause any difficult so long as we remember that radian measure is understood when not otherwis specified. Note that there is no ambiguity in the following examples.

EXAMPLE: $\sec 0° = 1$ (from Section 3–3) and also $\sec 0 = 1$. ■

EXAMPLE: $\sin \pi/6 = \frac{1}{2}$ and also $\sin 30° = \frac{1}{2}$. ■

EXAMPLE: $\cot 135° = -1$ (Section 3–3) and also $\cot 3\pi/4 = -1$. ■

The six trigonometric functions share an important property, called *periodicit* as follows.

The function f is *periodic* if there is a positive number h such that

$$f(\theta + h) = f(\theta)$$

for all real numbers θ in the domain of f.

The smallest such positive number h is the *period.*

Each trigonometric function f satisfies the formula

$$f(\theta + 2\pi) = f(\theta)$$

and hence also

$$f(\theta + 2\pi k) = f(\theta)$$

for any integer k. Thus each trigonometric function is periodic.

EXAMPLE: $\sec(\theta + 2\pi) = \sec\theta$. ■

EXAMPLE: $\cos 7\pi/3 = \cos(7\pi/3 - 2\pi) = \cos\pi/3 = \frac{1}{2}$. ■

EXAMPLE: $\tan(1 + 10\pi) = \tan 1$. ■

We will see later that, for each of the functions sine, cosine, secant, and cosecant, 2π is the smallest positive h with $f(\theta + h) = f(\theta)$ for all θ. The period of each of these functions is therefore 2π.

However, it turns out that $\tan(\theta + \pi) = \tan\theta$, and similarly for the cotangent function. The period of each of these is π.

In evaluating the trigonometric functions of the real number θ, recall (Section 3–3) that the process begins as follows:

1. Place an angle with radian measure θ in standard position.
2. Choose a point (x, y), other than the origin, on the terminal ray.

Often the most convenient choice for (x, y) is the point at which the terminal ray intersects the unit circle $x^2 + y^2 = 1$. This point is called the *trigonometric point* associated with θ; its symbol is $T(\theta)$. See Figure 3–25.

Because $T(\theta)$ is on the unit circle, its distance from the origin is 1. Substituting $r = 1$ into the definitions given in Section 3–3, we have, if $(x, y) = T(\theta)$,

$$\sin\theta = y$$

$$\cos\theta = x$$

$$\tan\theta = \frac{y}{x}, \quad x \neq 0$$

$$\cot\theta = \frac{x}{y}, \quad y \neq 0$$

$$\sec\theta = \frac{1}{x}, \quad x \neq 0$$

$$\csc\theta = \frac{1}{y},\ y \neq 0$$

Thus $T(\theta)$ is the point $(\cos\theta, \sin\theta)$ in the xy-plane.

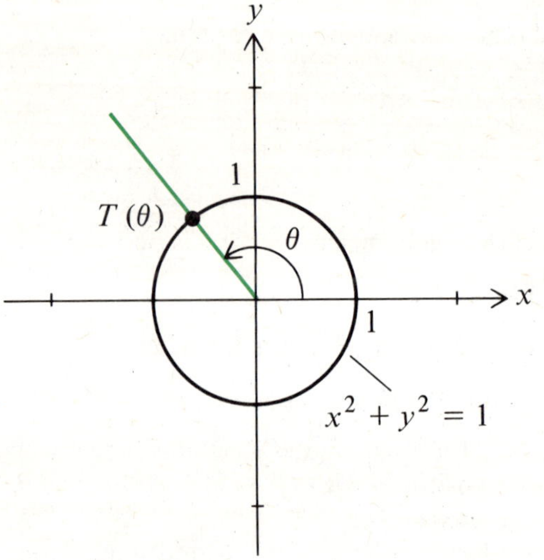

Figure 3–25
The trigonometric point $T(\theta)$

EXAMPLE: Find $T(\pi/3)$.

We know (Section 3-2) that $\cos \pi/3 = \frac{1}{2}$ and $\sin \pi/3 = \frac{1}{2}\sqrt{3}$. Thus

$$T\left(\frac{\pi}{3}\right) = \left(\cos\frac{\pi}{3}, \sin\frac{\pi}{3}\right) = \left(\frac{1}{2}, \frac{1}{2}\sqrt{3}\right)$$

$T(\pi/3)$, as shown in Figure 3–26, is the intersection point of the unit circle with the terminal ray of an angle of measure $\pi/3$ in standard position. ■

EXAMPLE: Find $T(-\pi)$.

We have $\cos(-\pi) = -1$ and $\sin(-\pi) = 0$. Therefore $T(-\pi) = (-1, 0)$. See Figure 3–27. ■

EXAMPLE: Find $T(0)$.

Because $\cos 0 = 1$ and $\sin 0 = 0$, $T(0) = (1, 0)$. ■

The trigonometric point $T(\theta)$ is defined for any real number θ. However, $\tan\theta$ and $\sec\theta$ are undefined when $x = 0$, while $\cot\theta$ and $\csc\theta$ are not defined when $y = 0$.

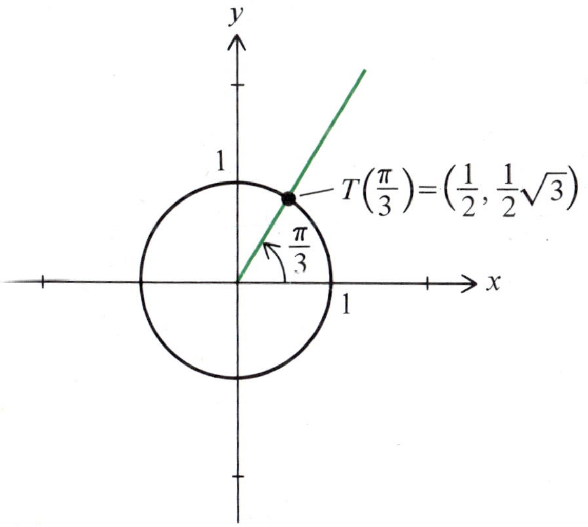

Figure 3–26

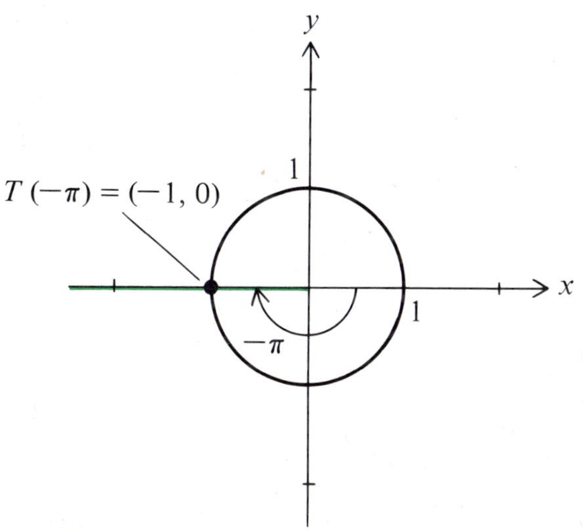

Figure 3–27

When $x = 0$, $T(\theta)$ must be the y-axis. This can happen in two ways. We may have $T(\theta) = (0, 1)$, in which case $\theta = \pi/2 + 2k\pi$ for some integer k. The other possibility is $T(\theta) = (0, -1)$, whence $\theta = 3\pi/2 + 2k\pi$ for some $k \in J$. Hence $x = 0$ if and only if $\theta = \pi/2 + k\pi$ for some $k \in J$.

Similarly, if $y = 0$, then $T(\theta)$ must be either $(1, 0)$ or $(-1, 0)$. It follows that $\theta = k\pi$ for some integer k.

We can therefore describe the domains of the trigonometric functions as follows.

The domain of sin and of cos is R.
The domain of tan and sec is $\{\theta \in R \mid \theta \neq \pi/2 + k\pi,\ k \in J\}$.
The domain of cot and csc is $\{\theta \in R \mid \theta \neq k\pi,\ k \in J\}$.

EXAMPLE: If $\theta = 5\pi/2$, which trigonometric functions of θ are not defined? Because $5\pi/2 = \pi/2 + 2\pi$, tan θ and sec θ are undefined. ■

EXAMPLE: csc (-11π) is undefined. ■

Looking at matters from a fairly sophisticated point of view, we may think of T as a function that associates to the real number θ the point $T(\theta)$ in the xy-plane. The domain of T is the set of all real numbers. Its range is the set of points on the unit circle—that is,

$$\{(x, y) \mid x^2 + y^2 = 1\}.$$

From the periodicity of the cosine and sine functions, it follows that

$$T(\theta + 2\pi) = T(\theta)$$

We may therefore think of T as another example of a periodic function. Its period is 2π.

3–4 *Exercises*

Evaluate.

1. sin $-\pi/2$
2. cos -3π
3. cot $9\pi/4$
4. tan $-7\pi/4$
5. sin $13\pi/3$
6. sin $13\pi/6$

Express in the form $f(\theta)$, where $0 \leq \theta < 2\pi$.

7. sin $5\pi/4$
8. tan $-21\pi/8$
9. sec $31\pi/6$
10. cot $-33\pi/7$

Locate on a graph.

11. $T(2\pi)$
12. $T(-4\pi)$
13. $T(\pi/4)$
14. $T(-\pi/2)$
15. $T(-\pi/3)$
16. $T(3\pi/2)$

Which of the following are undefined?

17. $\sin \pi$
18. $\tan 5\pi/2$
19. $\cot 7\pi$
20. $\csc 3\pi/4$

21. For what values of θ is $\sin \theta$ undefined? Why?
22. For what values of θ is $\tan \theta$ undefined? Why?

Locate approximately the trigonometric point $T(\theta)$ and use your graph to estimate $\sin \theta$ and $\cos \theta$.

23. $\theta = 1$
24. $\theta = -1$
25. $\theta = -3$
26. $\theta = 2$

27. Show on a graph how $T(\theta)$ and $T(-\theta)$ are related geometrically. How are their x-coordinates related? The y-coordinates?
28. The x-coordinate of $T(\theta)$ is $\cos \theta$, and the x-coordinate of $T(-\theta)$ is $\cos(-\theta)$. Use this to guess a formula for $\cos(-\theta)$.
29. The y-coordinate of $T(\theta)$ is $\sin \theta$, and the y-coordinate of $T(-\theta)$ is $\sin(-\theta)$. Use this to guess a formula for $\sin(-\theta)$.

3-5 Evaluation of trigonometric functions

At this point, we know (from Section 2-3) how to use the table to find approximate values of the trigonometric functions for any acute angle θ. If θ is a quadrantal angle, we worked out its trigonometric functions in Section 3-3.

Now we turn our attention to angles in the other quadrants. For this, it is easiest to use the *reference angle*, defined as follows.

If θ is an angle in standard position, then the reference angle of θ is the acute angle between the terminal ray of θ and the horizontal axis. The symbol for the reference angle of θ is θ^*.

When θ is a second-quadrant angle, then θ and its reference angle θ^* are as shown in Figure 3–28. The situation for third- and fourth-quadrant angles is pictured in Figures 3–29 and 3–30, respectively.

From Figures 3–28, 3–29, and 3–30, we can read off formulas for θ^*, tabulated in Figure 3–31.

EXAMPLE: Find θ^* for $\theta = 125°$.

Because 125° represents a second-quadrant angle, we have $\theta^* = 180° - \theta = 180° - 125° = 55°$. ■

EXAMPLE: If $\theta = 340°$, then θ is in Quadrant IV. Hence $\theta^* = 360° - \theta = 360° - 340° = 20°$. ■

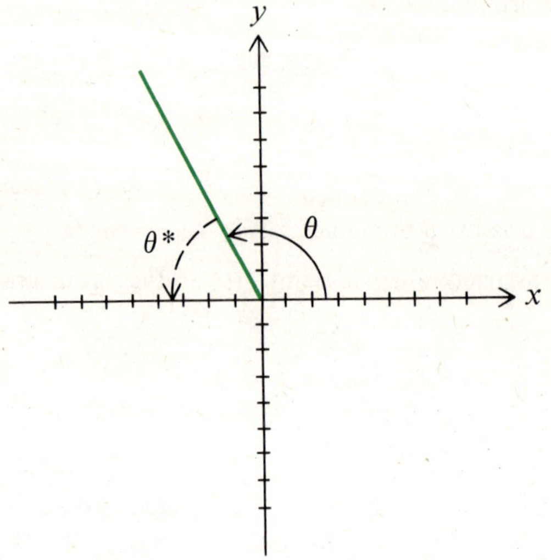

Figure 3–28
The reference angle for θ in Quadrant II
$\theta + \theta^* = 180°$

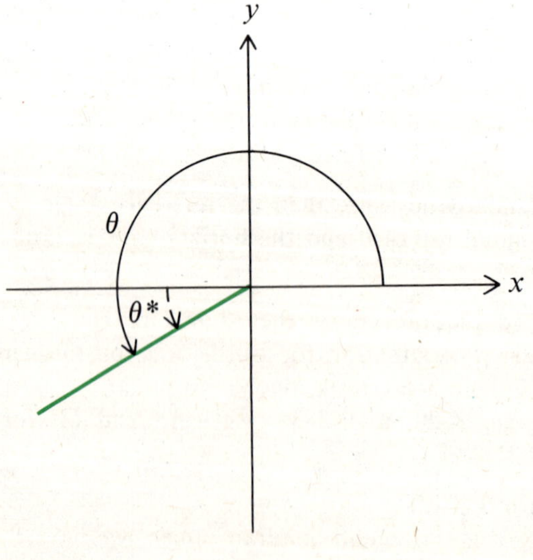

Figure 3–29
The reference angle for θ in Quadrant III
$180° + \theta^* = \theta$

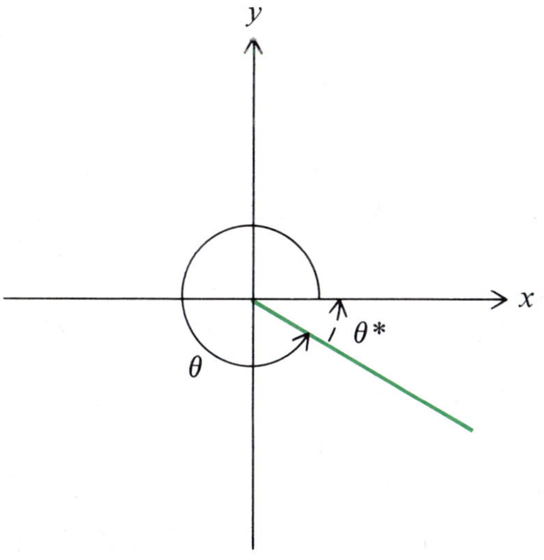

Figure 3–30
The reference angle for θ in Quadrant IV
$\theta + \theta^* = 360°$

Quadrant of θ	θ^*, in degrees	θ^*, in radians
I	$\theta^* = \theta$	$\theta^* = \theta$
II	$\theta^* = 180° - \theta$	$\theta^* = \pi - \theta$
III	$\theta^* = \theta - 180°$	$\theta^* = \theta - \pi$
IV	$\theta^* = 360° - \theta$	$\theta^* = 2\pi - \theta$

Figure 3–31
Formulas for the reference angle

EXAMPLE: If θ has radian measure $7\pi/6$, then θ is in Quadrant III. Hence $\theta^* = \theta - \pi = 7\pi/6 - \pi = \pi/6$. ■

Here is the way the reference angle is used to evaluate $f(\theta)$, where f is one of the trigonometric functions.

If θ^* is the reference angle for θ, and f is a trigonometric function, then $f(\theta) = \pm f(\theta^*)$. The sign is chosen according to the quadrant of θ.

EXAMPLE: Express sin 125° as a function of an acute angle.

We have already observed that for $\theta = 125°$, the reference angle is 55°. Thus $\sin 125° = \pm\sin 55°$. Because θ is in Quadrant II, where the sine function is positive, we have $\sin 125° = \sin 55°$. ■

EXAMPLE: We find tan 340°

For $\theta = 340°$, we have $\theta^* = 20°$, and θ is in Quadrant IV. For angles in Quadrant IV, the tangent is negative. Therefore, $\tan 340° = -\tan 20°$. From the table, $\tan 20° = 0.3640$. Hence

$$\tan 340° = -0.3640.$$ ■

EXAMPLE: Find cos $7\pi/6$.

As noted above, $7\pi/6$ is a third-quadrant angle with reference angle of $\pi/6$. We know $\cos \pi/6 = \sqrt{3}/2$. Because cosine is negative for an angle in Quadrant III, we have $\cos 7\pi/6 = -\sqrt{3}/2$. ■

Sometimes it is necessary to add or subtract one or more complete revolutions before evaluating $f(\theta)$. According to the periodicity of the trigonometric functions, this leaves $f(\theta)$ unchanged.

EXAMPLE: We find tan 1000° by the following steps.

$$\begin{aligned}\tan 1000° &= \tan(1000° - 2 \cdot 360°)\\ &= \tan 280°\\ &= -\tan 80°\\ &= -0.1763\end{aligned}$$

We have used that tangent is negative for a fourth-quadrant angle. ■

EXAMPLE: Find csc $(-13\pi/6)$.

$$\begin{aligned}\csc(-13\pi/6) &= \csc\left(-\frac{13\pi}{6} + 4\pi\right)\\ &= \csc\frac{11\pi}{6}\\ &= -\csc\left(2\pi - \frac{11\pi}{6}\right)\\ &= -\csc\frac{\pi}{6}\\ &= -2\end{aligned}$$

Here we used that cosecant is negative for a fourth-quadrant angle. ■

We now wish to explain the reasoning behind the formula

$$f(\theta) = \pm f(\theta^*)$$

where f is one of the trigonometric functions. For this, we should compare the trigonometric points $T(\theta)$ and $T(\theta^*)$.

Let us first consider the case where θ is in the second quadrant. The trigonometric points are then as in Figure 3–32. Let (x, y) be the coordinates of $T(\theta^*)$.

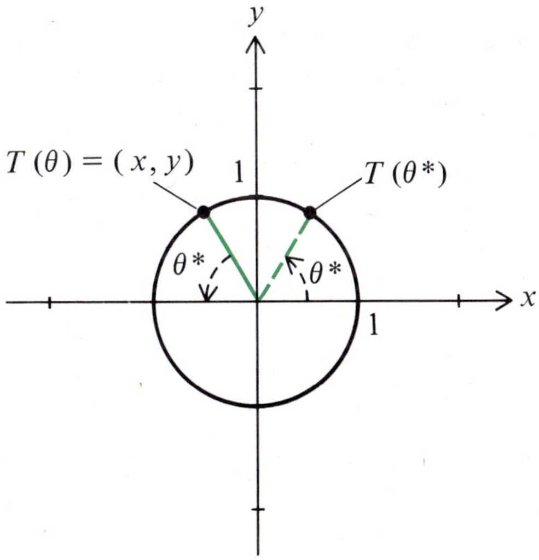

Figure 3–32
$T(\theta)$ and $T(\theta^*)$ for θ in Quadrant II

Then, from Figure 3–32, we see that $T(\theta)$ is the point $(-x, y)$. Thus, for θ in Quadrant II,

$$\sin\theta = y = \sin\theta^*$$
$$\cos\theta = -x = -\cos\theta^*$$
$$\tan\theta = -\frac{y}{x} = -\tan\theta^*$$
$$\cot\theta = -\frac{x}{y} = -\cot\theta^*$$
$$\sec\theta = -\frac{1}{x} = -\sec\theta^*$$
$$\csc\theta = \frac{1}{y} = \csc\theta^*$$

When θ is in the third quadrant, the situation is as shown in Figure 3–33. In this case we have $T(\theta) = (-x, -y)$ if $T(\theta^*) = (x, y)$. In Exercises 29 through 34 you are asked to use this to work out the trigonometric functions of θ in terms of those of θ^*.

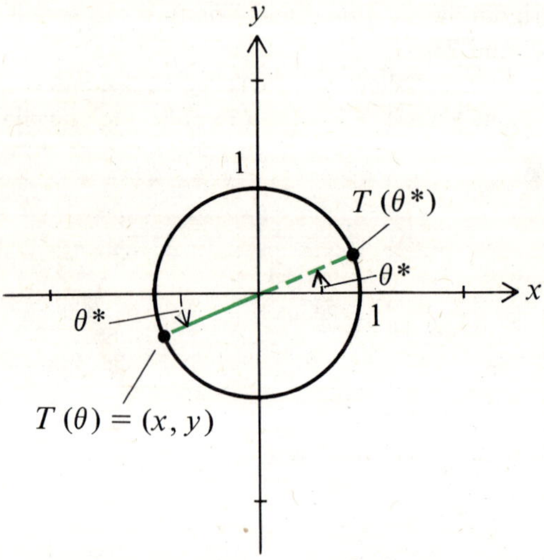

Figure 3–33
$T(\theta)$ and $T(\theta^*)$ for θ in Quadrant III

For θ a fourth-quadrant angle, we refer to Figure 3–34. Now, if $T(\theta^*) = (x, y)$, we have $T(\theta) = (x, -y)$. From that, it is possible to work out the trigonometric functions of θ in terms of those of θ^*.

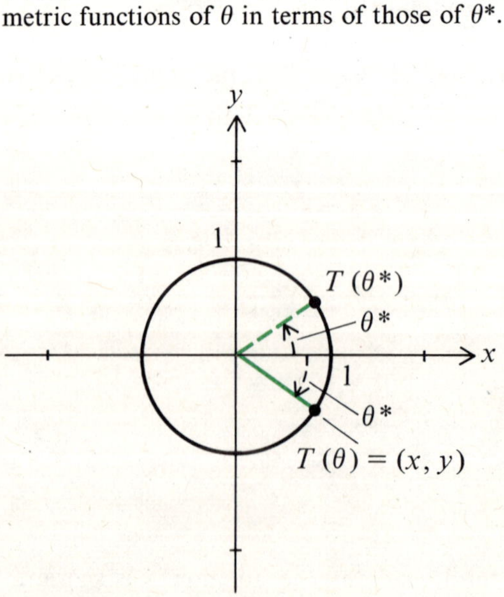

Figure 3–34
$T(\theta)$ and $T(\theta^*)$ for θ in Quadrant IV

3–5 *Exercises*

Find the reference θ^* for θ as given.

1. 160°
2. 221°
3. 303°
4. 262°
5. $5\pi/4$
6. $9\pi/8$
7. $-51°$
8. $-5\pi/6$
9. $12\pi/5$
10. 479°

Express in terms of a function of an acute angle.

11. sin 143°
12. cos $6\pi/5$
13. tan $15\pi/8$
14. cot $-318°$
15. csc 480°
16. sec $-110°$

Evaluate.

17. sin 210°
18. cot 240°
19. cos $5\pi/4$
20. tan $-3\pi/2$
21. sec 7π
22. csc $7\pi/4$
23. cos 173°
24. cot 312°
25. sin $-80°$
26. cos $-147°$
27. sec 500°
28. csc $-500°$

Let θ be a third-quadrant angle. Then, if $T(\theta^*) = (x, y)$, we have $T(\theta) = (-x, -y)$. Express the following first in terms of x and/or y, and then in terms of θ^*.

29. $\sin\theta$
30. $\cos\theta$
31. $\tan\theta$
32. $\cot\theta$
33. $\sec\theta$
34. $\csc\theta$

3-6 Graphs of the sine and cosine functions

Our object now is to sketch the graph of the sine function $\{(\theta, \sin\theta)\}$ and the cosine function $\{(\theta, \cos\theta)\}$. We will draw these graphs in the usual coordinate plane, using the variables x and y. Thus these graphs will be the graphs of the equations $y = \sin x$ and $y = \cos x$, respectively.

We begin with some known values of these functions. In Figure 3–35 we have tabulated the values for both $\sin\theta$ and $\cos\theta$, where θ is a quadrantal angle, a special angle for which we have previously found $\sin\theta$ and $\cos\theta$, or an angle with such a reference angle.

We turn our attention to the sine function. We would like to plot the points $(\theta, \sin\theta)$ given in the table of Figure 3–35. We need decimal approximations for the irrational numbers in the table. We use the following approximate values

θ, in radians	θ, in degrees	$\sin\theta$	$\cos\theta$
0	$0°$	0	1
$\frac{\pi}{6}$	$30°$	$\frac{1}{2}$	$\frac{\sqrt{3}}{2}$
$\frac{\pi}{4}$	$45°$	$\frac{\sqrt{2}}{2}$	$\frac{\sqrt{2}}{2}$
$\frac{\pi}{3}$	$60°$	$\frac{\sqrt{3}}{2}$	$\frac{1}{2}$
$\frac{\pi}{2}$	$90°$	1	0
$\frac{2\pi}{3}$	$120°$	$\frac{\sqrt{3}}{2}$	$-\frac{1}{2}$
$\frac{3\pi}{4}$	$135°$	$\frac{\sqrt{2}}{2}$	$-\frac{\sqrt{2}}{2}$
$\frac{5\pi}{6}$	$150°$	$\frac{1}{2}$	$-\frac{\sqrt{3}}{2}$
π	$180°$	0	-1
$\frac{7\pi}{6}$	$210°$	$-\frac{1}{2}$	$-\frac{\sqrt{3}}{2}$
$\frac{5\pi}{4}$	$225°$	$-\frac{\sqrt{2}}{2}$	$-\frac{\sqrt{2}}{2}$
$\frac{4\pi}{3}$	$240°$	$-\frac{\sqrt{3}}{2}$	$-\frac{1}{2}$
$\frac{3\pi}{2}$	$270°$	-1	0
$\frac{5\pi}{3}$	$300°$	$-\frac{\sqrt{3}}{2}$	$\frac{1}{2}$
$\frac{7\pi}{4}$	$315°$	$-\frac{\sqrt{2}}{2}$	$\frac{\sqrt{2}}{2}$
$\frac{11\pi}{6}$	$330°$	$-\frac{1}{2}$	$\frac{\sqrt{3}}{2}$
2π	$360°$	0	1

Figure 3–35

$$\sqrt{2} = 1.414$$
$$\sqrt{3} = 1.732$$
$$\pi = 3.142$$

These give us values

$$\frac{\sqrt{2}}{2} = 0.707$$

$$\frac{\sqrt{3}}{2} = 0.866$$

$$\frac{\pi}{6} = 0.524$$

$$\frac{\pi}{4} = 0.785$$

$$\frac{\pi}{3} = 1.047$$

$$\frac{\pi}{2} = 1.571$$

and so on. Now the points $(\theta, \sin \theta)$ can be plotted in the xy-plane (see Figure 3–36). These are points which lie on the graph of $y = \sin x$.

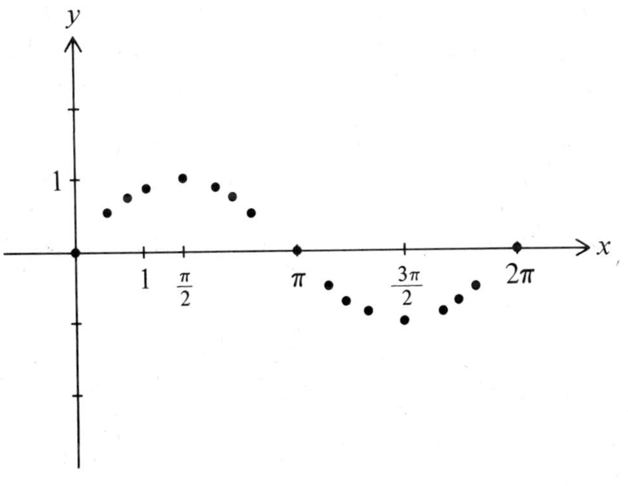

Figure 3–36
Some points on the graph of $y = \sin x$

Next, these points are connected by a smooth curve, as in Figure 3–37. This gives the graph of $y = \sin x$, where $0 \leq x \leq 2\pi$. This portion of the graph is referred to as one *cycle* of (Figure 3–37) the graph of the sine function.

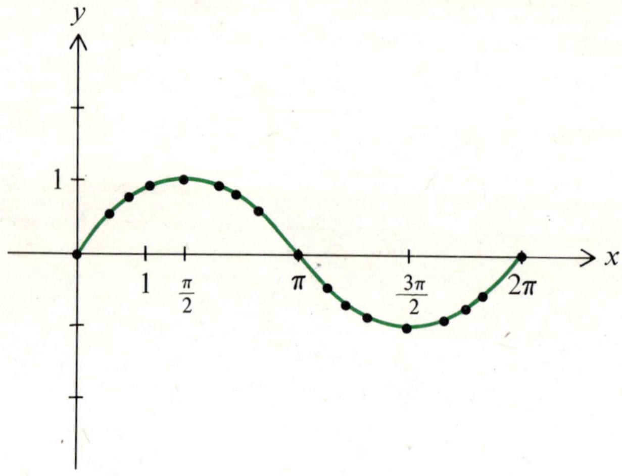

Figure 3–37
$\{(x, y) | y = \sin x, 0 \le x \le 2\pi\}$

Because of the periodicity relationship $\sin(\theta + 2\pi) = \sin \theta$, this complete cycle is repeated every 2π units on the x-axis. The domain of the sine function is the set of all real numbers. The graph thus extends indefinitely, but repeats itself every 2π units, as shown by Figure 3–38.

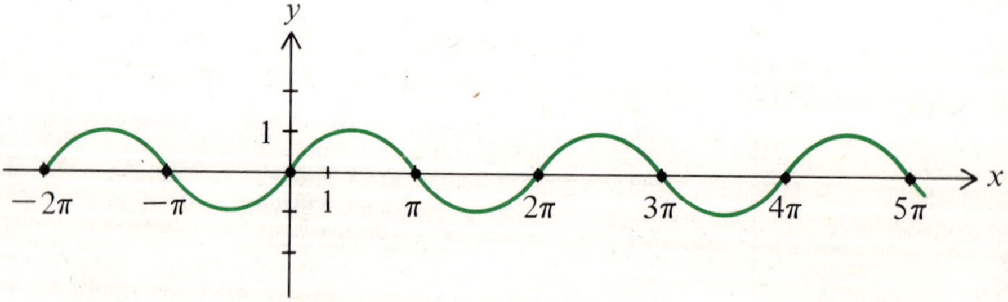

Figure 3–38
$\{(x, y) | y = \sin x\}$

The graph of the cosine function is obtained similarly. We plot the points $(\theta, \cos \theta)$ given by Figure 3–35 in the xy-plane and connect them with a smooth curve. The resulting graph (Figure 3–39) is one cycle of the graph of $y = \cos x$.

The complete graph is determined by the periodicity relation $\cos(\theta + 2\pi) = \cos \theta$. The domain of the cosine function is the set of real numbers. The graph extends indefinitely, but repeats itself every 2π units, as shown by Figure 3–40.

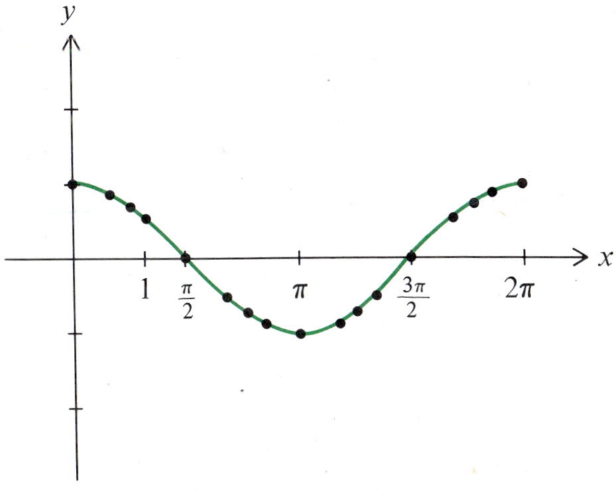

Figure 3–39
$\{(x, y) \mid y = \cos x,\ 0 \leq x \leq 2\pi\}$

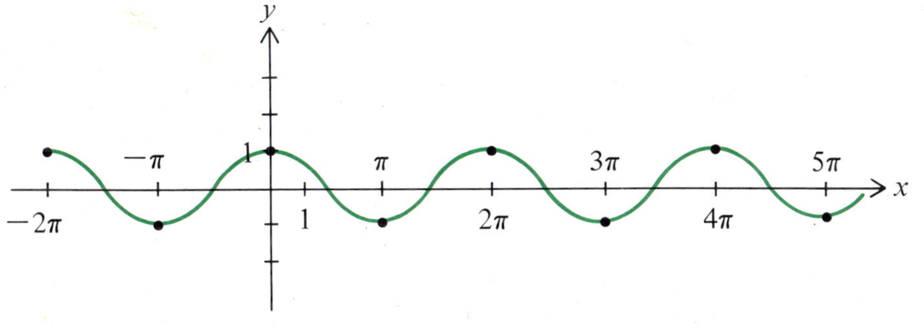

Figure 3–40
$\{(x, y) \mid y = \cos x\}$

3–6 *Exercises*

Use the graphs of $y = \sin x$ and $y = \cos x$ to find the solution set of the given equation.

1. $\sin x = 1$
2. $\sin x = 0$
3. $\sin x = -1$
4. $\cos x = 1$
5. $\cos x = 0$
6. $\cos x = -1$

7. (a) Draw the graphs of $y = \sin x$ and $y = \cos x$ on the same set of axes.
   (b) Use your graph to determine the values of x for which $\sin x = \cos x$.
8. Use the graph of $y = \sin x$ to read off the sign of $\sin x$ in the various quadrants.
9. Use the graph of $y = \cos x$ to read off the sign of $\cos x$ in the various quadrants.
10. Use the graph of $y = \cos x$ to guess a formula for $\cos(-x)$.
11. Use the graph of $y = \sin x$ to guess a formula for $\sin(-x)$.

12. Draw the graph of $y = \sin x$, but with x measured in degrees.
13. Draw the graph of $y = \cos x$, but with x measured in degrees.

Use the graphs from Exercises 12 and 13 to estimate the following. Check your estimate by referring to the table.

14. $\sin 20°$
15. $\cos 20°$
16. $\sin 65°$
17. $\cos 65°$
18. $\sin 140°$
19. $\cos 140°$
20. $\sin 200°$
21. $\cos 200°$
22. $\sin 305°$
23. $\cos 305°$
24. $\sin -25°$
25. $\cos -25°$

For each equation, construct a table of values for x and y, similar to the table given in Figure 3–35 for the equations $y = \sin x$ and $y = \cos x$. Plot the points (x, y) so obtained, and try to sketch the graph.

26. $y = 2 \sin x$
27. $y = 3 \cos x$
28. $y = -\sin x$
29. $y = -\cos x$
30. $y = \sin x + \cos x$

Review exercises (Chapter 3)

Explain the meaning.

1. unit circle
2. initial ray
3. terminal ray
4. standard position
5. reference angle
6. quadrantal angle
7. coterminal angles

Find the indicated distance.

8. between $(-2, -5)$ and the origin
9. between $(3, -2)$ and $(-1, 4)$

Write the equation for the circle indicated.

10. the unit circle
11. circle of radius 6 with center the origin

12. Express $\frac{3}{4}$ revolution in degrees and in radians.

13. Find the trigonometric ratios for an angle θ in standard position such that $(-2, 7)$ is on its terminal ray.

14. Find the trigonometric functions $f(\theta)$, given that the trigonometric point $T(\theta) = (-3/4, \sqrt{7}/4)$.

15. What can you say about the measure, in degrees, of coterminal angles?

16. What can you say about the trigonometric ratios for angles which have the same reference angle?

17. If the reference angle for θ is $\pi/6$, show on a sketch the possibilities for $T(\theta)$.

18. Sketch one complete cycle of the graph of the function defined by $y = \cos x$, starting with an x-value of $-\pi$.

Geometery

Theorms
Postulates
axioms
Systems

4

Trigonometric identities and equations

In advanced work, especially in calculus, the identities satisfied by trigonometric functions turn out to be most useful. These identities often permit us to rewrite complicated trigonometric expressions in much simpler form.

For our purposes, identities will be used to facilitate the solution of the trigonometric equations and to provide justification for the reference angle method for evaluation of trigonometric functions.

4-1 The Pythagorean identities

In Section 2-2 we noted the Pythagorean identities, valid for any acute angle $\angle A$

$$\cos^2 A + \sin^2 A = 1$$
$$1 + \tan^2 A = \sec^2 A$$
$$1 + \cot^2 A = \csc^2 A$$

The proof of these identities made use of the Pythagorean theorem in the right triangle $\triangle ABC$.

The same identities are valid for any real number θ, as well as for any angle

These, then, are the *Pythagorean identities* for the trigonometric functions:

If θ is any real number, then

$$\cos^2 \theta + \sin^2 \theta = 1$$
$$1 + \tan^2 \theta = \sec^2 \theta$$
$$1 + \cot^2 \theta = \csc^2 \theta$$

EXAMPLE: $\cos^2 4\pi/3 + \sin^2 4\pi/3 = 1$. ■

EXAMPLE: $1 + \tan^2 216° = \sec^2 216°$. ■

EXAMPLE: $1 + \cot^2 2 = \csc^2 2$. ■

EXAMPLE: $\cos^2 (-14°) + \sin^2 (-14°) = 1$. ■

To verify the Pythagorean identities, we look at the interpretation of trigonometric functions using the trigonometric point $T(\theta)$ (see Section 3-4).

We know that $T(\theta)$ is a certain point (x, y) on the unit circle $x^2 + y^2 = 1$, as in Figure 4–1. The trigonometric functions of θ are then given by:

$$\sin \theta = y$$

$$\cos \theta = x$$

$$\tan \theta = \frac{y}{x} \text{ if } x \neq 0$$

$$\cot \theta = \frac{x}{y} \text{ if } y \neq 0$$

$$\sec \theta = \frac{1}{x} \text{ if } x \neq 0$$

$$\csc \theta = \frac{1}{y} \text{ if } y \neq 0$$

We have $x^2 + y^2 = 1$ for $T(\theta) = (x, y)$, and we have $x = \cos \theta$, $y = \sin \theta$. Substituting, we obtain

$$(\cos \theta)^2 + (\sin \theta)^2 = 1$$

or

$$\cos^2 \theta + \sin^2 \theta = 1$$

and the first identity is proved.

The steps to justify the second identity are as follows:

$$1 + \tan^2 \theta = 1 + \left(\frac{y}{x}\right)^2$$

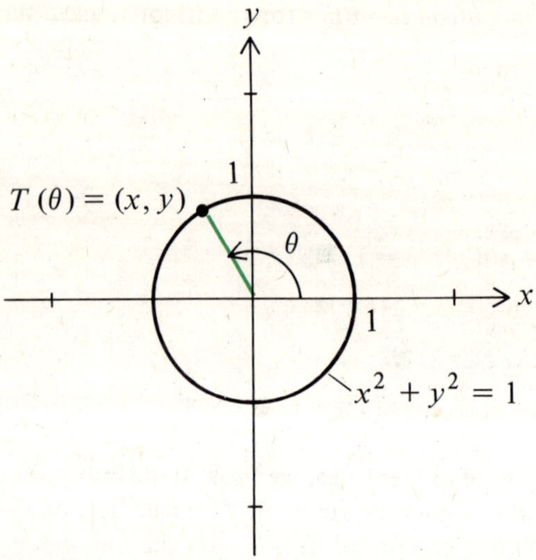

Figure 4–1
$T(\theta) \in \{(x, y) | x^2 + y^2 = 1\}$

$$= 1 + \frac{y^2}{x^2}$$

$$= \frac{x^2 + y^2}{x^2}$$

$$= \frac{1}{x^2}$$

$$= \left(\frac{1}{x}\right)^2$$

$$= \sec^2 \theta$$

In the exercises you are asked to verify the third of the Pythagorean identities. The following examples illustrate ways the Pythagorean identities can be used.

EXAMPLE: Find $\cos \theta$, given that $\sin \theta = 1/4$.

Because $\cos^2 \theta + \sin^2 \theta = 1$, we have

$$\cos^2 \theta = 1 - \sin^2 \theta = 1 - \tfrac{1}{16} = \tfrac{15}{16}$$

Therefore

$$\cos \theta = \pm \sqrt{\frac{15}{16}} = \pm \frac{\sqrt{15}}{4}$$

From the information given, it is not possible to determine the correct sign. ■

EXAMPLE: Find $\cos\theta$, given that θ is a second-quadrant angle such that $\sin\theta -$ $1/4$.

From the preceding example, $\cos\theta = \pm\sqrt{15}/4$. Because θ is in Quadrant II, $\cos\theta$ is negative. Therefore $\cos\theta = -\sqrt{15}/4$. ■

EXAMPLE: If $\tan\theta = -2$ and θ is in Quadrant IV, find $\sec\theta$.

We have $\sec^2\theta = 1 + \tan^2\theta = 1 + 4 = 5$. Therefore $\sec\theta = \pm\sqrt{5}$. Because θ is in Quadrant IV, $\sec\theta$ is positive. Therefore $\sec\theta = \sqrt{5}$. ■

EXAMPLE: Express $\cot\theta$ in terms of $\csc\theta$ for third-quadrant angles.

From $1 + \cot^2\theta = \csc^2\theta$, we obtain

$$\cot^2\theta = \csc^2\theta - 1$$

or

$$\cot\theta = \pm\sqrt{\csc^2\theta - 1}.$$

In Quadrant III, $\cot\theta > 0$, and so

$$\cot\theta = \sqrt{\csc^2\theta - 1}$$ ■

4–1 *Exercises*

Find the indicated function of θ using the information given.

1. $\sin\theta = 3/5$, θ in Quadrant I, $\cos\theta = ?$
2. $\sin\theta = 3/5$, θ in Quadrant II, $\cos\theta = ?$
3. $\sec\theta = -13/5$, θ in Quadrant II, $\tan\theta = ?$
4. $\sec\theta = -13/5$, θ in Quadrant III, $\tan\theta = ?$
5. $\cos\theta = 1/2$, θ in Quadrant I, $\sin\theta = ?$
6. $\cos\theta = 1/2$, θ in Quadrant IV, $\sin\theta = ?$
7. $\cot\theta = -5/3$, θ in Quadrant II, $\csc\theta = ?$
8. $\cot\theta = -5/3$, θ in Quadrant IV, $\csc\theta = ?$

Find general formulas as indicated.

9. For $\cos\theta$ in terms of $\sin\theta$, assuming θ in Quadrant I.
10. For $\sin\theta$ in terms of $\cos\theta$, assuming θ in Quadrant II.
11. For $\tan\theta$ in terms of $\sec\theta$, assuming θ in Quadrant III.
12. For $\sec\theta$ in terms of $\tan\theta$, assuming θ in Quadrant IV.
13. For $\cot\theta$ in terms of $\csc\theta$, assuming θ in Quadrant II.
14. For $\csc\theta$ in terms of $\cot\theta$, assuming θ in Quadrant IV.
15. For $\sin^4\theta$ in terms of $\cos\theta$.
16. For $\cos^4\theta$ in terms of $\sin\theta$.

17. Give the steps to justify the identity $1 + \cot^2\theta = \csc^2\theta$.
18. Find the steps to justify the equation $\sin^2\theta = (1 + \cos\theta)(1 - \cos\theta)$.

Verify the following simplification identities, valid for all real numbers θ, excep when division by 0 would be involved.

19. $\tan\theta = (\sin\theta)/\cos\theta$

20. $\cot\theta = (\cos\theta)/\sin\theta$

21. $\sec\theta = 1/\cos\theta$

22. $\csc\theta = 1/\sin\theta$

4-2 The cosine of a difference

We are going to use geometric reasoning and a little algebra to derive a fundamenta formula. This identity gives a formula for the cosine of the difference of two angle (or numbers).

In Figure 4–2 are shown the trigonometric points $T(\alpha) = (\cos\alpha, \sin\alpha)$ an $T(\beta) = (\cos\beta, \sin\beta)$. The angle between the terminal rays of α and β is $\alpha - \beta$

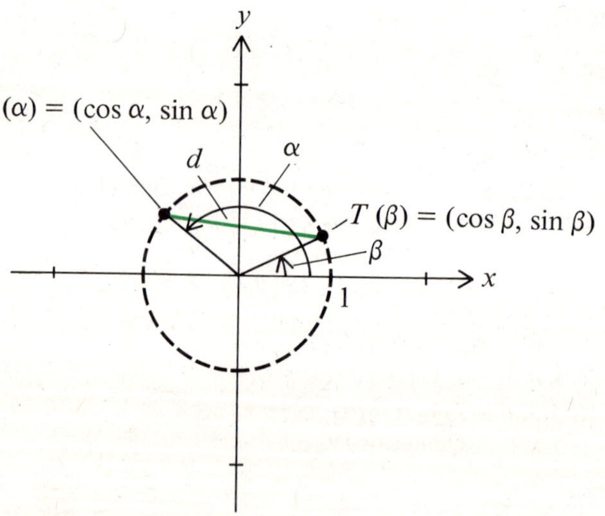

Figure 4–2
The distance between $T(\alpha)$ and $T(\beta)$ is d

The distance between $T(\alpha)$ and $T(\beta)$ is denoted by d.

According to the distance formula (Section 3-1), we have

$$d = \sqrt{(\cos\alpha - \cos\beta)^2 + (\sin\alpha - \sin\beta)^2}$$

or

$$\begin{aligned} d^2 &= (\cos\alpha - \cos\beta)^2 + (\sin\alpha - \sin\beta)^2 \\ &= \cos^2\alpha - 2\cos\alpha\cos\beta + \cos^2\beta + \sin^2\alpha - 2\sin\alpha\sin\beta + \sin^2\beta \end{aligned}$$

Using that $\sin^2 \alpha + \cos^2 \alpha = \sin^2 \beta + \cos^2 \beta = 1$, we have

$$d^2 = 2 - 2 \cos \alpha \cos \beta - 2 \sin \alpha \sin \beta$$

If we rotate Figure 4–2 so that the terminal ray of β is brought back to the x-axis, the result is Figure 4–3. By the rotation, $T(\beta)$ is moved to (1, 0) and $T(\alpha)$ moves to $T(\alpha - \beta) = (\cos(\alpha - \beta), \sin(\alpha - \beta))$. The distance between $T(\alpha - \beta)$ and (1, 0) is still d, the same as in Figure 4–2.

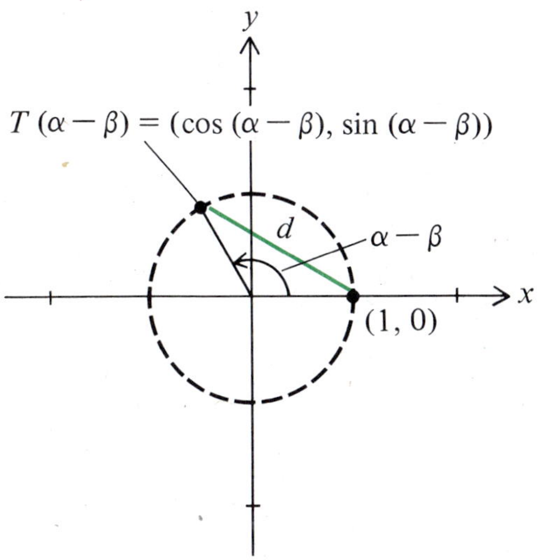

Figure 4–3
d is also the distance between $T(\alpha - \beta)$ and (1, 0)

Using the distance formula in Figure 4–3, we have

$$d = \sqrt{[\cos(\alpha - \beta) - 1]^2 + [\sin(\alpha - \beta) - 0]^2}$$

or

$$d^2 = \cos^2(\alpha - \beta) - 2\cos(\alpha - \beta) + 1 + \sin^2(\alpha - \beta)$$

Because $\cos^2(\alpha - \beta) + \sin^2(\alpha - \beta) = 1$, this gives

$$d^2 = 2 - 2\cos(\alpha - \beta)$$

The two values obtained for d^2 from Figures 4–2 and 4–3 must be the same. Setting them equal, we have

$$2 - 2\cos \alpha \cos \beta - 2 \sin \alpha \sin \beta = 2 - 2\cos(\alpha - \beta)$$

Simplifying, we obtain

$$\cos(\alpha - \beta) = \cos \alpha \cos \beta + \sin \alpha \sin \beta$$

This last is the formula we were seeking. Stated formally, we have the following:

If α and β are any angles, then

$$\cos(\alpha - \beta) = \cos\alpha\cos\beta + \sin\alpha\sin\beta$$

Or, restated in terms of real numbers:

If x and y are any real numbers, then

$$\cos(x - y) = \cos x\cos y + \sin x\sin y$$

EXAMPLE: $\cos(\theta - 20°) = \cos\theta\cos 20° + \sin\theta\sin 20°$. ■

EXAMPLE: $\cos(x - \pi/3) = \cos x\cos\pi/3 + \sin x\sin\pi/3$. ■

EXAMPLE:
$$\begin{aligned}\cos 15° &= \cos(45° - 30°)\\ &= \cos 45°\cos 30° + \sin 45°\sin 30°\\ &= \frac{\sqrt{2}}{2}\cdot\frac{\sqrt{3}}{2} + \frac{\sqrt{2}}{2}\cdot\frac{1}{2} = \frac{\sqrt{6}}{4} + \frac{\sqrt{2}}{4} = \frac{(\sqrt{6}+\sqrt{2})}{4}\end{aligned}$$ ■

Some further important identities can be derived easily from the formula for $\cos(x - y)$. We present a few of these in the following examples.

EXAMPLE: Find a formula for $\cos(-x)$.

We have

$$\begin{aligned}\cos(-x) &= \cos(0 - x)\\ &= \cos 0\cos x + \sin 0\sin x\\ &= 1\cdot\cos x + 0\cdot\sin x = \cos x\end{aligned}$$

Therefore, $\cos(-x) = \cos x$. ■

EXAMPLE: $\cos -78° = \cos 78°$, because $\cos(-x) = \cos x$. ■

EXAMPLE: $\cos -9\pi/8 = \cos 9\pi/8$. ■

EXAMPLE: We verify the cofunction identity $\cos(90° - \theta) = \sin\theta$.

We have

$$\begin{aligned}\cos(90° - \theta) &= \cos 90°\cos\theta + \sin 90°\sin\theta\\ &= 0\cdot\cos\theta + 1\cdot\sin\theta\\ &= \sin\theta\end{aligned}$$ ■

EXAMPLE: We derive a formula for $\cos(x - \pi)$ as follows:

$$\begin{aligned}\cos(x - \pi) &= \cos x\cos\pi + \sin x\sin\pi\\ &= (\cos x)\cdot(-1) + (\sin x)\cdot(0)\\ &= -\cos x\end{aligned}$$

and so

$$\cos(x - \pi) = -\cos x$$ ■

Among the results of this section thus far, the most important are these:

$$\cos(x - y) = \cos x \cos y + \sin x \sin y$$
$$\cos(-x) = \cos x$$
$$\cos\left(\frac{\pi}{2} - x\right) = \sin x$$

All three of these can be used to derive one further key formula:

$$\sin(-x) = -\sin x$$

EXAMPLE: $\sin(-78°) = -\sin 78°$. ■

EXAMPLE: $\sin(-9\pi/8) = -\sin 9\pi/8$

Here are the steps for the proof.

$$\begin{aligned}
\sin(-x) &= \cos\left[\frac{\pi}{2} - (-x)\right] \\
&= \cos\left(\frac{\pi}{2} + x\right) \\
&= \cos -\left(\frac{\pi}{2} + x\right) \\
&= \cos\left(-\frac{\pi}{2} - x\right) \\
&= \cos\left(-\frac{\pi}{2}\right)\cos x + \sin\left(-\frac{\pi}{2}\right)\sin x \\
&= 0 \cdot \cos x + (-1) \cdot \sin x \\
&= -\sin x
\end{aligned}$$ ■

–2 *Exercises*

Give a formula for each.

1. $\cos(x - y)$
2. $\cos(\pi/2 - x)$
3. $\cos(\theta - 180°)$
4. $\cos(\theta - 90°)$
5. $\cos(x + \pi/2)$
6. $\cos(x + \pi)$
7. $\cos(\theta - 30°)$
8. $\cos(\theta - 45°)$
9. $\cos(\theta + 45°)$
10. $\cos(\theta + 60°)$

Verify the following formulas

11. $\sin(\pi/2 - x) = \cos x$
12. $\sin(90° - \theta) = \cos\theta$

Find without use of the table.

13. $\cos(-15°)$
14. $\sin 75°$
15. $\cos 75°$
16. $\sin 15°$

17. $\cos(-\pi/12)$

18. $\sin 7\pi/12$

19. $\cos 7\pi/12$

20. $\sin 285°$

21. $\cos 285°$

22. $\sin -30°$

23. $\sin -\pi/4$

24. $\sin -75°$

25. $\sin -15°$

26. $\sin -7\pi/12$

27. $\sin -285°$

4-3 The addition formulas

We are now ready to obtain the *addition formulas*

$$\cos(x+y) = \cos x \cos y - \sin x \sin y$$
$$\sin(x+y) = \sin x \cos y + \cos x \sin y$$

as well as a number of other useful identities.

Let us first recall the main formulas from the previous section:

$$\cos(x-y) = \cos x \cos y + \sin x \sin y$$
$$\cos(-x) = \cos x$$
$$\cos\left(\frac{\pi}{2} - x\right) = \sin x$$
$$\sin(-x) = -\sin x$$

From these four results, the proof of the addition formulas is not difficult. Here are the steps.

$$\begin{aligned}\cos(x+y) &= \cos[x-(-y)] \\ &= \cos x \cos(-y) + \sin x \sin(-y) \\ &= \cos x \cos y + (\sin x)\cdot(-\sin y) \\ &= \cos x \cos y - \sin x \sin y\end{aligned}$$

and

$$\begin{aligned}\sin(x+y) &= \cos\left[\frac{\pi}{2} - (x+y)\right] \\ &= \cos\left[\left(\frac{\pi}{2} - x\right) - y\right] \\ &= \cos\left(\frac{\pi}{2} - x\right)\cos y + \sin\left(\frac{\pi}{2} - x\right)\sin y \\ &= \sin x \cos y + \cos\left[\frac{\pi}{2} - \left(\frac{\pi}{2} - x\right)\right]\sin y \\ &= \sin x \cos y + \cos x \sin y\end{aligned}$$

The examples that follow include some specific applications and also some new general formulas. In the next section we summarize the important identities of trigonometry.

EXAMPLE:
$$\begin{aligned}\cos 75^\circ &= \cos(30^\circ + 45^\circ)\\ &= \cos 30^\circ \cos 45^\circ - \sin 30^\circ \sin 45^\circ\\ &= \frac{\sqrt{3}}{2}\cdot\frac{\sqrt{2}}{2} - \frac{1}{2}\cdot\frac{\sqrt{2}}{2}\\ &= \frac{\sqrt{6}}{4} - \frac{\sqrt{2}}{4} = \frac{(\sqrt{6}-\sqrt{2})}{4}\end{aligned}$$ ■

EXAMPLE:
$$\begin{aligned}\sin 75^\circ &= \sin(30^\circ + 45^\circ)\\ &= \sin 30^\circ \cos 45^\circ + \cos 30^\circ \sin 45^\circ\\ &= \frac{1}{2}\cdot\frac{\sqrt{2}}{2} + \frac{\sqrt{3}}{2}\cdot\frac{\sqrt{2}}{2}\\ &= \frac{\sqrt{2}}{4} + \frac{\sqrt{6}}{4} = \frac{(\sqrt{2}+\sqrt{6})}{4}\end{aligned}$$ ■

EXAMPLE: We find a formula for $\sin(x - y)$.

$$\begin{aligned}\sin(x - y) &= \sin[x + (-y)]\\ &= \sin x \cos(-y) + \cos x \sin(-y)\\ &= \sin x \cos y + (\cos x)\cdot(-\sin y)\end{aligned}$$

Therefore

$$\sin(x - y) = \sin x \cos y - \cos x \sin y$$ ■

EXAMPLE: We find a formula for $\sin(\theta - 180^\circ)$ as follows

$$\begin{aligned}\sin(\theta - 180^\circ) &= \sin\theta \cos 180^\circ - \cos\theta \sin 180^\circ\\ &= (\sin\theta)\cdot(-1) - (\cos\theta)\cdot(0)\\ &= -\sin\theta\end{aligned}$$

and so

$$\sin(\theta - 180^\circ) = -\sin\theta$$ ■

EXAMPLE: $\sin(190^\circ - 180^\circ) = -\sin 190^\circ$, by the preceding example. Therefore,

$$\sin 10^\circ = -\sin 190^\circ$$

or

$$\sin 190^\circ = -\sin 10^\circ$$ ■

EXAMPLE: We find a formula for $\sin(x + \pi/6)$ as follows.

$$\begin{aligned}\sin\left(x+\frac{\pi}{6}\right) &= \sin x \cos\frac{\pi}{6} + \cos x \sin\frac{\pi}{6} \\ &= (\sin x)\cdot\left(\frac{\sqrt{3}}{2}\right) + (\cos x)\cdot\left(\frac{1}{2}\right) \\ &= \frac{\sqrt{3}}{2}\sin x + \frac{1}{2}\cos x\end{aligned}$$

and so

$$\sin\left(x+\frac{\pi}{6}\right) = \frac{\sqrt{3}}{2}\sin x + \frac{1}{2}\cos x \;■$$

4–3 *Exercises*

Obtain formulas for the following

1. $\cos(\theta - 180°)$

2. $\cos(x + \pi)$

3. $\cos(x + \pi/6)$

4. $\cos(\theta - 30°)$

5. $\sin(\theta + 60°)$

6. $\sin(x - \pi/3)$

7. $\sin(x + \pi/2)$

8. $\cos(\theta + 90°)$

Derive the following.

9. $\cos 2\theta = \cos^2\theta - \sin^2\theta$

10. $\sin 2\theta = 2\sin\theta\cos\theta$

Find formulas for

11. $\cos 3\theta$

12. $\sin 3\theta$

To verify the four *product-to-sum identities*, use substitutions chosen from

$$\begin{aligned}\sin(x+y) &= \sin x\cos y + \cos x \sin y \\ \sin(x-y) &= \sin x\cos y - \cos x \sin y \\ \cos(x+y) &= \cos x\cos y - \sin x \sin y \\ \cos(x-y) &= \cos x\cos y + \sin x \sin y\end{aligned}$$

13. $\sin x\cos y = \frac{1}{2}[\sin(x+y) + \sin(x-y)]$

14. $\cos x\sin y = \frac{1}{2}[\sin(x+y) - \sin(x-y)]$

15. $\cos x\cos y = \frac{1}{2}[\cos(x+y) + \cos(x-y)]$

16. $\sin x\sin y = -\frac{1}{2}[\cos(x+y) - \cos(x-y)]$

In the formulas of Exercises 13 through 16, use the substitution $A = x + y$ and $B = x - y$ to derive the following *sum-to-product identities*.

17. $\sin A + \sin B = 2\sin\frac{1}{2}(A+B)\cos\frac{1}{2}(A-B)$

18. $\sin A - \sin B = 2\cos\frac{1}{2}(A+B)\sin\frac{1}{2}(A-B)$

19. $\cos A + \cos B = 2 \cos \frac{1}{2}(A + B) \cos \frac{1}{2}(A - B)$
20. $\cos A - \cos B = -2 \sin \frac{1}{2}(A + B) \sin \frac{1}{2}(A - B)$

Use an identity from one of Exercises 13 through 20 to evaluate or simplify the following.
21. $\sin 15° \cos 45°$
22. $\cos 22\frac{1}{2}° \sin 67\frac{1}{2}°$
23. $\sin 97\frac{1}{2}° \sin 217\frac{1}{2}°$
24. $\cos 105° - \cos 135°$
25. $\cos 67\frac{1}{2}° + \cos 112\frac{1}{2}°$
26. $\sin 7\frac{1}{2}° + \sin 37\frac{1}{2}°$

-4 Trigonometric identities

This is a convenient moment to review the most important identities obtained thus far.

Simplification identities:

$$\tan x = \frac{\sin x}{\cos x} \qquad \text{if } \cos x \neq 0$$

$$\cot x = \frac{\cos x}{\sin x} \qquad \text{if } \sin x \neq 0$$

$$\sec x = \frac{1}{\cos x} \qquad \text{if } \cos x \neq 0$$

$$\csc x = \frac{1}{\sin x} \qquad \text{if } \sin x \neq 0$$

Pythagorean identities:

$$\sin^2 x + \cos^2 x = 1$$
$$1 + \tan^2 x = \sec^2 x$$
$$1 + \cot^2 x = \csc^2 x$$

Addition and subtraction formulas

$$\cos (x + y) = \cos x \cos y - \sin x \sin y$$
$$\cos (x - y) = \cos x \cos y + \sin x \sin y$$
$$\sin (x + y) = \sin x \cos y + \cos x \sin y$$
$$\sin (x - y) = \sin x \cos y - \cos x \sin y$$

The identities in the following groups have been developed in the text and/or exercises. All can be recovered quickly from the addition and subtraction formulas.

Cofunction identities:

$$\sin\left(\frac{\pi}{2} - x\right) = \cos x$$

$$\cos\left(\frac{\pi}{2} - x\right) = \sin x$$

Reference identities:

$$\sin(\pi - x) = \sin x$$
$$\cos(\pi - x) = -\cos x$$
$$\sin(\pi + x) = -\sin x$$
$$\cos(\pi + x) = -\cos x$$
$$\sin(2\pi - x) = \sin(-x) = -\sin x$$
$$\cos(2\pi - x) = \cos(-x) = \cos x$$

Dougle-angle formulas:

$$\sin 2x = 2 \sin x \cos x$$
$$\cos 2x = \cos^2 x - \sin^2 x$$

Many more identities can be derived from these. In the following examples we verify some others which turn out to be useful in applications of trigonometry.

EXAMPLE: Verify that $\cos 2x = 2\cos^2 x - 1$.

We have

$$\begin{aligned}\cos 2x &= \cos^2 x - \sin^2 x \\ &= \cos^2 x - (1 - \cos^2 x) \\ &= 2\cos^2 x - 1 \;\blacksquare\end{aligned}$$

In Exercise 8 you are asked to check a closely-related identity:

$$\cos 2x = 1 - 2\sin^2 x$$

EXAMPLE: Verify the cofunction identity $\tan(\pi/2 - x) = \cot x$.

We have

$$\begin{aligned}\tan\left(\frac{\pi}{2} - x\right) &= \frac{\sin(\pi/2 - x)}{\cos(\pi/2 - x)} \\ &= \frac{\cos x}{\sin x} \\ &= \cot x \qquad (\text{provided } \sin x \neq 0) \;\blacksquare\end{aligned}$$

EXAMPLE: Verify the formula

$$\tan(x + y) = \frac{\tan x + \tan y}{1 - \tan x \tan y}$$

We have

$$\begin{aligned}
\tan(x+y) &= \frac{\sin(x+y)}{\cos(x+y)} \\
&= \frac{\sin x \cos y + \cos x \sin y}{\cos x \cos y - \sin x \sin y} \\
&= \frac{\dfrac{\sin x \cos y}{\cos x \cos y} + \dfrac{\cos x \sin y}{\cos x \cos y}}{\dfrac{\cos x \cos y}{\cos x \cos y} - \dfrac{\sin x \sin y}{\cos x \cos y}} \\
&= \frac{\dfrac{\sin x}{\cos x} + \dfrac{\sin y}{\cos y}}{1 - \dfrac{\sin x}{\cos x} \cdot \dfrac{\sin y}{\cos y}} \\
&= \frac{\tan x + \tan y}{1 - \tan x \tan y}
\end{aligned}$$

provided no division by zero is involved. ■

EXAMPLE: Verify the half-angle formula $\cos \frac{1}{2}x = \pm\sqrt{\frac{1}{2}(1 + \cos x)}$ (The sign depends on the quadrant of $\frac{1}{2}x$).

We have, from an example earlier in this section,

$$\cos 2x = 2\cos^2 x - 1$$

We solve for $\cos x$ as follows.

$$\begin{aligned}
\cos 2x &= 2\cos^2 x - 1 \\
2\cos^2 x &= 1 + \cos 2x \\
\cos^2 x &= \tfrac{1}{2}(1 + \cos 2x) \\
\cos x &= \pm\sqrt{\tfrac{1}{2}(1 + \cos 2x)}
\end{aligned}$$

We now let $y = 2x$, and so $x = \frac{1}{2}y$. Making this substitution, we obtain

$$\cos \tfrac{1}{2}y = \pm\sqrt{\tfrac{1}{2}(1 + \cos y)}$$

or, equivalently,

$$\cos \tfrac{1}{2}x = \pm\sqrt{\tfrac{1}{2}(1 + \cos x)}$$ ■

In Exercise 12 you are asked to check the half-angle formula for $\sin \frac{1}{2}x$:

$$\sin \tfrac{1}{2}x = \pm\sqrt{\tfrac{1}{2}(1 - \cos x)}$$

Finally, we note two identities which are often useful in calculus:

$$\sin^2 x = \tfrac{1}{2}(1 - \cos 2x)$$
$$\cos^2 x = \tfrac{1}{2}(1 + \cos 2x)$$

The first of these is obtained from the identity

$$\cos 2x = 1 - 2\sin^2 x$$

as follows.

$$\begin{aligned} \cos 2x &= 1 - 2\sin^2 x \\ 2\sin^2 x &= 1 - \cos 2x \\ \sin^2 x &= \tfrac{1}{2}(1 - \cos 2x) \end{aligned}$$

The second identity is gotten similarly from the identity

$$\cos 2x = 2\cos^2 x - 1$$

4–4 *Exercises*

Verify the identities.

1. $\tan(-x) = -\tan x$
2. $\cot(-x) = -\cot x$
3. $\sec(-x) = \sec x$
4. $\csc(-x) = -\csc x$
5. $\cot(\pi/2 - x) = \tan x$
6. $\sec(\pi/2 - x) = \csc x$
7. $\csc(\pi/2 - x) = \sec x$
8. $\cos 2x = 1 - 2\sin^2 x$
9. $\tan(x - y) = \dfrac{\tan x - \tan y}{1 + \tan x \tan y}$
10. $\cot(x + y) = \dfrac{\cot x \cot y - 1}{\cot x + \cot y}$
11. $\cot(x - y) = \dfrac{1 + \cot x \cot y}{\cot y - \cot x}$
12. $\sin \tfrac{1}{2}x = \pm\sqrt{\tfrac{1}{2}(1 - \cos x)}$
13. $\tan \tfrac{1}{2}x = \pm\sqrt{(1 - \cos x)/(1 + \cos x)}$
14. $\cot \tfrac{1}{2}x = \pm\sqrt{(1 + \cos x)/(1 - \cos x)}$

Use the half-angle formulas to find the following.

15. $\sin 22\tfrac{1}{2}°$
16. $\cos 22\tfrac{1}{2}°$
17. $\tan 22\tfrac{1}{2}°$
18. $\sin 15°$
19. $\cos 15°$
20. $\sin 7\tfrac{1}{2}°$
21. $\cos 7\tfrac{1}{2}°$

22. For what angles can you find the trigonometric functions without use of table? Explain.

4-5 Verification of identities

A *trigonometric equation* is an equation containing one or more trigonometric functions of the same variable. The equation may be an identity—that is, true for all values of the variable—or it may be a conditional equation—that is, true for only certain values of the variable. For now, we will concentrate on determining whether a given trigonometric equation is an identity or not. Solution of conditional trigonometric equations will be discussed in the next section. In this section, we will show how to recognize identities.

First, we will show how to determine if the equation is not an identity. To show that an equation is not an identity and thus a conditional equation, it is enough to find one value of the variable for which the equation is not true.

EXAMPLE: Show that the equation

$$\cos 2x = 2 \cos x$$

is not an identity.

We try $x = 0$. We have

$$\cos 2x = \cos 2 \cdot 0 = \cos 0 = 1$$

but

$$2 \cos x = 2 \cos 0 = 2$$

Therefore the equation is false for $x = 0$, and so it is a conditional equation, not an identity. ■

A trigonometric equation in two variables is an identity only if it is satisfied by all ordered pairs in the replacement set.

EXAMPLE: Show that the equation

$$\sin (x + y) = \sin x + \sin y$$

is not an identity.

We try the ordered pair $(\pi, \pi/2)$. We then have

$$\sin (x + y) = \sin \left(\pi + \frac{\pi}{2}\right) = \sin 3\pi/2 = -1$$

but

$$\sin x + \sin y = \sin \pi + \sin \frac{\pi}{2} = 0 + 1 = 1.$$

The equation is false for $(x, y) = (\pi, \pi/2)$ and so is not an identity. ■

Now, we will show how to determine if the equation is an identity. To do this, it is desirable to gain skill in verifying identities—that is, to be able to prove that a given equation is indeed an identity. This skill comes through practice and knowing the proper technique to use. It demands familiarity with the many identities alread developed. The most important of these identities are summarized in the previou section, and again in Appendix B. However, in some of the more challengin exercises of this section, you will need to use your ingenuity to select identities fror past examples and exercises. Good luck!

The actual technique for proving an identity is straightforward. We begin wit the equation that is asserted to be an identity and proceed in several steps. At eac step, a new, but equivalent, equation is obtained by substitution or some othe algebraic transformation. We examine each of these equations and if we arrive at a equation that is obviously an identity, it follows that the original equation is als an identity.

The following examples illustrate the general method. Throughout, it is assume that the replacement set is the set of all real numbers for which both sides of th equation are defined.

EXAMPLE: Verify that the following trigonometric equation is an identity

$$\csc x \tan x = \sec x$$

It often proves to be most convenient to use the simplification identities s that only the sine and cosine functions appear. In this case, the steps are as follow:

$$\csc x \tan x = \sec x$$

$$\frac{1}{\sin x} \cdot \frac{\sin x}{\cos x} = \frac{1}{\cos x}$$

$$\frac{1}{\cos x} = \frac{1}{\cos x}$$

This last is an identity, and therefore the original equation is also an identity. ■

EXAMPLE: Verify the identity

$$\frac{\sin x}{\sec x} = \frac{1}{\tan x + \cot x}$$

Cross-multiplication yields the equivalent equation

$$(\sin x)(\tan x + \cot x) = \sec x$$

or

$$\sin x \tan x + \sin x \cot x = \sec x$$

Now the simplification identities give us

$$(\sin x) \cdot \frac{\sin x}{\cos x} + (\sin x) \cdot \frac{\cos x}{\sin x} = \frac{1}{\cos x}$$

$$\frac{\sin^2 x}{\cos x} + \cos x = \frac{1}{\cos x}$$

Multiplying each side by cos x, we have

$$\sin^2 x + \cos^2 x = 1$$

This is a Pythagorean identity, and so the verification that our equation is an identity is complete. ■

EXAMPLE: Verify that

$$\sec x \csc x = 2 \csc 2x$$

is an identity.

We have

$$\frac{1}{\cos x} \cdot \frac{1}{\sin x} = \frac{2}{\sin 2x}$$

By the double-angle formula, this is

$$\frac{1}{\cos x} \cdot \frac{1}{\sin x} = \frac{2}{2 \sin x \cos x}$$

This last is obviously an identity. ■

EXAMPLE: Show that

$$\tan \frac{x}{2} = \frac{1 - \cos x}{\sin x}$$

is an identity.

We make the substitution $y = \frac{1}{2}x$, and so $x = 2y$, in order to rewrite this equation as

$$\tan y = \frac{1 - \cos 2y}{\sin 2y}$$

By the double-angle formulas, this is equivalent to

$$\tan y = \frac{1 - (1 - 2 \sin^2 y)}{2 \sin y \cos y}$$

Proceeding, we have

$$\tan y = \frac{2 \sin^2 y}{2 \sin y \cos y}$$

$$\tan y = \frac{\sin y}{\cos y}$$

$$\tan y = \tan y$$

and the identity is proved. ■

EXAMPLE: Verify the identity

$\cos^4 x + \sin^4 x = 1 - \frac{1}{2}\sin^2 2x$

Here is one possible sequence of steps to show that we have an identity (ther are many other possible sequences):

$$\begin{aligned}
\cos^4 x + \sin^4 x &= 1 - \tfrac{1}{2}\sin^2 2x \\
\cos^4 x + \sin^4 x &= 1 - \tfrac{1}{2}(2\sin x \cos x)^2 \\
\cos^4 x + \sin^4 x &= 1 - \tfrac{1}{2}(4\sin^2 x \cos^2 x) \\
\cos^4 x + \sin^4 x &= 1 - 2\sin^2 x \cos^2 x \\
\cos^4 x + \sin^4 x &= 1 - 2(\sin^2 x)(1 - \sin^2 x) \\
\cos^4 x + \sin^4 x &= 1 - 2\sin^2 x + 2\sin^4 x \\
(\cos^2 x)^2 + \sin^4 x &= 1 - 2\sin^2 x + 2\sin^4 x \\
(1 - \sin^2 x)^2 + \sin^4 x &= 1 - 2\sin^2 x + 2\sin^4 x \\
1 - 2\sin^2 x + \sin^4 x + \sin^4 x &= 1 - 2\sin^2 x + 2\sin^4 x \\
1 - 2\sin^2 x + 2\sin^4 x &= 1 - 2\sin^2 x + 2\sin^4 x
\end{aligned}$$

Thus, since both sides are equal we have an identity. ■

◇4–5 *Exercises*

Verify that each is not an identity.

1. $\sin 2x = 2\sin x$

2. $\cos(-x) = -\cos x$

3. $\sin(x - y) = \sin x - \sin y$

4. $\cos(x + y) = \cos x + \cos y$

Verify that each is an identity.

5. $\sin x \cot x = \cos x$

6. $\cos(x + \pi) - \cos(x - \pi) = 0$

7. $\cos^2 x = \frac{1}{2}(1 + \cos 2x)$

8. $\sin^2 x = \frac{1}{2}(1 - \cos 2x)$

9. $\dfrac{\tan x}{\sec x - 1} = \dfrac{\sec x + 1}{\tan x}$

10. $\dfrac{\sin x}{\csc x} + \dfrac{\cos x}{\sec x} = 1$

11. $\dfrac{\sin x - \cos x + 1}{\sin x + \cos x - 1} = \dfrac{\sin x + 1}{\cos x}$

12. $\dfrac{\sec x - \csc x}{\sec x + \csc x} = \dfrac{\tan x - 1}{\tan x + 1}$

13. $\dfrac{\sin x + \cos x}{\sin x \cos x} = \sec x + \csc x$

14. $\dfrac{\csc x + 1}{\csc x - 1} = \dfrac{1 + \sin x}{1 - \sin x}$

15. $\csc x = \dfrac{\sec x + \csc x}{1 + \tan x}$

16. $\dfrac{1}{\sec x - \tan x} = \sec x + \tan x$

17. $\sin 2x = \dfrac{2 \tan x}{1 + \tan^2 x}$

18. $\tan x = \dfrac{\sin 2x}{1 + \cos 2x}$

19. $\tan \dfrac{x}{2} = \dfrac{\sin x}{1 + \cos x}$

20. $\tan x + \cot x = \sec x \csc x$

21. $\dfrac{(1 + \sin x)(1 - \sin x)}{\cos x} = \cos x$

22. $(1 - \tan x)(1 - \cot x) = 2 - \sec x \csc x$
23. $\cot \frac{1}{2}x = \csc x + \cot x$
24. $\tan \frac{1}{2}x = \csc x - \cot x$
25. $\sec^2 x(1 - \sin^2 x) = 1$
26. $\sec^2 x + \csc^2 x = \sec^2 x \csc^2 x$
27. $\tan^2 x - \sin^2 x = \tan^2 x \sin^2 x$

28. $\dfrac{1 + \cos x}{1 - \cos x} - \dfrac{1 - \cos x}{1 + \cos x} = 4 \cot x \csc x$

29. $\cot^2 x - \cos^2 x = \cot^2 x \cos^2 x$
30. $\tan x (\sin x + \cot x \cos x) = \sec x$
31. $\sin^2 x (2 + \tan^2 x) = \sec^2 x - \cos^2 x$
32. $\cos 4x = 1 - 8 \sin^2 x \cos^2 x$
33. $\sin^2 x(2 + \tan^2 x) = \sin^2 x + \tan^2 x$
34. $\tan^4 x + \tan^2 x = \sec^4 x - \sec^2 x$
35. $\cos 2x = \cos^4 x - \sin^4 x$
36. $\sin^3 x = \frac{1}{4}(3 \sin x - \sin 3x)$
37. $\cos^3 x = \frac{1}{4}(\cos 3x + 3 \cos x)$
38. $\sin^4 x = \frac{3}{8} - \frac{1}{2} \cos 2x + \frac{1}{8} \cos 4x$
39. $\cos^4 x = \frac{3}{8} + \frac{1}{2} \cos 2x + \frac{1}{8} \cos 4x$

Home work

40. $\dfrac{\sin x - \sin y}{\sin x + \sin y} = \dfrac{\tan \frac{1}{2}(x - y)}{\tan \frac{1}{2}(x + y)}$

41. $\dfrac{\cos x + \cos y}{\cos x - \cos y} = -\cot \frac{1}{2}(x - y) \cot \frac{1}{2}(x + y)$

42. $\dfrac{\sin 4x + \sin 2x}{\cos 4x + \cos 2x} = \tan 3x$

43. $\sin^6 x + \cos^6 x = 1 - 3 \sin^2 x \cos^2 x$

44. $\sec^6 x - \tan^6 x = 1 + 3 \sec^2 x \tan^2 x$

◇4-6 Conditional trigonometric equations

A trigonometric equation that is not an identity is a conditional trigonometri equation. Our task in this section is to find the solution set of such an equation.

As with identifying trigonometric identities, skill for finding the solution set o conditional trigonometric equations comes with practice. Various techniques ar illustrated in the examples, but there are no set rules for determining which sub stitutions or simplifications will work best. Most often, however, it is wise to us trigonometric identities for substitution into the given equation so that an equatio is obtained where only one trigonometric function appears. We then solve for tha trigonometric function, and finally obtain the set of values of the variable for whic the equation is valid.

Because the trigonometric functions are periodic, it is typical for the solutio set to be infinite. We follow the customary practice and restrict the replacement se to be $\{x \mid 0 \le x < 2\pi\}$. Hence we consider only solutions between 0 and 2π (o between 0° and 360°, when we work in degrees).

As always, the results should be checked in the original equation so that error and extraneous solutions can be weeded out.

EXAMPLE: Solve

$$2 \sin x + 1 = 0$$

Our strategy is to solve for $\sin x$, and then to determine x. The steps are:

$$\begin{aligned} 2 \sin x + 1 &= 0 \\ 2 \sin x &= -1 \\ \sin x &= -\tfrac{1}{2} \end{aligned}$$

Because $\sin \pi/6 = \frac{1}{2}$, the reference angle for x is $\pi/6$. However, x must be in Quad rant III or Quadrant IV. Hence the solution set is $\{7\pi/6, 11\pi/6\}$. ■

EXAMPLE: Solve

$$\sin x + \cos^2 x + 1 = 0$$

Our plan is to use the substitution

$\cos^2 x = 1 - \sin^2 x$ (from a Pythagorean identity)

This will give us an equation involving $\sin x$ and $\sin^2 x$—that is, a quadratic equation in $\sin x$. We solve the quadratic for $\sin x$, and then determine x, as follows.

$$\begin{aligned} \sin x + \cos^2 x + 1 &= 0 \\ \sin x + (1 - \sin^2 x) + 1 &= 0 \\ \sin^2 x - \sin x - 2 &= 0 \\ (\sin x - 2)(\sin x + 1) &= 0 \\ \sin x = 2 \qquad \text{or} \qquad \sin x &= -1 \end{aligned}$$

Because $-1 \leq \sin x \leq 1$ for any real number x, no solutions result from $\sin x = 2$. From $\sin x = -1$, we obtain the solution set $\{3\pi/2\}$. ■

EXAMPLE: We solve

$3 \tan^2 x + 5 \sec x + 1 = 0$

The solution will be worked out by two different methods.

The first plan of attack is to use the simplification identities and then to clear the resulting equation of fractions. The resulting equation involves $\sin^2 x$, $\cos^2 x$, and $\cos x$. A Pythagorean substitution ($\sin^2 x = 1 - \cos^2 x$) reduces this to a quadratic in $\cos x$. The quadratic is solved for $\cos x$, and then x is determined. Here are the steps.

$$3 \tan^2 x + 5 \sec x + 1 = 0$$

$$3 \frac{\sin^2 x}{\cos^2 x} + 5 \frac{1}{\cos x} + 1 = 0$$

$$\begin{aligned} 3 \sin^2 x + 5 \cos x + \cos^2 x &= 0 \\ 3(1 - \cos^2 x) + 5 \cos x + \cos^2 x &= 0 \\ 2 \cos^2 x - 5 \cos x - 3 &= 0 \\ (2 \cos x + 1)(\cos x - 3) &= 0 \end{aligned}$$

$$2 \cos x + 1 = 0 \qquad \text{or} \qquad \cos x - 3 = 0$$
$$\cos x = -\tfrac{1}{2} \qquad \text{or} \qquad \cos x = 3$$

No solutions result from $\cos x = 3$. From $\cos x = -1/2$, we obtain $x = 2\pi/3$ or $x = 4\pi/3$.

We should check each of these solutions. If $x = 2\pi/3$, then $\tan x = -\sqrt{3}$ and $\sec x = -2$. We find that

$3(-\sqrt{3})^2 + (5) \cdot (-2) + 1 = 0$

as desired.

If $x = 4\pi/3$, then $\tan x = \sqrt{3}$ and $\sec x = -2$. This gives

$3(\sqrt{3})^2 + (5) \cdot (-2) + 1 = 0$

which is also valid. Thus both solutions check and the solution set is $\{2\pi/3, 4\pi/3\}$.

An alternate solution is based on the Pythagorean identity $1 + \tan^2 x = \sec^2 x$. Here are the steps:

$$\begin{aligned} 3\tan^2 x + 5\sec x + 1 &= 0 \\ 3(\sec^2 x - 1) + 5\sec x + 1 &= 0 \\ 3\sec^2 x - 3 + 5\sec x + 1 &= 0 \\ 3\sec^2 x + 5\sec x - 2 &= 0 \\ (3\sec x - 1)(\sec x + 2) &= 0 \end{aligned}$$

$$3\sec x - 1 = 0 \quad \text{or} \quad \sec x + 2 = 0$$

$$\sec x = \tfrac{1}{3} \quad \text{or} \quad \sec x = -2$$

Because $|\sec x| \geq 1$ for all real numbers x, no solutions will result from $\sec x = 1/3$. From $\sec x = -2$, we obtain the solution set $\{2\pi/3, 4\pi/3\}$, which agrees with the solution we found previously. ■

EXAMPLE: We solve

$$2\sin 2x - 3\sin x = 0$$

We are going to make a substitution

$$\sin 2x = 2\sin x\cos x$$

according to the double-angle formula. The resulting equation involves $\sin x$ and $\cos x$.

An equation involving $\sin x$ and $\cos x$ can be handled by a substitution

$$\sin x = \pm\sqrt{1 - \cos^2 x}$$

or

$$\cos x = \pm\sqrt{1 - \sin^2 x}$$

but easier methods may be available. Sometimes division by $\sin x$ or $\cos x$ leads to a simple equation in $\tan x$ or $\cot x$. In this particular example, factoring works out well. Here now are the steps involved in the solution.

$$\begin{aligned} 2\sin 2x - 3\sin x &= 0 \\ 2(2\sin x\cos x) - 3\sin x &= 0 \\ 4\sin x\cos x - 3\sin x &= 0 \\ (\sin x)(4\cos x - 3) &= 0 \end{aligned}$$

$$\sin x = 0 \quad \text{or} \quad \cos x = \tfrac{3}{4}$$

From $\sin x = 0$, we obtain $x = 0$ or $x = \pi$. To obtain a value for x when $\cos x = \frac{3}{4} = 0{\cdot}7500$, we use the table of Trigonometric Ratios and get the approximate solution $x = 41°$. Another possibility is the fourth-quadrant angle $360° - 41° = 319°$. A check verifies that each of these angles is valid. Expressing the solutions all in degrees, we have a solution set of $\{0°, 41°, 180°, 319°\}$. ■

Sometimes extra care is needed to obtain all the solutions, as in the following example.

EXAMPLE: Solve $\tan 3\theta + 1 = 0$, where θ is measured in degrees.

We have $\tan 3\theta = -1$, hence $3\theta = 135°$ or $3\theta = 315°$. This gives $\theta = 45°$ or $\theta = 105°$. However, these are not the only solutions!

If we wish to consider all possible θ from 0° to 360°, we must allow 3θ to go from 0° to $3 \cdot 360° = 1080°$. Or, said in another way, if we add 120° to θ, we will add $3 \cdot 120° = 360°$ to 3θ so as not to change the value of $\tan 3\theta$.

The values for 3θ which must be considered are then $3\theta = 135°$, 315°, 495°, 675°, 855°, and 1035°. Thus, the solution set for θ is {45°, 105°, 165°, 225°, 285°, 345°}. ■

This last example is typical of a general situation that we will elaborate on. First, we review what we did in the last example. We had obtained a value for $\tan 3\theta$. To get all solutions θ for $0° \leq \theta < 360°$, it was necessary for us to consider all possible 3θ satisfying the inequality $0° \leq 3\theta < 3 \cdot 360°$.

More generally, suppose the solution process yields a value of $f(k\theta)$, where f is a trigonometric function and k is an integer. To get all solutions θ for $0° \leq \theta < 360°$, we must consider all possible $k\theta$ satisfying $0° \leq k\theta < k \cdot 360°$. Or, if we are working in radians, to get all solutions for $0 \leq \theta < 2\pi$, we consider all θ satisfying $0 \leq k\theta < 2k\pi$. Let us consider an example.

EXAMPLE: We solve

$$\cos 3x \cos x + \sin 3x \sin x = 0$$

for all x such that $0 \leq x < 2\pi$.

By the formula for cosine of a difference, we have

$$\begin{aligned}\cos 3x \cos x + \sin 3x \sin x &= \cos (3x - x)\\ &= \cos 2x\end{aligned}$$

Thus our equation is

$$\cos 2x = 0$$

For $2x$ in the interval $0 \leq 2x < 2 \cdot 2\pi = 4\pi$, we have

$$2x = \pi/2,\ 3\pi/2,\ 5\pi/2, \text{ or } 7\pi/2$$

or

$$x = \pi/4,\ 3\pi/4,\ 5\pi/4, \text{ or } 7\pi/4$$

Hence the solution set is $\{\pi/4, 3\pi/4, 5\pi/4, 7\pi/4\}$. ■

◇4-6 *Exercises*

Find the solution set. The replacement set is $\{x|0 \leq x < 2\pi\}$.

1. $\sin x = \cos x$

2. $2 \cos x - 1 = 0$

3. $\sin 2x + \cos x = 0$

4. $4 \sin^2 x - 3 = 0$

5. $\sin 4x = 1$

6. $\cos 2x + \sin 2x = 0$

7. $\tan x + 2 \cos x = \sec x$

8. $\tan 2x + \cot 2x = 2$

9. $\sin^2 x - \cos^2 x = 1$

10. $3 + 3 \cos x = 2 \sin x$

Find all solutions with $0° \leq \theta < 360°$.

11. $2 \cos^2 \theta + 3 \sin \theta = 0$

12. $2 \sin \theta \cos \theta - \sin \theta - 2 \cos \theta + 1 = 0$

13. $3 \cot \theta - 1 = 0$

14. $\sin \theta + \cos \theta \tan \theta = 3$

15. $3 \sin \theta + \cos \theta = 0$

16. $2 \sin \theta \tan \theta + \tan \theta - 2 \sin \theta - 1 = 0$

17. $\tan \theta + 3 \cot \theta - 4 = 0$

18. $\sin \theta \cos \theta = \frac{1}{3}$

19. $\sin \theta + \cos \theta = 1$

20. $3 \sec^2 \theta + \cot^2 \theta - 7 = 0$

21. $\sin 3\theta \cos \theta - \cos 3\theta \sin \theta = 0$

22. $\sin 2\theta \cos \theta + \cos 2\theta \sin \theta = -\frac{1}{2}$

Review exercises (Chapter 4)

Complete the identities.

1. $\sin^2 x + \cos^2 x = ?$

2. $1 + \tan^2 x = ?$

3. $1 + \cot^2 x = ?$

4. $\sin(-x) = ?$

5. $\cos(-x) = ?$

6. $2 \sin x \cos x = ?$

7. $\cos^2 x - \sin^2 x = ?$

8. $2 \cos^2 x - 1 = ?$

9. $1 - 2 \sin^2 x = ?$

10. $\sin(x + y) = ?$

11. $\cos(x + y) = ?$

Given that B is a first-quadrant angle with $\sin B = \frac{4}{5}$, find:

12. $\cos B$

13. a formula for $\sin(A - B)$

14. a formula for $\cos(A - B)$

15. a formula for $\tan(A + B)$

16. $\sin 2B$

17. $\tan 2B$

18. $\cos \frac{1}{2}B$

19. $\cot \frac{1}{2}B$

Verify the cofunction identities:

20. $\tan(\pi/2 - \theta) = \cot \theta$

21. $\sec(\pi/2 - \theta) = \csc \theta$

◇**22.** Show that $\tan 2x = 2 \tan x$ is not an identity.

Verify the identities.

◇**23.** $\csc x - \cos x \cot x = \sin x$

◇**24.** $\sec^4 x - \tan^4 x = 1 + 2\tan^2 x$

◇**25.** $\csc x = \dfrac{\cot x + 1}{\sin x + \cos x}$

Find all solutions such that $0° \le \theta < 360°$.

◇**26.** $\cos^2 2\theta = \cos^2 \theta$

◇**27.** $3\cos^2 \theta + 11\sin \theta = 9$

5

Trigonometry of arbitrary triangles

The traditional applications of trigonometry, like measuring and surveying, demand the abilit to solve any type of triangle, not just a right triangle. In this chapter we develo that skill. Two key formulas, the law of sines and the law of cosines, are the basi of the computations.

5-1 The law of cosines

We turn our attention to the trigonometry of arbitrary triangles, not just righ triangles. Our practice will still be to use A, B, and C for the vertices of the triangle with a, b, and c the sides, lettered as in Figure 5–1. We no longer assume $\angle C$ t be a right triangle.

For a right triangle, the Pythagorean formula

$$c^2 = a^2 + b^2$$

relates the lengths of the sides. However, if $\triangle ABC$ is not a right triangle, thi formula is not valid. (Consider, for example, equilateral $\triangle ABC$ with $a = b = c = 1$.

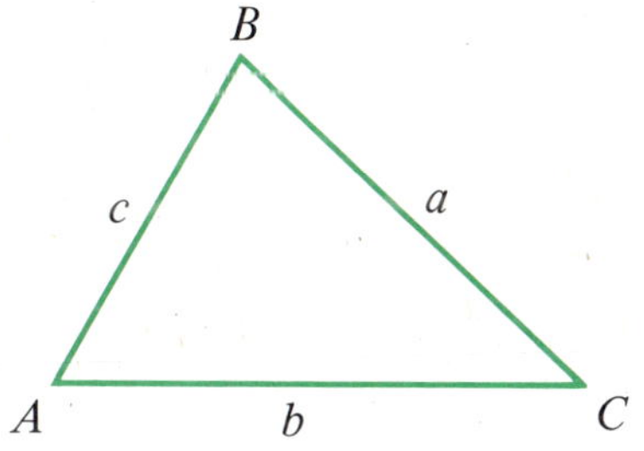

Figure 5–1
$\triangle ABC$

Still, we would like to have a formula relating the lengths of the sides of any triangle. In a right triangle, the measure of one angle is 90°. For a formula valid in any triangle, the additional ingredient needed is the measure of one angle, as in the following *law of cosines.*

In any triangle $\triangle ABC$, we have

$$c^2 = a^2 + b^2 - 2ab \cos C$$

Thus the law of cosines expresses the length of one side in terms of the lengths of the other two sides *and* the cosine of the angle included between them. Using $\angle A$ or $\angle B$ as the included angle, we get the alternate forms

$$a^2 = b^2 + c^2 - 2bc \cos A$$

and

$$b^2 = a^2 + c^2 - 2ac \cos B$$

Because $\cos 90° = 0$, for a right triangle with right angle at C we have

$$c^2 = a^2 + b^2 - 2ab \cos 90°$$

or simply

$$c^2 = a^2 + b^2$$

Thus the law of cosines includes the Pythagorean theorem as a special case.

The law of cosines involves both addition and subtraction. For that reason it is not efficient to use logarithms when the computations are based on the law of cosines. Formulas better suited for logarithmic calculation are presented in Appendix A.

Before proving the law of cosines, we give some examples.

EXAMPLE: In $\triangle ABC$, shown in Figure 5-2, we have $a = 9$ inches, $b = 4$ inches, and $\angle C = 60°$. Find the length of side c.

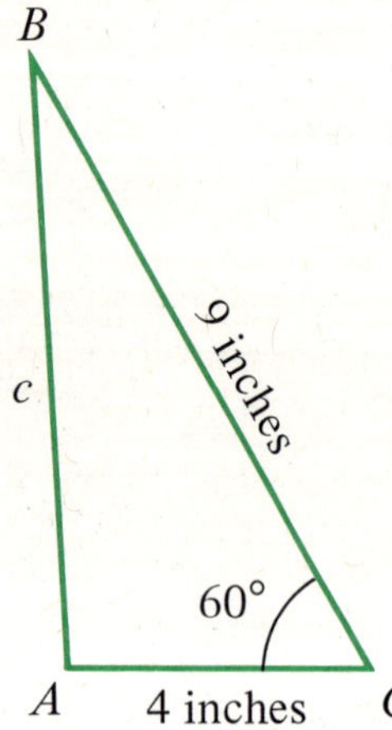

Figure 5–2

We have

$$\begin{aligned} c^2 &= a^2 + b^2 - 2ab \cos C \\ &= 9^2 + 4^2 - 2 \cdot 9 \cdot 4 \cos 60^\circ \\ &= 81 + 16 - 72 \cdot \tfrac{1}{2} \\ &= 61 \end{aligned}$$

Therefore $c = \sqrt{61}$, or approximately 7.810 inches. ■

In the preceding example we used the law of cosines to find the measure of the third side of a triangle, given two sides and the angle included between them. The next two examples illustrate how to find the measure of any angle, given all three sides.

EXAMPLE: In the triangle $\triangle ABC$ shown in Figure 5–3, we are given $a = 8$ units, $b = 10$, and $c = 7$. Find the measure of $\angle A$.

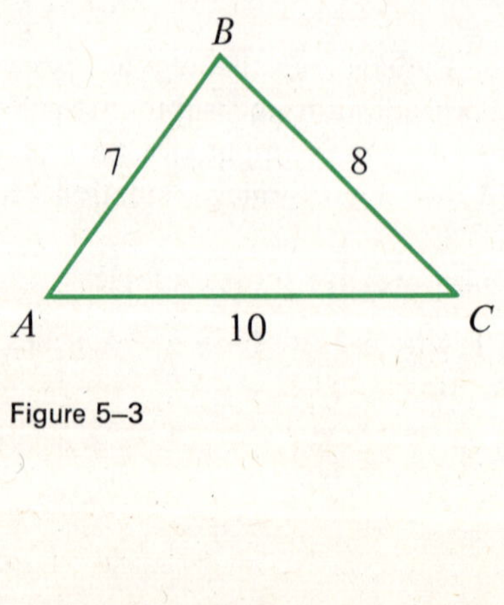

Figure 5–3

Because we wish to find $\angle A$, we use the formula

$$a^2 = b^2 + c^2 - 2bc \cos A$$

and have

$$8^2 = 10^2 + 7^2 - 2 \cdot 10 \cdot 7 \cos A$$

Therefore

$$140 \cos A = 100 + 49 - 64 = 85$$

and

$$\cos A = 85/140$$

or $\cos A$ is approximately 0.6071. From the table of Trigonometric Ratios, we find that $\angle A$ is approximately 37°. ■

EXAMPLE: Find the measure of $\angle B$ in $\triangle ABC$, given that $a = 12$ feet, $b = 19$ feet, and $c = 10$ feet (see Figure 5–4).

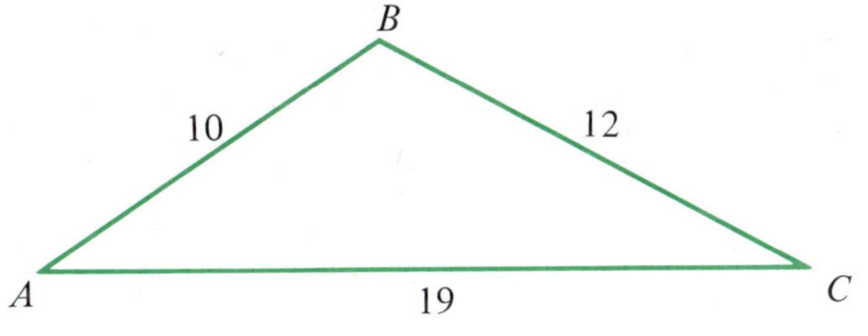

Figure 5–4

To find $\angle B$, we use the formula

$$b^2 = a^2 + c^2 - 2ac \cos B$$

Thus we have

$$19^2 = 12^2 + 10^2 - 2 \cdot 12 \cdot 10 \cos B$$

and so

$$240 \cos B = 144 + 100 - 361 = -117$$

Hence

$$\cos B = \frac{-117}{240} = -0.4875$$

It follows that B is an obtuse angle.[1] From the table of Trigonometric Ratios, the reference angle B^* is found to be 61°, and thus

$$\angle B = 180° - 61° = 119° \blacksquare$$

Geometrically, it is apparent that the sum of any two sides of a triangle must exceed the third side. (The shortest distance between two points is a straight line!) Otherwise there is no triangle with the three given sides. The difficulty will show up when we use the law of cosines.

EXAMPLE: Find the measure of $\angle C$ in ΔABC, given $a = 6$, $b = 2$, and $c = 3$.

We have

$$\begin{aligned} c^2 &= a^2 + b^2 - 2ab \cos C \\ 3^2 &= 6^2 + 2^2 - 2 \cdot 6 \cdot 2 \cos C \\ 9 &= 36 + 4 - 24 \cos C \\ 24 \cos C &= 31 \\ \cos C &= \frac{31}{24} \end{aligned}$$

This is impossible. For any angle $\angle C$, we must have $|\cos C| \leq 1$. There is no such triangle, which should come as no surprise, because $b + c \not> a$. $\blacksquare$

For the proof of the law of cosines, we refer to Figure 5–5. The altitude h from B meets AC at D. The lengths of AD and DC are denoted by u and v, respectively.

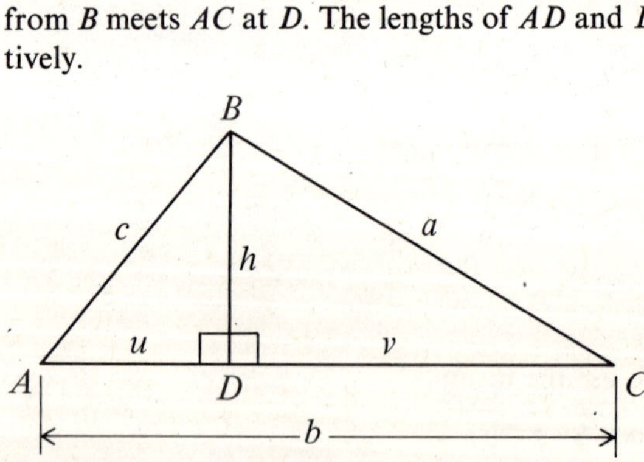

Figure 5–5

From right ΔADB, we have

$$c^2 = h^2 + u^2$$

From right ΔBDC, we have

$$\frac{h}{a} = \sin C$$

[1] An *obtuse* angle is an angle with measure between 90° and 180°.

and so

$$h = a \sin C$$

We also have

$$\frac{v}{a} = \cos C$$

and so

$$v = a \cos C$$

We further observe that

$$u = b - v$$

or

$$u = b - a \cos C$$

Putting this all together, we have

$$\begin{aligned} c^2 &= h^2 + u^2 \\ &= (a \sin C)^2 + (b - a \cos C)^2 \\ &= a^2 \sin^2 C + b^2 - 2ab \cos C + a^2 \cos^2 C \\ &= a^2(\sin^2 C + \cos^2 C) + b^2 - 2ab \cos C \\ &= a^2 + b^2 - 2ab \cos C \end{aligned}$$

because $\sin^2 C + \cos^2 C = 1$.

In this proof we have actually assumed that $\angle C$ is acute, because we drew the altitude AD inside $\triangle ABC$. The proof for $\angle C > 90°$ is left to you in Exercise 15.

5–1 *Exercises*

Find the measure of the indicated side of $\triangle ABC$.

1. $a = 4$, $b = 11$, $C = 60°$, $c = ?$
2. $a = 7$ feet, $c = 5$ feet, $B = 120°$, $b = ?$
3. $b = 6$ meters, $c = 9$ meters, $A = 90°$, $a = ?$
4. $a = 1$ inch, $b = 3$ inches, $C = \pi/4$, $c = ?$
5. $a = 8$, $c = 5$, $B = 28°$, $b = ?$
6. $b = 10$, $c = 20$, $A = 41°$, $a = ?$
7. $a = 100$ feet, $b = 50$ feet, $C = 63°$, $c = ?$
8. $a = 12$ meters, $c = 10$ meters, $B = 110°$, $b = ?$

Find the measures of the indicated angle of $\triangle ABC$, given that $a = 4$, $b = 5$, and $c = 6$.

9. $\angle A = ?$
10. $\angle B = ?$
11. $\angle C = ?$

Find the measures of the indicated angle of $\triangle ABC$, given that $a = 5$ feet, $b = 6$ feet, and $c = 9$ feet.

12. $\angle A = ?$

13. $\angle B = ?$

14. $\angle C = ?$

15. Prove the formula $c^2 = a^2 + b^2 - 2ab \cos C$ for the case $\angle C > 90°$. Use the figure shown and note that $u/a = \cos(180° - C)$ and $h/a = \sin(180° - C)$.

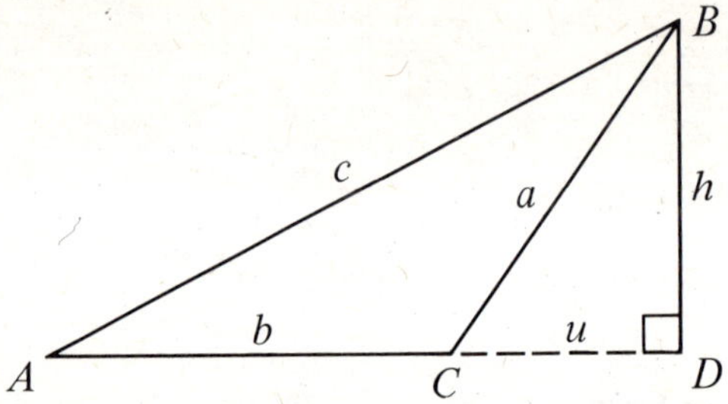

16. For $\angle A$ an angle in a triangle, we cannot have either $\cos A = 1$ or $\cos A = -1$. Why not?

The law of cosines can be applied to the triangle illustrated to derive the identity for the cosine of a difference. The steps are outlined in Exercises 17 through 19.

17. Let d be the distance between the points $(\cos\alpha, \sin\alpha)$ and $(\cos\beta, \sin\beta)$, as shown in the figure. What is the value for d^2 given by the distance formula?

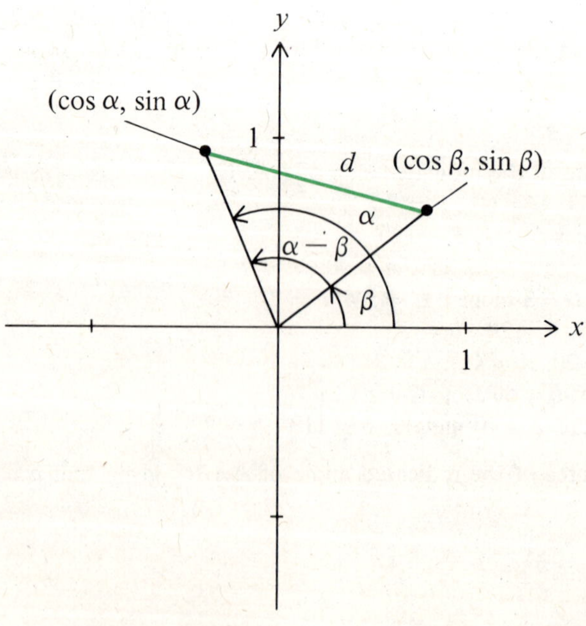

18. Use the law of cosines to get another value for d^2 (Note that $(\cos \alpha, \sin \alpha)$ and $(\cos \beta, \sin \beta)$ are each 1 unit away from the origin.)
19. Equate the values for d^2 obtained in Exercises 17 and 18 and simplify to derive the formula for $\cos(\alpha - \beta)$.

5-2 The law of sines

The law of cosines has proved helpful when all three sides of a triangle are known, or when we know two sides and the angle included between them. However, the law of cosines cannot be used when the known data consists of two angles and a side or two sides and the angle opposite one of them.

These latter cases can be handled with the following formula, the *law of sines*.

In any triangle ABC, we have

$$\frac{\sin A}{a} = \frac{\sin B}{b} = \frac{\sin C}{c}$$

or, equivalently,

$$\frac{a}{\sin A} = \frac{b}{\sin B} = \frac{c}{\sin C}$$

Before proceeding with the proof, we give some illustrations. Our first example shows how the law of sines is used to find the measure of an unknown side of a triangle, given the measures of two angles and another side.

EXAMPLE: In $\triangle ABC$, shown in Figure 5–6, we are given $a = 13.6$, $\angle A = 41°$, and $\angle C = 72°$. Find c.

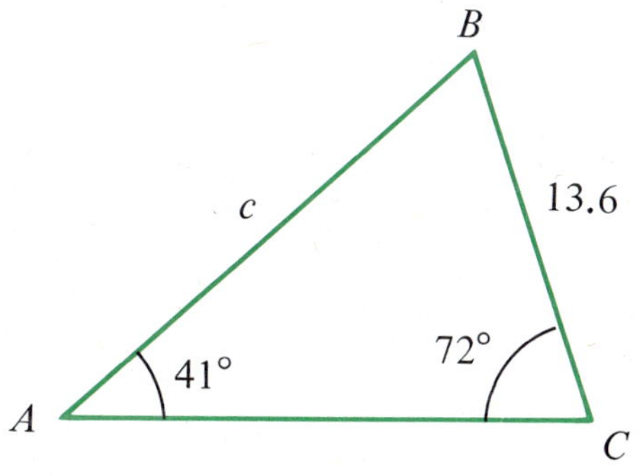

Figure 5–6

By the law of sines, in the form

$$\frac{a}{\sin A} = \frac{c}{\sin C}$$

we have

$$\frac{13.6}{\sin 41^\circ} = \frac{c}{\sin 72^\circ}$$

Hence

$$c = \frac{13.6 \sin 72^\circ}{\sin 41^\circ}$$

or

$$c = \frac{(13.6) \cdot (0.9511)}{0.6561} = 19.7 \text{ (approximately)}$$ ■

When we are given the measures of two sides and the angle opposite one of them, the law of sines can be used to find the measure of another of the angles. However, there are some pitfalls, as the following examples will show.

EXAMPLE: In $\triangle ABC$, we are given $\angle A = 35^\circ$, $a = 9$, and $b = 10$. Find $\angle B$.

By the law of sines, in the form

$(\sin A)/a = (\sin B)/b$

we have

$(\sin 35^\circ)/9 = (\sin B)/10$

and so

$$\begin{aligned} \sin B &= (10/9) \sin 35^\circ \\ &= (10/9) \cdot 0.5736 \\ &= 0.6373, \text{ approximately} \end{aligned}$$

Knowing that $\sin B = 0.6373$, we refer to the table of Trigonometric Ratios and find that $\angle B = 40^\circ$, assuming B to be an acute angle.

However, we cannot be sure that B is an acute angle! There is also a second-quadrant angle θ for which the value of $\sin \theta$ is 0.6373, because $\sin(180^\circ - \theta) = \sin \theta$. Thus the other possibility for $\angle B$ is the obtuse angle $180^\circ - 40^\circ = 140^\circ$. ■

Here is how to determine whether both possibilities in the preceding example are valid. Use the given value of $\angle A$ and the possibilities for $\angle B$ to calculate $\angle C$. Any solution that does not lead to a negative value for $\angle C$ is valid.

EXAMPLE: In the preceding example, we were given $\angle A = 35^\circ$. Which of the possibilities $\angle B = 40^\circ$ and $\angle B = 140^\circ$ are valid?

If $\angle B = 40^\circ$, then $\angle C = 180^\circ - 35^\circ - 40^\circ = 105^\circ$
If $\angle B = 140^\circ$, then $\angle C = 180^\circ - 35^\circ - 140^\circ = 5^\circ$

Both of these possibilities are acceptable, and there are actually two different triangles which are consistent with the given data. These two triangles are shown in Figure 5–7. Because of the possibility of two solutions, this situation (given two sides and the angle opposite one of them) is known as the *ambiguous case.* ■

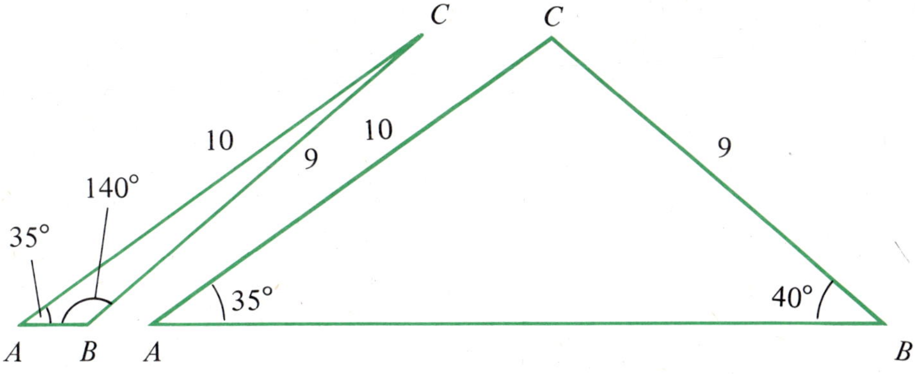

Figure 5–7
The two possibilities for $\triangle ABC$

EXAMPLE: Find $\angle C$ in $\triangle ABC$, given $\angle B = 23°$, $b = 11$, and $c = 8$.

By the law of sines, in the form

$$\frac{\sin B}{b} = \frac{\sin C}{c}$$

we have

$$\frac{\sin 23°}{11} = \frac{\sin C}{8}$$

and so

$$\begin{aligned}\sin C &= \frac{8}{11} \sin 23° \\ &= \frac{8}{11} \cdot 0.3907 \\ &= 0.2841\end{aligned}$$

Thus $\angle C = 17°$ or $\angle C = 180° - 17° = 163°$

If $\angle C = 17°$, then $\angle A = 180° - 23° - 17° = 140°$, which is acceptable.

However, if $\angle C = 163°$, then $\angle A = 180° - 23° - 163° = -6°$, which is impossible.

Therefore we must have $\angle C = 17°$, and the triangle is as shown in Figure 5–8. ■

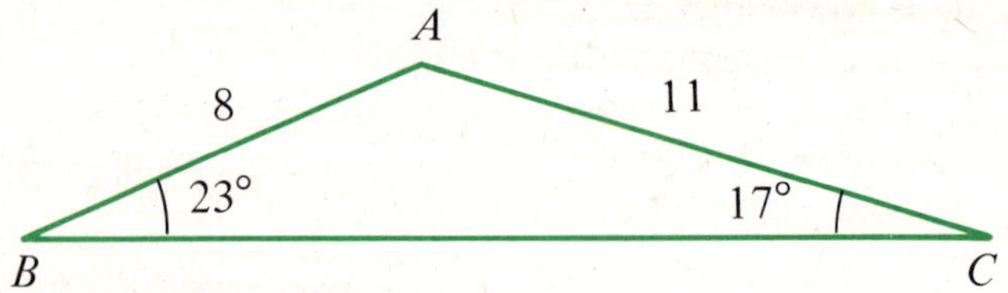

Figure 5–8

From the last two examples, we see that, given two sides and the angle opposite one of them, there may be one triangle or two triangles satisfying the given data. There may be no solution at all, as in the following example.

EXAMPLE: Find $\angle A$ in $\triangle ABC$, given $\angle C = 33°$, $c = 21$, and $a = 53$.

We use the law of sines in the form

$$\frac{\sin A}{a} = \frac{\sin C}{c}$$

Thus

$$\frac{\sin A}{53} = \frac{\sin 33°}{21}$$

and so

$$\begin{aligned}\sin A &= \frac{53}{21} \sin 33° \\ &= \frac{53}{21} \cdot 0.5446 \\ &= 1.374\end{aligned}$$

But there is no angle $\angle A$ such that $\sin A = 1.374$, because $-1 \leq \sin A \leq 1$ for every angle $\angle A$. Thus, there is no solution. ■

To understand the different possibilities in the ambiguous case, we look at the situation geometrically. Let us try to construct $\triangle ABC$, given $\angle A$ and sides a and b. We place $\angle A$ in standard position and measure off side b. This gives us the diagram in Figure 5–9. Then we try to complete the triangle by drawing an arc of radius a centered at C. The point where the arc meets the initial ray of $\angle A$ will be the vertex B.

The various possibilities are shown in Figures 5–10 through 5–13. If a is too short, there may be no point of intersection, and hence no triangle, as in Figure 5–10. There is a value for a that leads to one right triangle, as in Figure 5–11. For certain values of a, the arc meets the initial ray of $\angle A$ in two different points, and two triangles result, as in Figure 5–12. Finally, if a is large enough there is again only one point of intersection, and hence just one triangle, as shown by Figure 5–13:

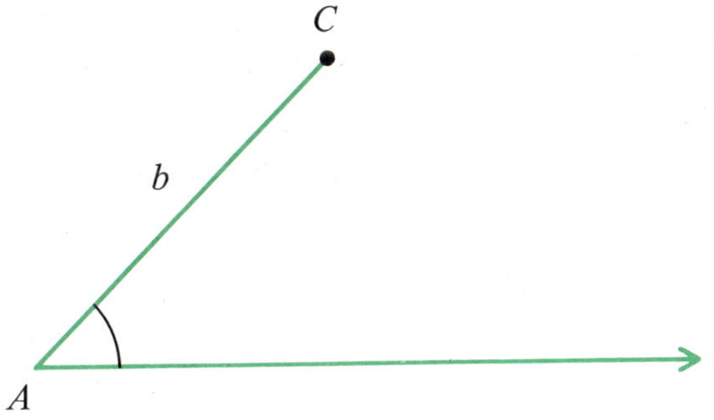

Figure 5–9

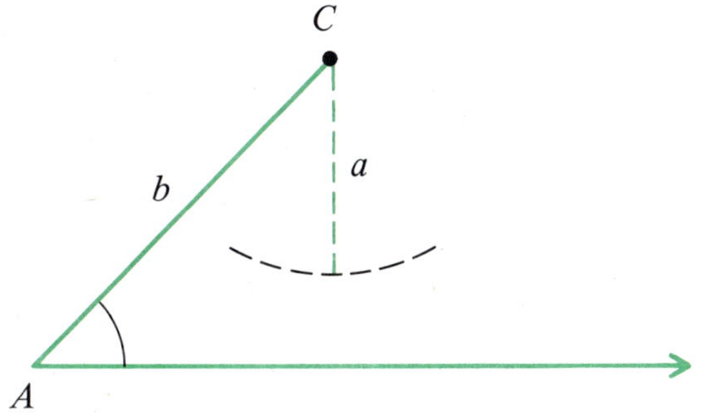

Figure 5–10
No triangle

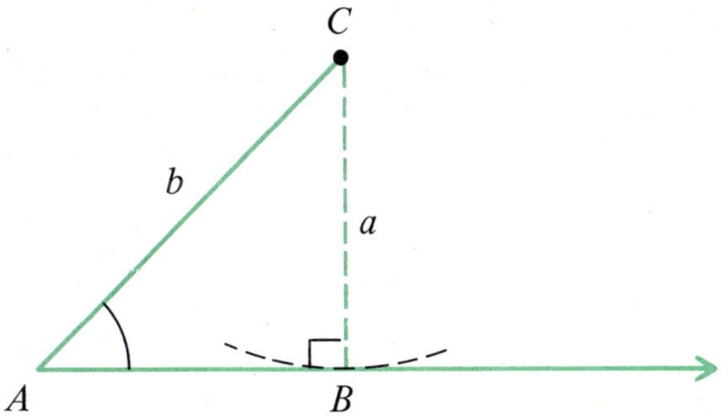

Figure 5–11
One right triangle

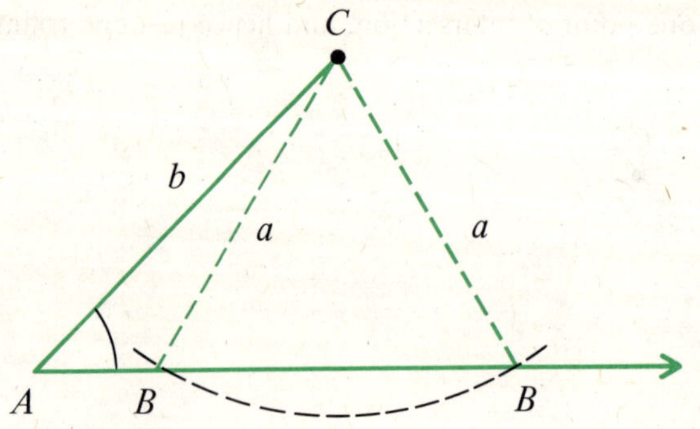

Figure 5–12
Two triangles

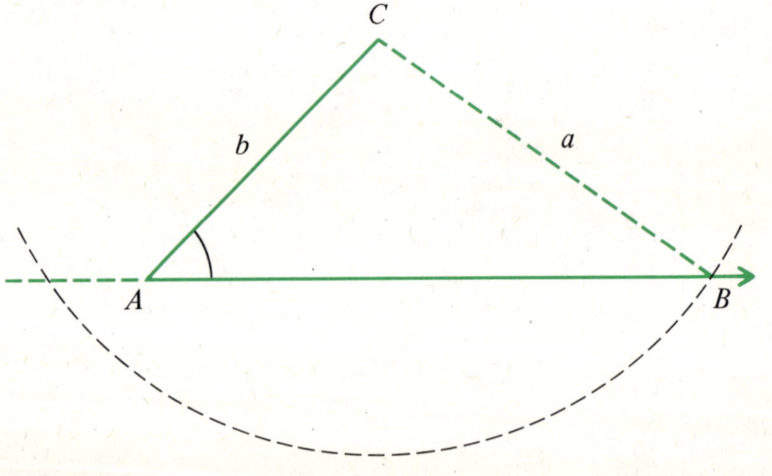

Figure 5–13
One triangle

We now turn to the proof of the law of sines. Figure 5–14 shows $\triangle ABC$, together with the altitude h drawn from B to AC. This altitude meets AC at D. In right $\triangle ADB$, we have $h/c = \sin A$, and so $h = c \sin A$. From right $\triangle BDC$, we have $h/a = \sin C$, and so $h = a \sin C$. We equate these two expressions for h and obtain

$$c \sin A = a \sin C$$

Dividing by ac, we get

$$\frac{\sin A}{a} = \frac{\sin C}{c}$$

A similar proof shows

$$\frac{\sin A}{a} = \frac{\sin B}{b}$$

and thus

$$\frac{\sin B}{b} = \frac{\sin C}{c}$$

also.

In this proof we have assumed that $\angle C$ is acute, because the altitude BD was drawn inside the triangle. You are asked to check the other case, for C an obtuse angle, in Exercise 14.

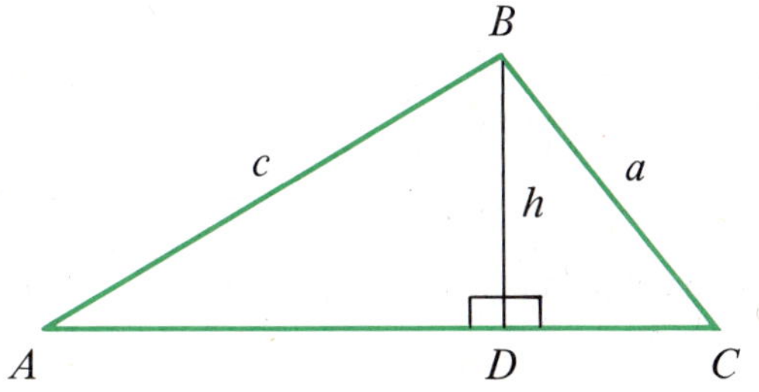

Figure 5–14

At this point it is convenient to use Figure 5–14 to derive a formula for the area of a triangle. The general principle is:

$$\text{Area} = \tfrac{1}{2}(\text{base}) \cdot (\text{altitude})$$

In Figure 5–14, the base is b. We found the altitude h to be $h = a \sin C$ and also $h = c \sin A$. Thus the area K is:

$$K = \tfrac{1}{2}bh = \tfrac{1}{2}bc \sin A$$

or

$$K = \tfrac{1}{2}bh = \tfrac{1}{2}b(a \sin C) = \tfrac{1}{2}ab \sin C$$

A similar proof gives also $K = \tfrac{1}{2}ac \sin B$. To summarize:

The area K of $\triangle ABC$ is:

$$\begin{aligned} K &= \tfrac{1}{2}bc \sin A \\ &= \tfrac{1}{2}ab \sin C \\ &= \tfrac{1}{2}ac \sin B \end{aligned}$$

Thus, as for a right triangle (Section 2-4), the area is one-half the product of two sides and the sine of the angle included between them.

EXAMPLE: Find the area of $\triangle ABC$, given $\angle A = 46°$, $b = 13$, and $c = 22$.
The formula for the area K is

$$K = \tfrac{1}{2}bc \sin A$$

Thus

$$\begin{aligned} K &= \tfrac{1}{2} \cdot 13 \cdot 22 \cdot \sin 46° \\ &= \tfrac{1}{2} \cdot 13 \cdot 22 \cdot 0.7193 \\ &= 103 \text{ square units} \end{aligned}$$ ■

5–2 *Exercises*

Find the measure of the indicated side of $\triangle ABC$.

1. $\angle A = 38°$, $\angle B = 17°$, $a = 4$, $b = ?$
2. $\angle A = 38°$, $\angle B = 17°$, $a = 4$, $c = ?$
3. $\angle A = 112°$, $\angle C = 36°$, $a = 23$ feet, $b = ?$
4. $\angle A = 112°$, $\angle C = 36°$, $a = 23$ feet, $c = ?$
5. $\angle B = 76°$, $\angle C = 66°$, $a = 18$ meters, $b = ?$
6. $\angle B = 76°$, $\angle C = 66°$, $a = 18$ meters, $c = ?$

Find the measure of the indicated angle of $\triangle ABC$.

7. $\angle A = 80°$, $a = 40$, $b = 30$, $\angle B = ?$
8. $\angle A = 22°$; $a = 90$, $c = 150$, $\angle C = ?$
9. $\angle B = 43°$, $a = 50$ feet, $b = 10$ feet, $\angle A = ?$
10. $\angle B = 99°$, $b = 100$ feet, $c = 130$ feet, $\angle C = ?$
11. $\angle C = 104°$, $a = 35$ meters, $c = 50$ meters, $\angle A = ?$
12. $\angle C = 48°$, $b = 60$ meters, $c = 70$ meters, $\angle B = ?$

13. Check that the law of sines remains valid when $\triangle ABC$ is a right triangle.
14. Prove the law of sines for $\triangle ABC$ when C is an obtuse angle. Use the figure shown. Note that $h/c = \sin A$ and $h/a = \sin (180° - C)$.

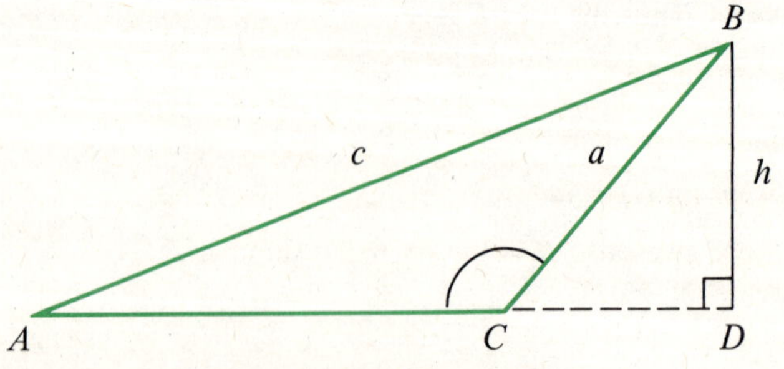

Find the area of $\triangle ABC$ from the given data.

15. $a = 28$ feet, $b = 34$ feet, $\angle C = 76°$
16. $a = 11$ inches, $c = 43$ inches, $\angle B = 20°$
17. $b = 100$ meters, $c = 40$ meters, $\angle A = 125°$

Use the formulas for K given in this section and the law of sines to derive the following additional formulas for the area K of $\triangle ABC$.

18. $K = \frac{1}{2}c^2 \dfrac{\sin A \sin B}{\sin C}$

19. $K = \frac{1}{2}a^2 \dfrac{\sin B \sin C}{\sin A}$

20. $K = \frac{1}{2}b^2 \dfrac{\sin A \sin C}{\sin B}$

Use the formulas from Exercises 18 through 20 to find the area of $\triangle ABC$ from the given data.

21. $\angle A = 13°$, $\angle B = 80°$, $c = 25$ feet

22. $\angle A = 76°$, $\angle B = 39°$, $a = 72$ meters

23. $\angle A = 26°$, $\angle B = 74°$, $b = 33$ inches.

Suppose we are given the measures of an acute angle $\angle A$ and sides a and b in $\triangle ABC$. Refer to Figures 5–10 through 5–13 to verify the following.

24. If $a < b \sin A$, there is no triangle satisfying the given data.

25. If $a = b \sin A$, there is one right triangle which fits the data.

26. If $b \sin A < a < b$, there are two triangles which fit the data.

27. If $b \leq a$, there is only one triangle which fits the data.

5-3 The solution of triangles

When enough data is given, we can use the law of sines and/or the law of cosines to solve a triangle—that is, to find the measures of the unknown sides and/or angles.

The exact procedure depends on the type of data given. The various cases are as follows.

Case 1. Given two angles and one side. Because the sum of the three angles is 180°, the third angle is found by subtracting the sum of the two given angles from 180°. The measures of the remaining sides can be calculated by the law of sines.

Case 2. Given the three sides. One angle is found by the law of cosines. Another is gotten by another application of the law of cosines or by the law of sines[2]. The

[2] When the law of sines is used here, an ambiguity may arise. It is sometimes resolved by noting that the larger angle is always opposite the longer side—that is, $a > b$ whenever $\angle A > \angle B$ in $\triangle ABC$. It may be necessary to compute the third angle as well, and/or to use that $\angle A + \angle B + \angle C = 180°$.

third is then available by subtraction from 180°.

Case 3. Given two sides and the angle included between them. The third side is gotten by the law of cosines. We then proceed as in Case 2.

Case 4. Given two sides and the angle opposite one of them. This is the ambiguou case. The possibilities for the angle opposite the other given side are computed b the law of sines. The third angle is found by subtraction, and any inconsisten possibility is eliminated. Finally, the remaining side is gotten by either the law o sines or the law of cosines.

The following examples illustrate each of these four cases.

EXAMPLE: In $\triangle ABC$, we have $\angle A = 46°$, $\angle B = 79°$, and $a = 31.2$ units. Solv the triangle.

This is Case 1 since we are given two angles and one side. The solution is a follows.

First,

$$\angle C = 180° - \angle A - \angle B = 180° - 46° - 79° = 55°$$

Next, from the law of sines, we get

$$b = \frac{a \sin B}{\sin A} = \frac{31.2 \sin 79°}{\sin 46°}$$

$$= (31.2) \cdot (0.9816)/0.7193$$
$$= 42.58$$

Also from the law of sines, we have

$$c = \frac{a \sin C}{\sin A} = \frac{31.2 \sin 55°}{\sin 46°}$$

$$= \frac{(31.2) \cdot (0.8192)}{0.7193}$$

$$= 35.53$$ ■

EXAMPLE: Solve $\triangle ABC$, where $a = 83.7$, $b = 76.2$, and $c = 88.4$.

This is Case 2 since all three sides are given. The process is as follows.

By the law of cosines,

$$c^2 = a^2 + b^2 - 2ab \cos C$$

Therefore

$$\cos C = \frac{a^2 + b^2 - c^2}{2ab}$$

$$= \frac{83.7^2 + 76.2^2 - 88.4^2}{2 \cdot (83.7) \cdot (76.2)}$$

$$= 0.3918$$

Therefore the value for $\angle C$ is 67°.

Similarly,

$$a^2 = b^2 + c^2 - 2bc \cos A$$

and so

$$\cos A = \frac{b^2 + c^2 - a^2}{2bc}$$

$$= \frac{76.2^2 + 88.4^2 - 83.7^2}{2 \cdot (76.2) \cdot (88.4)}$$

$$= 0.4910$$

Hence

$$\angle A = 61°$$

Finally,

$$\angle B = 180° - 67° - 61° = 52°$$ ■

EXAMPLE: Solve $\triangle ABC$, given $a = 8.27$ centimeters, $c = 5.38$ centimeters, and $\angle B = 28°$.

This example fits Case 3 since we are given two sides and the angle included between them.

From the law of cosines,

$$\begin{aligned} b^2 &= a^2 + c^2 - 2ac \cos B \\ &= 8.27^2 + 5.38^2 - 2 \cdot (8.27) \cdot (5.38) \cdot (0.8829) \\ &= 18.77 \end{aligned}$$

and

$$b = \sqrt{18.77} = 4.33 \text{ centimeters}$$

Then, from the law of sines,

$$\sin A = \frac{a \sin B}{b}$$

$$= \frac{8.27 \sin 28°}{4.33}$$

$$\begin{aligned} &= (8.27) \cdot (0.4695)/4.33 \\ &= 0.8967 \end{aligned}$$

Thus either

$$\angle A = 64°$$

or[3]

[3] This ambiguity would not arise if $\angle A$ were found by the law of cosines.

$$\angle A = 180° - 64° = 116°.$$

If

$$\angle A = 64°$$

then

$$\angle C = 180° - 28° - 64° = 88°$$

This is impossible, because, from the given data, $a > c$. Therefore $\angle A > \angle C$, bu $64° \not> 88°$. Thus, it follows that

$$\angle A = 116°$$

and so

$$\angle C = 180° - 116° - 28° = 36° \blacksquare$$

EXAMPLE: Solve $\triangle ABC$, given $\angle A = 19°$, $a = 6.81$, and $b = 10.9$.

These data fit the pattern of Case 4 since the two sides and the angle opposit one of them are given. Here is the solution.

By the law of sines,

$$\sin B = \frac{b \sin A}{a}$$

$$= \frac{10.9 \sin 19°}{6.81}$$

$$= \frac{(10.9) \cdot (0.3256)}{6.81}$$

$$= 0.5212$$

Therefore,

$$\angle B = 31°$$

or

$$\angle B = 180° - 31° = 149°$$

Both possibilities are valid. If $\angle B = 31°$, then

$$\angle C = 180° - 19° - 31° = 130°$$

Then

$$c = \frac{a \sin C}{\sin A}$$

$$= \frac{6.81 \sin 130°}{\sin 19°}$$

$$= \frac{(6.81) \cdot (0.7660)}{0.3256}$$

$$= 16.0$$

If

$$\angle B = 149^\circ$$

then

$$\angle C = 180^\circ - 19^\circ - 149^\circ = 12^\circ.$$

In this case

$$c = \frac{a \sin C}{\sin A}$$

$$= \frac{6.81 \sin 12^\circ}{\sin 19^\circ}$$

$$= \frac{(6.81) \cdot (0.2079)}{0.3256}$$

$$= 4.35$$

The two different solutions are shown in Figure 5–15. ■

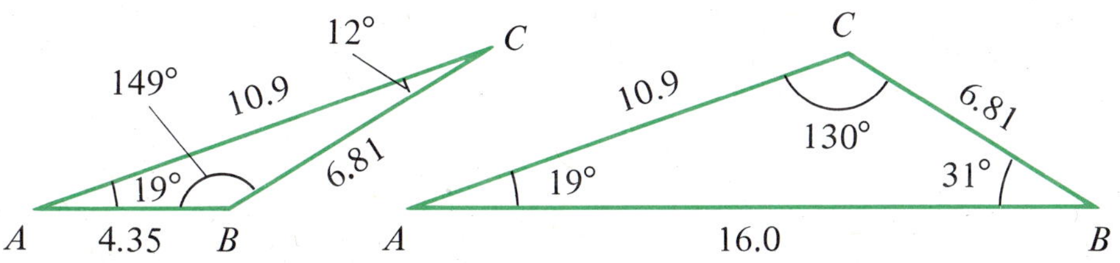

Figure 5–15

5–3 *Exercises*

Solve $\triangle ABC$ from the given data, and, where requested, find the area K.

1. $\angle A = 67^\circ$, $\angle B = 13^\circ$, $c = 97.3$
2. $\angle B = 42^\circ$, $\angle C = 70^\circ$, $a = 13.1$ meters
3. $\angle A = 98^\circ$, $\angle B = 53^\circ$, $b = 80.6$ feet, $K = ?$
4. $\angle A = 29^\circ$, $\angle C = 80^\circ$, $c = 41$ kilometers, $K = ?$
5. $a = 8$, $b = 9$, $c = 10$, $K = ?$
6. $a = 47$ meters, $b = 38$ meters, $c = 42$ meters, $K = ?$
7. $a = 6$ inches, $b = 2$ inches, $c = 5$ inches
8. $a = 19.2$ miles, $b = 17.3$ miles, $c = 14.8$ miles
9. $\angle A = 63^\circ$, $b = 46$ feet, $c = 15$ feet, $K = ?$
10. $\angle B = 89^\circ$, $a = 33$ meters, $c = 50$ meters, $K = ?$

11. $\angle C = 118°$, $a = 77.3$, $b = 89.3$, $K = ?$
12. $\angle A = 24°$, $b = 7$ kilometers, $c = 10$ kilometers, $K = ?$
13. $\angle A = 20°$, $a = 30$ miles, $c = 22$ miles
14. $\angle B = 34°$, $a = 47.1$, $b = 44.3$
15. $\angle C = 32°$, $a = 12$ feet, $c = 11$ feet
16. $\angle A = 80°$, $a = 75$ meters, $b = 70$ meters

17. Try to solve $\triangle ABC$, with $a = 1$, $b = 2$, and $c = 5$. What happens? Why?
18. Try to solve $\triangle ABC$ with $\angle B = 60°$, $b = 3$, and $c = 8$. What happens? Why?

5-4 Word problems

Practical problems in surveying, navigation, and physics often require the solutio of a triangle by trigonometry. Usually the first step should be to draw the picture

EXAMPLE: Two trees are on opposite sides of a pond. From a marker on th shore George sights first to one tree, then to the other, and then measures th angle between the lines of sight to be 86°. The distance from the marker to or tree is 53 feet, and to the other is 38 feet. How far apart are the trees?

First we draw the picture and label it, as illustrated in Figure 5–16. We need t find the length of side a, given that $\angle A = 86°$, $b = 38$ feet, and $c = 53$ feet. Th law of cosines affords the solution as follows:

$$\begin{aligned} a^2 &= b^2 + c^2 - 2bc \cos A \\ &= 38^2 + 53^2 - 2 \cdot 38 \cdot 53 \cos 86° \\ &= 1444 + 2809 - (4028) \cdot (0.0698) \\ &= 3972 \\ a &= \sqrt{3972} = 63.0 \end{aligned}$$

Thus the trees are 63 feet apart. ■

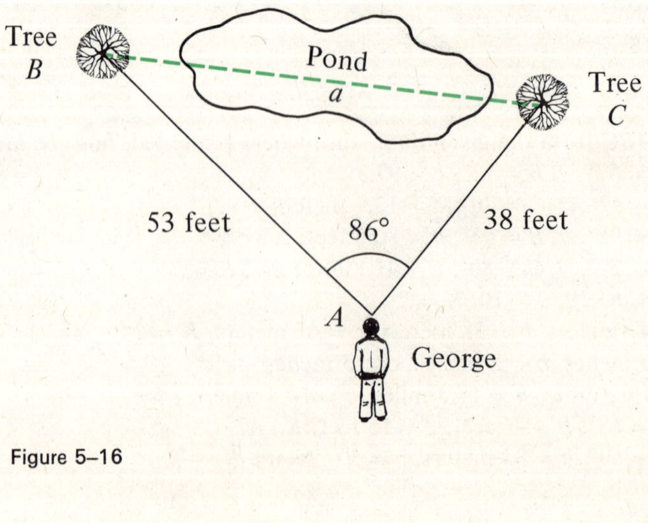

Figure 5–16

EXAMPLE: The pilot of a plane flying north at a speed of 600 miles per hour observes the bearing of a distant peak to be 45°. How far is the plane from the peak 1 minute later, if the bearing is then 55°?

The picture is as shown in Figure 5–17. In $\triangle ABC$, $\angle B = 180° - 55° = 125°$ and $\angle A = 180° - 125° - 45° = 10°$. At 600 miles per hour the plane flies 10 miles in 1 minute, and so $a = 10$. We find c by the law of sines:

$$c = (a \sin C)/\sin A$$

$$= \frac{10 \cdot \sin 45°}{\sin 10°}$$

$$= \frac{(10) \cdot (0.7071)}{0.1736}$$

$$= 40.7$$

and so the peak is now about 41 miles away. ■

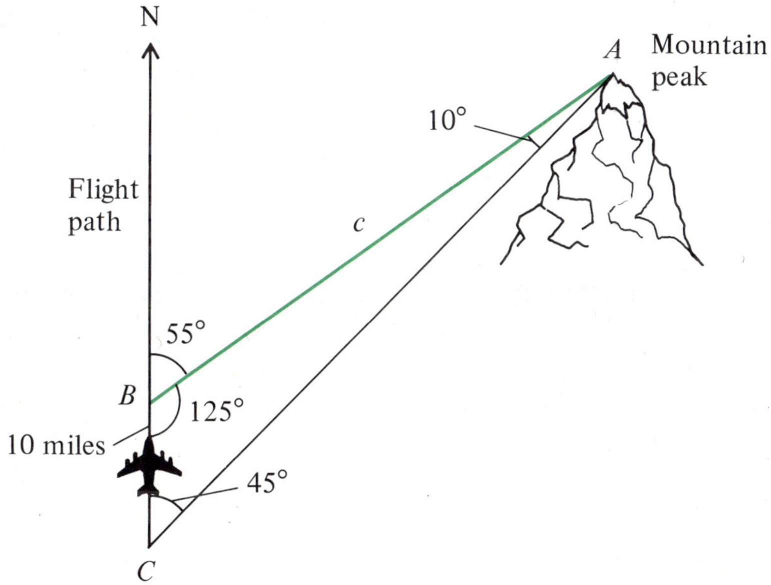

Figure 5–17

EXAMPLE: The Acme Realty offers for sale a triangular lot. Jennifer paces off the lengths of the sides to be roughly 63 feet, 143 feet, and 120 feet. What is the cost per square foot of the lot if the total price is $18,720?

The lot is shown in Figure 5–18. The area can be found by the formul $K = \frac{1}{2}bc \sin A$.

By the law of cosines,

$$a^2 = b^2 + c^2 - 2bc \cos A$$

and so

$$\begin{aligned}\cos A &= (b^2 + c^2 - a^2)/2bc \\ &= (63^2 + 143^2 - 120^2)/2 \cdot 63 \cdot 143 \\ &= 0.5560\end{aligned}$$

Hence

$$\sin A = \sqrt{1 - \cos^2 A} = \sqrt{1 - 0.5560^2} = 0.8312$$

and

$$K = \tfrac{1}{2}bc \sin A = \tfrac{1}{2} \cdot 63 \cdot 143 \cdot (0.8312) = 3744 \text{ square feet}$$

Therefore the cost per square foot is

$18{,}720 \div 3744 = \$5$ ■

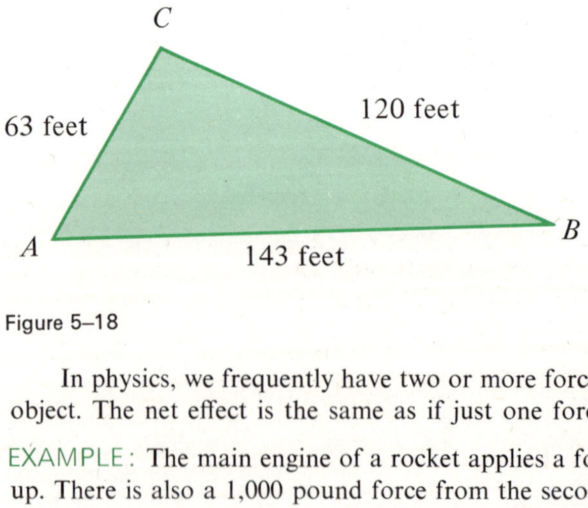

Figure 5–18

In physics, we frequently have two or more forces acting on the same body c object. The net effect is the same as if just one force, the resultant, were actin

EXAMPLE: The main engine of a rocket applies a force of 10,000 pounds straigh up. There is also a 1,000 pound force from the secondary, steering-control rock directed at an angle of 15° away from the vertical. Assuming these are the onl forces acting on the rocket, at what angle will it take off?

We must determine the direction of the resultant of the two forces. The situ ation is as shown in Figure 5–19. We need to find $\angle A$.

By the law of cosines,

$$\begin{aligned}c^2 &= a^2 + b^2 - 2ab \cos C \\ &= 10{,}000^2 + 1{,}000^2 - 2 \cdot 10{,}000 \cdot 1{,}000 \cos 165°\end{aligned}$$

from which we obtain

$c = 10{,}970$ pounds of force

From the law of sines, we have

$$\sin A = \frac{a \sin C}{c}$$

$$= \frac{1000 \sin 165°}{10970}$$

$$= 0.0236$$

and hence $\angle A = 1°$, approximately. ■

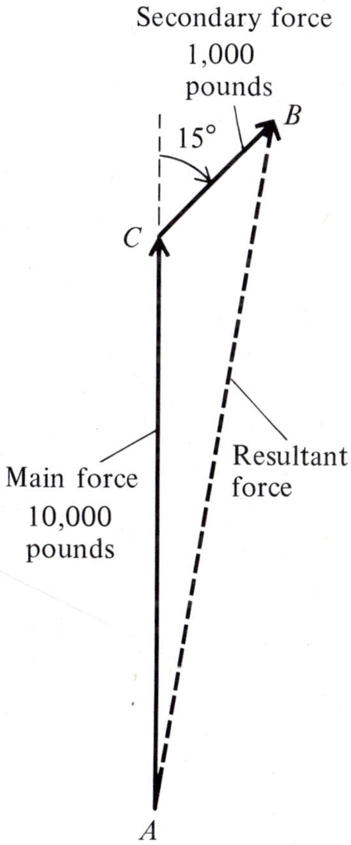

Figure 5–19

5–4 *Exercises*

1. To avoid sailing directly into the wind, a sailboat may have to travel a zig-zag course. To go to an island, the Zephyr started from port on a course of 13°. The Captain continued on this course for 6 miles, then set a new course of 93° and continued 8 miles until the Zephyr reached the island.

(a) How far is the island from the port?

(b) What is its bearing from port?

2. Thomas sets up his transit at the base of the lookout tower at the edge of the East Rim of a canyon, sights to the tree on the West Rim, then turns the transit through 94° to the rock 1000 yards away. Now, Thomas moves the transit to the rock, sights the lookout tower, and then turns it through 43° to sight the tree. How far is the tree from the lookout tower?

3. Russ has a triangular garden plot measuring 40 feet by 25 feet by 20 feet. How many bags of fertilizer should he purchase if each bag covers 75 square feet?

4. The State Recreation Commission is planning to buy a piece of forest land for a park. The land is in the shape of a triangle, 3 miles by 4 miles by 6 miles, and is valued at $600 per acre. How much should the state pay? (*Note:* 640 acres = 1 square mile.)

5. Because of the intervening mountain range, Linda cannot fly her twin-engine plane directly from Evansville to Folgerton. Instead, she leaves Evansville on a course of 280°, goes 170 miles, and changes course to 202°. She arrives at Folgerton after another 140 miles.

(a) How far apart are Evansville and Folgerton?

(b) What is the bearing of Folgerton from Evansville?

6. From Alberton, Bartlesville is 83 miles away at a bearing of 41°. The bearing of Centerburg from Alberton is 108°, and the distance from Bartlesville to Centerburg is 128 miles. How far apart are Alberton and Centerburg?

7. The gravitational acceleration vector on a falling object is 32 feet per second per second, pointing downward. The object is also acted upon by another acceleration vector of 14 feet per second per second, pointing upward at an angle of 26° from the horizontal. What angle does the resultant acceleration make with the horizontal?

8. The angle of elevation of the top of an apartment building, measured from the base of a nearby tower, is 33°. From the top of the tower, the angle of depression of the top of the apartment is 52°. The tower is 1000 feet high. How far is the top of the apartment from the top of the tower?

9. Home plate, first base, second base, and third base are the four corners of a square 90 feet on a side. The pitcher stands 60 feet, 6 inches from home plate on a line between home plate and second base.

(a) How far is the pitcher from second base?

(b) How far is the pitcher from first base?

10. Linda is flying to Georgetown, due North from Haroldsburg, but, because of the 35-mile-per-hour wind, she points the nose of her plane 36° east of north at an air speed of 200 miles per hour. Assuming Linda has figured properly, in what direction is the wind blowing?

11. On another flight due North from Haroldsburg to Georgetown, Linda is flying in a 45-mile-per-hour wind at an air speed of 200 miles per hour. If her ground speed is 190 miles per hour, what is the direction of the wind?

Review exercises (Chapter 5)

State the indicated formulas for triangle $\triangle ABC$.

1. Law of cosines: $c^2 = ?$
2. Law of sines: $(\sin A)/a = ?$
3. Area: $K = ?$

Solve the triangle from the given data.

4. $a = 29$, $b = 43$, $\angle C = 117°$
5. $a = 673$ meters, $\angle B = 48°$, $\angle C = 60°$
6. $a = 38$ feet, $b = 27$ feet, $c = 15$ feet
7. $\angle A = 32°$, $a = 65$ meters, $c = 100$ meters

Assume the given data are as follows. What restrictions on the data are needed to be sure that there is a triangle $\triangle ABC$ satisfying the given conditions? When can there be more than one solution?

8. Given sides a and b and angle $\angle C$.
9. Given side a and angles $\angle B$ and $\angle C$.
10. Given sides a and b and angle $\angle A$.
11. Given sides a, b, and c.
12. Given angles $\angle A$, $\angle B$, and $\angle C$.

13. A steamship moving due East at 8 knots notes the bearing of the lighthouse to be 62°. One hour later, the bearing is 43°. When the ship is closest to the lighthouse, determine how far apart they will be. (*Note:* 1 knot = 1 nautical mile per hour.)

6 Trigonometric graphs

We now extend our study of trigonometric graphs in two directions. First, we present the graph of the four remaining trigonometric functions, namely tangent, cotangent, secan and cosecant. Second, we will show some of the wave forms that arise as graphs c equations involving trigonometric functions.

Trigonometric graphs are used in physics to describe sound waves, electr magnetic waves, alternating current flow, vibrations of a string, and many othe periodic phenomena. A branch of higher mathematics called Fourier analysis dea with a closely related situation. There we start with a periodic function; the proble is to express that function in terms of trigonometric functions.

6-1 Graphs of trigonometric functions

The graphs of the sine function $\{(x, y) \mid y = \sin x\}$ and the cosine functio $\{(x, y) \mid y = \cos x\}$, obtained in Section 3-6, are shown in Figures 6–1 and 6– These two functions are periodic with period 2π. The domain of each is R, and th range is $\{y \mid -1 \leq y \leq 1\}$.

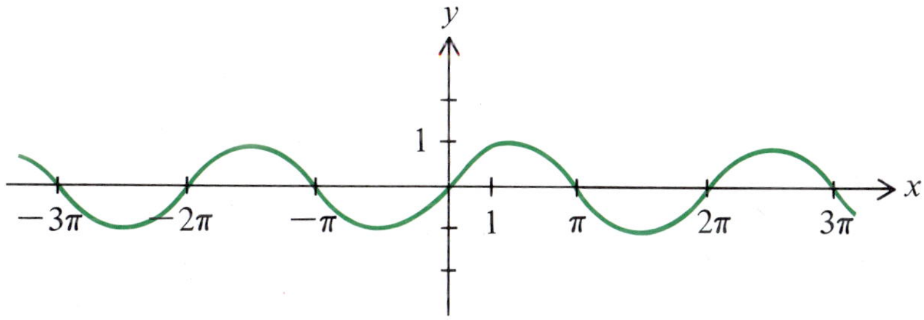

Figure 6–1
$\{(x, y) \mid y = \sin x\}$

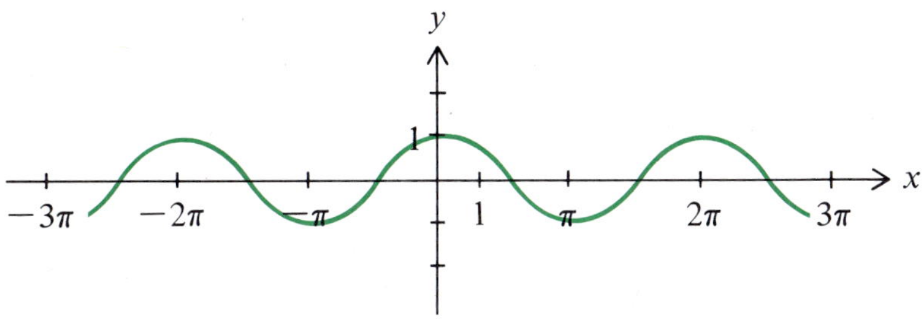

Figure 6–2
$\{(x, y) \mid y = \cos x\}$

Our aim now is to discuss the graphs of the tangent, cotangent, secant, and cosecant functions. Let us begin with the tangent function $\{(x, y) \mid y = \tan x\}$.

It would be possible for us to plot any number of points on the graph of $y = \tan x$. For each value of x, we can determine $\tan x$ from the table of Trigonometric Ratios (with the help of the reference angle, if necessary). Each time we do this, we have found a point $(x, \tan x)$ on the graph of $y = \tan x$. When the points so obtained are connected by smooth curves, the result, in Figure 6–3, is the graph of $y = \tan x$ for $0 \leq x \leq 2\pi$.

Because $\tan \pi/2$ and $\tan 3\pi/2$ are undefined, there are no points on the graph for $x = \pi/2$ or $x = 3\pi/2$. The graph approaches, but never reaches, the vertical lines $x = \pi/2$ and $x = 3\pi/2$, shown as dashed lines in Figure 6–3. These vertical lines are known as *asymptotes*.

From the formula

$$\tan(x + y) = \frac{\tan x + \tan y}{1 - \tan x \tan y}$$

we obtain $\tan(x + \pi) = \tan x$. It can be shown that the period of the tangent function is indeed π. The graph repeats after π units on the x-axis. One complete cycle is shown in Figure 6–4.

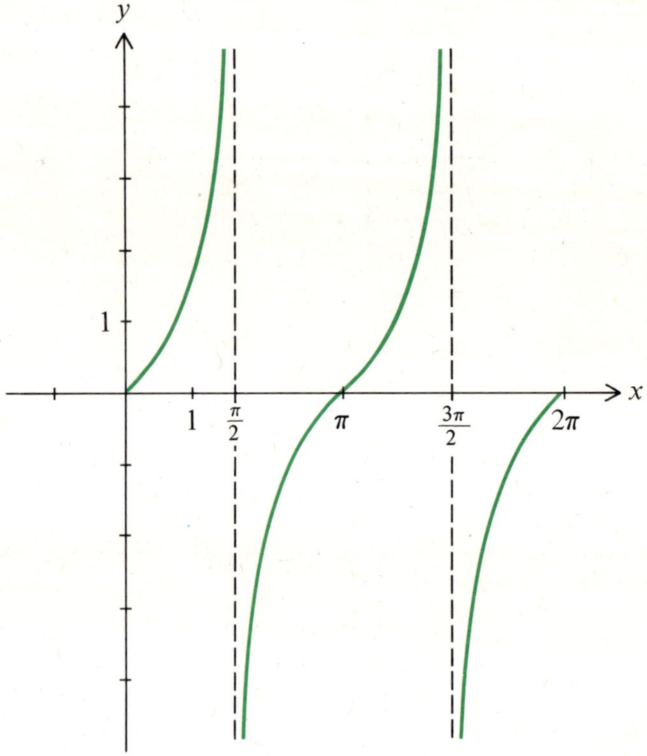

Figure 6–3
$\{(x, y) | y = \tan x, 0 \leq x \leq 2\pi\}$

By repeating this cycle every π units, we get the graph of $y = \tan x$, as in Figure 6–5. From the graph we make the following observations:

1. The period of the tangent function is π.
2. The domain of the tangent function is $\{x | x \neq \pi/2 + k\pi, k \in J\}$
3. The range of the tangent function is R.
4. The asymptotes to the graph of $y = \tan x$ are the lines $x = \pi/2 + k\pi$, $k \in J$

In the same way, by plotting points, it is feasible to sketch the graphs of the cotangent, secant, and cosecant functions. The results are as follows.

The graph of the function defined by $y = \cot x$ is as shown in Figure 6–6. From the graph, we note these facts:

1. The period of the cotangent function is π.
2. The domain of the cotangent function is $\{x | x \neq k\pi, k \in J\}$.
3. The range of the cotangent function is R.
4. The asymptotes to the graph of $y = \cot x$ are the lines $x = k\pi$, $k \in J$

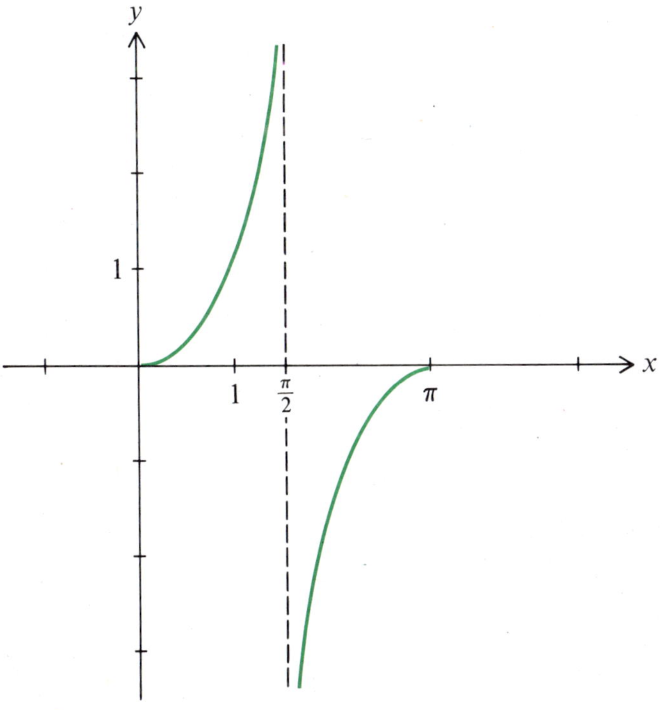

Figure 6–4
One complete cycle of the graph of $y = \tan x$

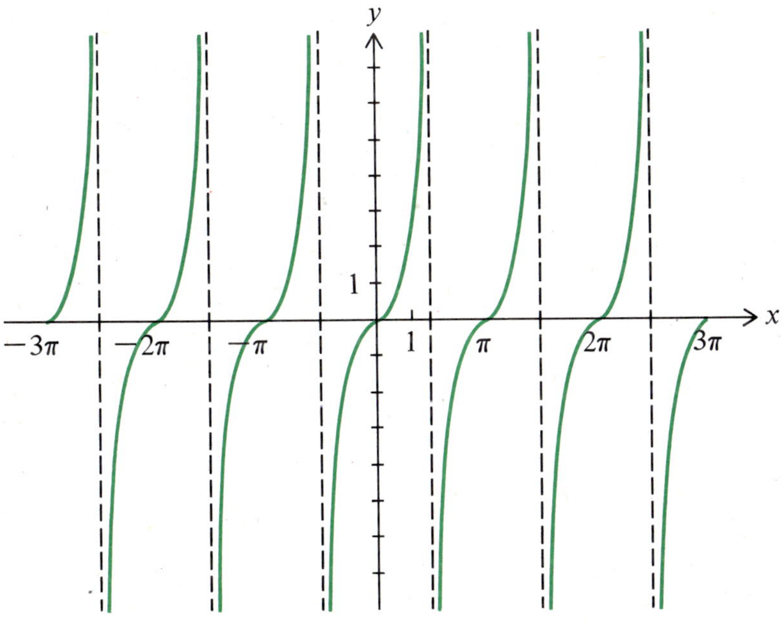

Figure 6–5
$\{(x, y) \mid y = \tan x\}$

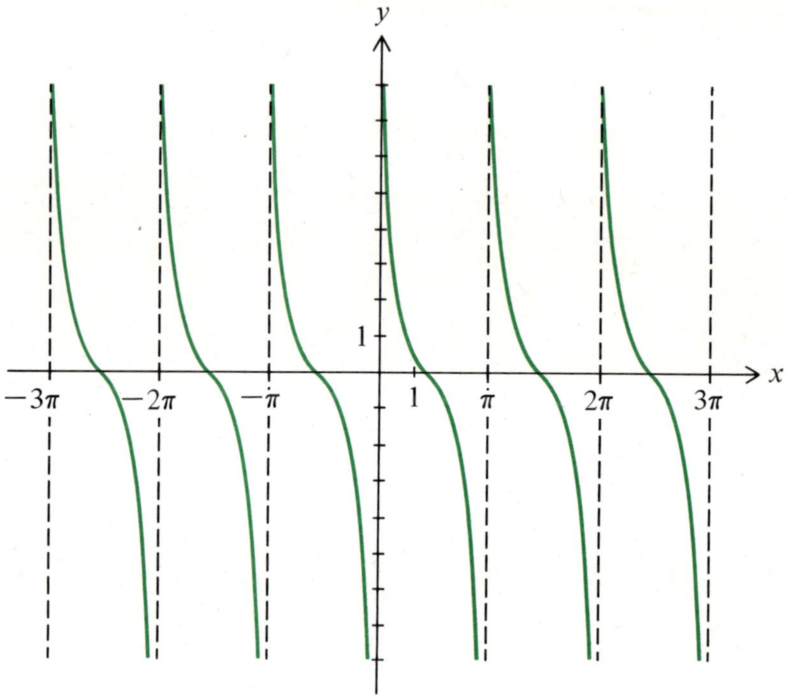

Figure 6–6
$\{(x, y) \mid y = \cot x\}$

In Figure 6–7 we have the graph of $y = \sec x$. These points should be observed

1. The period of the secant function is 2π.
2. The domain of the secant function is $\{x \mid x \neq \pi/2 + k\pi,\ k \in J\}$.
3. The range of the secant function is $\{y \mid |y| \geq 1\}$.
4. The asymptotes to the graph of $y = \sec x$ are the lines $x = \pi/2 + k\pi,\ k \in J$

Finally, the graph of the cosecant function is pictured in Figure 6–8. We note:

1. The period of the cosecant function is 2π.
2. The domain of the cosecant function is $\{x \mid x \neq k\pi,\ k \in J\}$.
3. The range of the cosecant function is $\{y \mid |y| \geq 1\}$.
4. The asymptotes to the graph of $y = \csc x$ are the lines $x = k\pi,\ k \in J$.

6–1 *Exercises*

Read off the solution set from the appropriate graph.

1. $\sin x = 0$

2. $\tan x = 0$

3. $\cot x = 0$

4. $\sec x = 0$

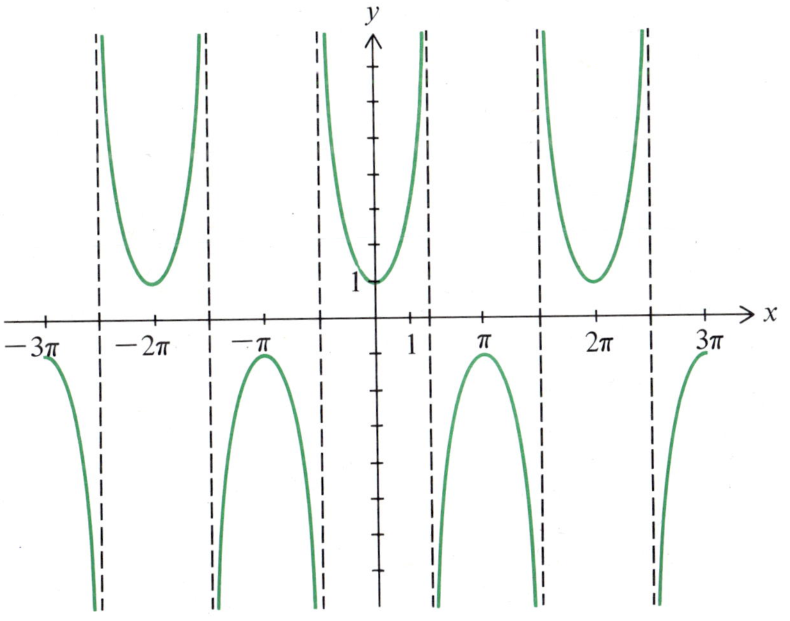

Figure 6–7
$\{(x, y) \mid y = \sec x\}$

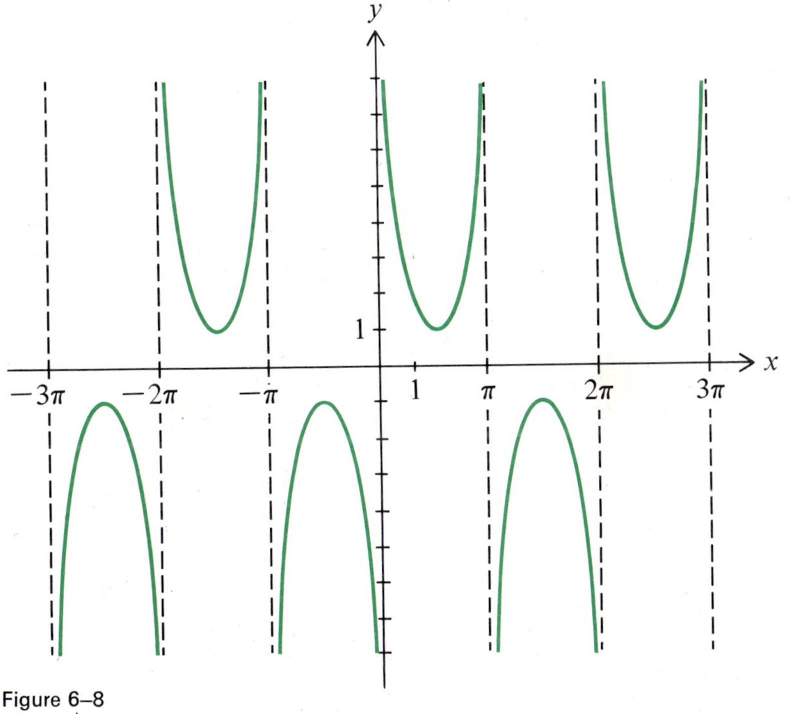

Figure 6–8
$\{(x, y) \mid y = \csc x\}$

5. $\cos x = 1$

6. $\tan x = 1$

7. $\csc x = 1$

8. $\sin x = -1$

9. $\cot x = -1$

10. $\sec x = -1$

Solve graphically. The replacement set is $\{x \mid 0 \leq x < 2\pi\}$.

11. $\tan x = \cot x$

12. $\sec x = \csc x$

13. $\cos x = \tan x$

14. $\sin x = \csc x$

15. The period of the cotangent function is π rather than 2π because $\cot(x + \pi) = \cot x$. Use the formula $\tan(x + \pi) = \tan x$ to prove that $\cot(x + \pi) = \cot x$.

Sketch the graphs, with x measured in degrees.

16. $y = \tan x$

17. $y = \csc x$

Give a graphical interpretation of the equation.

18. $\sin(x + 2\pi) = \sin x$

19. $\tan(x + \pi) = \tan x$

20. $\csc(x + 2\pi) = \csc x$

21. $\cot(-x) = -\cot x$

22. $\sec(-x) = \sec x$

23. $\csc(-x) = -\csc x$

24. $\cos(x - \pi/2) = \sin x$

25. $\sec(x - \pi/2) = \csc x$

6-2 Graphing $y = a \sin b(x - c)$

A function with equation $y = a \sin b(x - c)$ is known as a *sinusoidal function*. We assume here that a, b, and c are real numbers with $a \neq 0$ and $b > 0$.

EXAMPLE: The equation $y = \sin x$ defines a sinusoidal function. In this case we have $a = 1$, $b = 1$, and $c = 0$. ■

EXAMPLE: The equation $y = 2 \sin 3(x - \pi/4)$ defines a sinusoidal function, with $a = 2$, $b = 3$, and $c = \pi/4$. ■

EXAMPLE: The equation $y = -5 \sin 2(x + \pi)$ defines a sinusoidal function. ■

The graph of a sinusoidal function is called a *sine wave*. Perhaps the least complicated sine wave is the familiar graph of $y = \sin x$, shown again in Figure 6–9. We should note the following points:

1. The range is $\{y \mid |y| \leq 1\}$.
2. The period is 2π.
3. One complete cycle begins at (0, 0) and starts upward from there.

The general shape of any sine wave is similar. However, the numbers a, b, and c in the equation $y = a \sin b(x - c)$ affect the graph in the following ways.

1. The range is $\{y \mid |y| \leq |a|\}$.

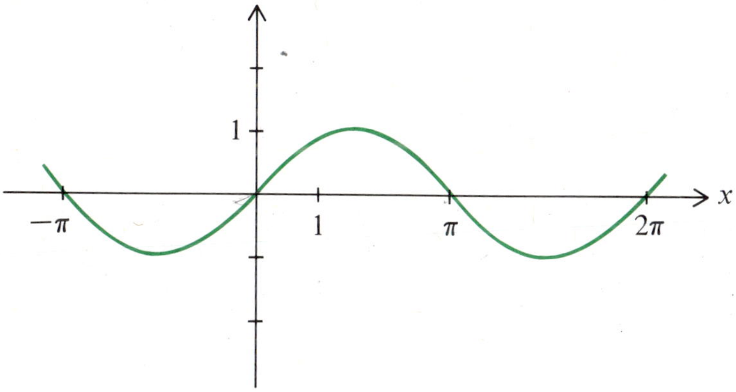

Figure 6–9
$\{(x, y) | y = \sin x\}$

2. The period is $2\pi/b$.
3. One complete cycle begins at $(c, 0)$ and starts upward from there if $a > 0$, but starts downward from $(c, 0)$ if a is negative.

The reason the cycle begins when $x = c$ is that the value $x = c$ makes the expression $b(x - c)$ equal to zero since $b(c - c) = b \cdot 0$. The first cycle is completed when we reach the value of x for which $b(x - c) = 2\pi$. Solving this for x, we have $x = c + 2\pi/b$. Therefore, in one complete cycle, the x-value goes from c to $c + 2\pi/b$. Thus the period is $(c + 2\pi/b) - c = 2\pi/b$.

In the equation $y = a \sin b(x - c)$, the number $|a|$ is referred to as the *amplitude* of the sine wave, and c is the *phase shift*.

EXAMPLE: Discuss and sketch the graph of $y = 4 \sin x$.

We observe:

1. The range is $\{y \,|\, |y| \leq 4\}$.
2. The period is 2π.
3. One complete cycle begins at $(0, 0)$ and starts upward from there.

The graph is, as shown in Figure 6–10, a sine wave with amplitude 4, period 2π, and phase shift 0. The cycle which begins at $(0, 0)$ ends at $(2\pi, 0)$, because the period is 2π. ■

EXAMPLE: Discuss and sketch the graph of $y = 2 \sin 3(x - \pi/4)$.

We have:

1. The range is $\{y \,|\, |y| \leq 2\}$.
2. The period is $2\pi/3$.
3. One complete cycle begins at $(\pi/4, 0)$, and the graph starts upward from there.

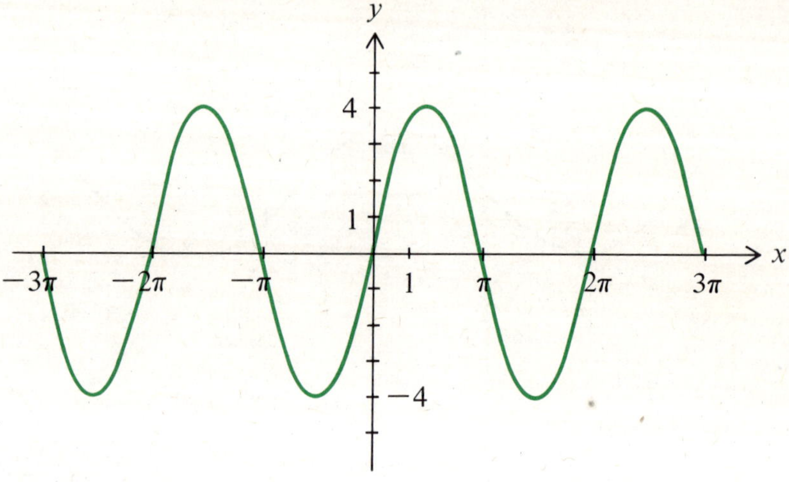

Figure 6–10
$\{(x, y) \mid y = 4 \sin x\}$

The graph, shown in Figure 6–11, is a sine wave with amplitude 2, period $2\pi/3$, and phase shift $\pi/4$. The cycle that begins at $(\pi/4, 0)$ ends at $(\pi/4 + 2\pi/3, 0) = (11\pi/12, 0)$ because the period is $2\pi/3$. ■

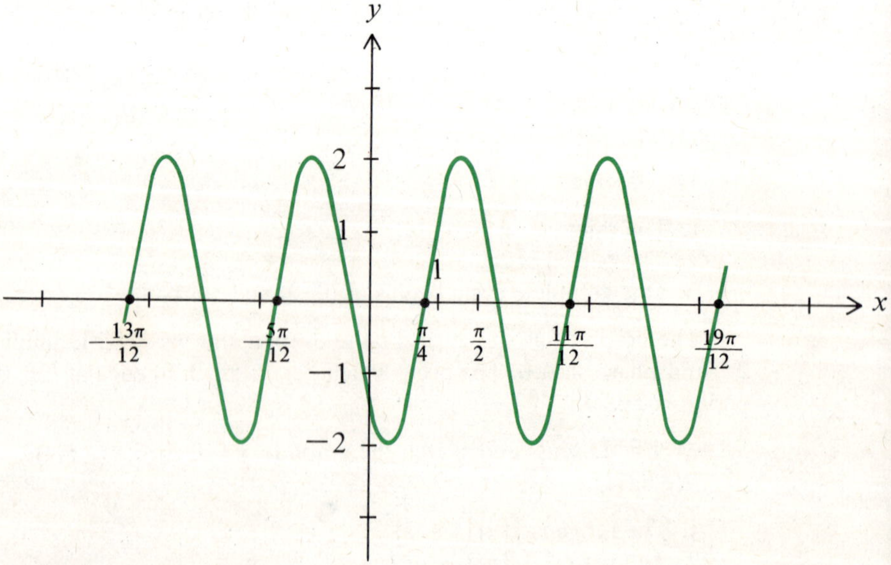

Figure 6–11
$\{(x, y) \mid y = 2 \sin 3(x - \pi/4)\}$

EXAMPLE: Discuss and sketch the graph of the function defined by $y = -5 \sin 2(x + \pi)$.

We note:

1. The range is $\{y \mid |y| \leq 5\}$.
2. The period is $2\pi/2 = \pi$.
3. One complete cycle begins at $(-\pi, 0)$ and starts downward from there.

As we see in Figure 6–12, the graph is a sine wave with amplitude 5, period π, and phase shift $-\pi$. ■

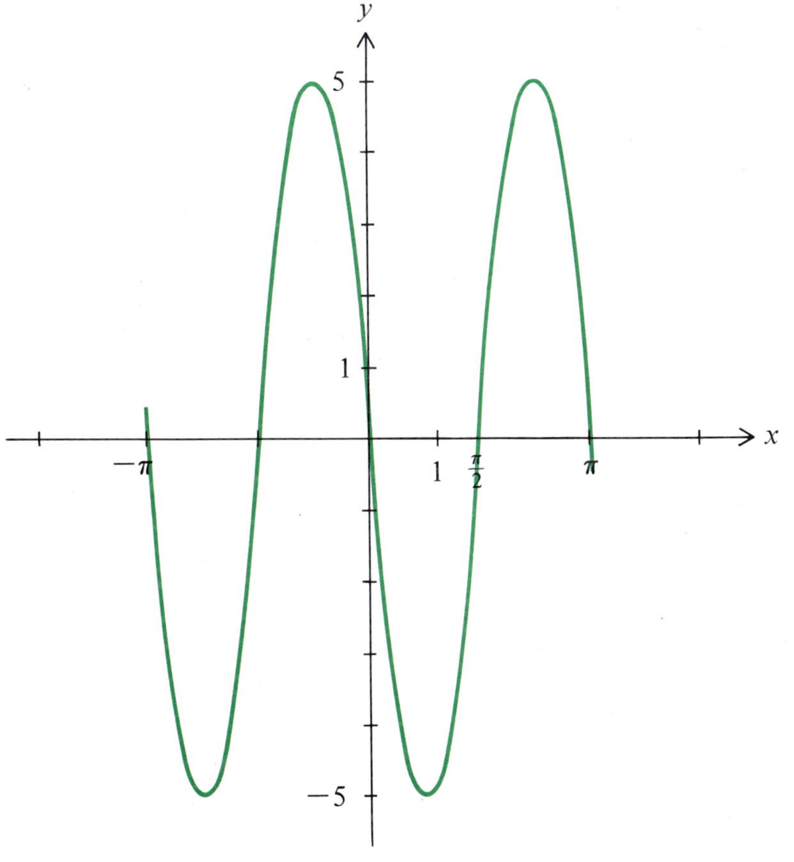

Figure 6–12
$\{(x, y) \mid y = -5 \sin 2(x + \pi)\}$

Warning: Be sure that the equation of the sinusoidal function is actually in the form $y = a \sin b(x - c)$ before interpreting the constants a, b, and c!

EXAMPLE: Discuss and sketch the graph of $y = 2 \sin (3x + \frac{1}{2}\pi)$.

It is necessary to rewrite the equation as $y = 2 \sin 3(x + \pi/6)$. We then rea off the amplitude as 2, the period as $2\pi/3$, and the phase shift as $-\pi/6$. Thus, th graph is as pictured in Figure 6–13. The complete cycle which begins at $(-\pi/6, 0$ ends at $(-\pi/6 + 2\pi/3, 0) = (\pi/2, 0)$. ■

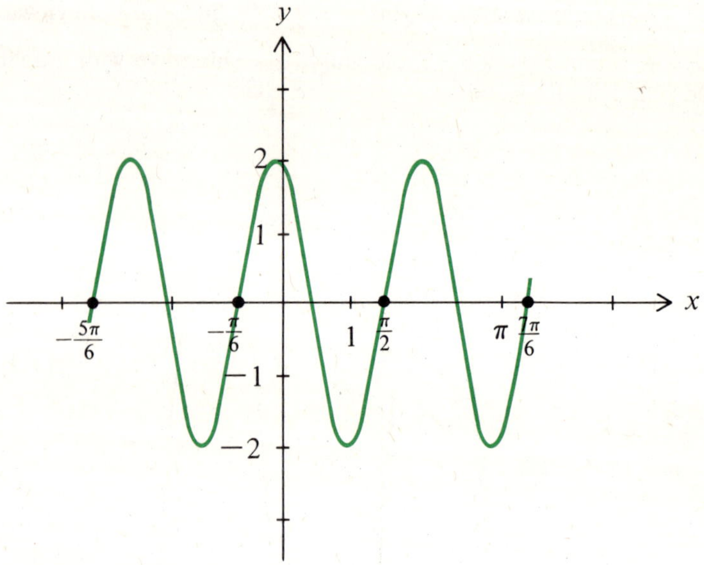

Figure 6–13
$\{(x, y) \mid y = 2 \sin 3(x + \pi/6)\}$

A function defined by an equation of the form

$$y = a \cos b(x - c)$$

is also a sinusoidal function. To see this formally, we use the identity

$$\cos x = \sin \left(x + \frac{\pi}{2}\right)$$

to rewrite $y = a \cos b(x - c)$ as follows

$$\begin{aligned} y &= a \cos b(x - c) \\ &= a \sin [b(x - c) + \pi/2] \\ &= a \sin b \left[x + \left(\frac{\pi}{2b} - c\right)\right] \\ &= a \sin b \left[x - \left(c - \frac{\pi}{2b}\right)\right] \end{aligned}$$

Therefore $y = a \cos b(x - c)$ defines a sinusoidal function with amplitude $|a|$, period $2\pi/b$, and phase shift $c - \pi/2b$.

However, in practice, there is a more convenient way to sketch the graph of $y = a \cos b(x - c)$. We draw a graph with the same general shape as that of $y = \cos x$ (Figure 6–14). The effect of the numbers a, b, and c on the graph of $y = a \cos b(x - c)$ is as follows.

1. The amplitude is $|a|$.
2. The period is $2\pi/b$.
3. One complete cycle begins at (c, a).

The reason a cycle begins at (c, a), rather than $(c, 0)$, is that $y = a \cos 0 = a$ if $x = c$. Because the period is $2\pi/b$, this cycle ends at $(c + 2\pi/b, a)$.

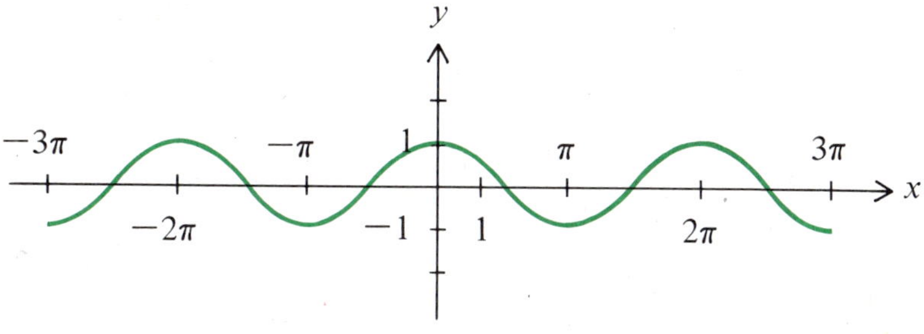

Figure 6–14
$\{(x, y) \mid y = \cos x\}$

EXAMPLE: We discuss and sketch the graph of $y = 2 \cos 3(x - \pi/4)$. We follow the pattern of the graph of $y = \cos x$ (Figure 6–14). The amplitude is 2, and a complete cycle begins at $(\pi/4, 2)$. The period is $2\pi/3$, and therefore this cycle ends at $(\pi/4 + 2\pi/3, 2) = (11\pi/12, 2)$. The graph is sketched in Figure 6–15. ■

–2 Exercises

Find the amplitude, period, and phase shift and sketch the graph.

1. $y = 2 \sin x$
2. $y = 8 \sin x$
3. $y = -2 \sin x$
4. $y = -8 \sin x$
5. $y = \sin 5x$
6. $y = \sin x/2$
7. $y = \sin (x - \pi/3)$
8. $y = \sin (x + \pi/3)$
9. $y = 3 \sin x/2$
10. $y = 2 \sin x/4$
11. $y = -\sin 5x$
12. $y = -3 \sin 3x$
13. $y = \frac{1}{2} \sin \frac{1}{2}(x - \pi)$
14. $y = \frac{1}{4} \sin 2(x + \pi/6)$
15. $y = -\pi \sin \pi(x - 1)$
16. $y = 3 \sin (2x + \pi)$
17. $y = -3 \sin (x/2 + \pi/3)$

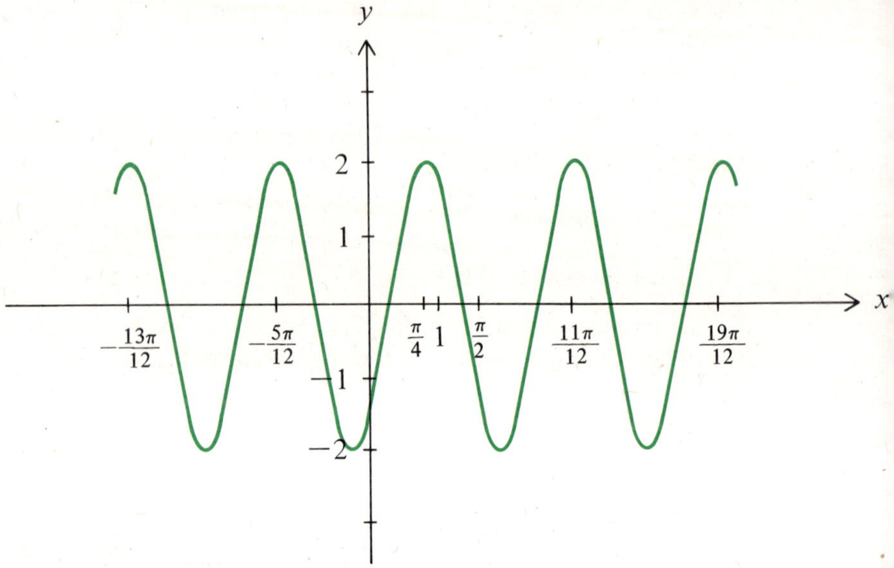

Figure 6–15
$\{(x, y) \mid y = 2 \cos 3(x - \pi/4)\}$

18. Discuss the graph of $y = a \sin b(x - c)$ when $a = 0$.
19. Discuss the graph of $y = \sin -2x$.
20. Discuss the graph of $y = a \sin b(x - c)$ for $b < 0$.

Sketch the graph.
21. $y = 2 \cos x$
22. $y = -8 \cos x$
23. $y = \cos 5x$
24. $y = \cos (x + \pi/3)$
25. $y = 3 \cos x/2$
26. $y = -3 \cos 3x$
27. $y = \frac{1}{2} \cos \frac{1}{2}(x - \pi)$
28. $y = 3 \cos (2x + \pi)$
29. $y = -3 \cos (x/2 + \pi/3)$

30. Discuss the graph of $y = a \cos b(x - c)$ for $b < 0$

Sketch the graph.
31. $y = 2 + 3 \sin 5x$
32. $y = -5 + 2 \cos 3x$

◇6-3 Sums of sinusoidal functions

The function defined by $y = \sin x + \sin 2x$ may be thought of as the sum of th two sinusoidal functions given by the equations $y = \sin x$ and $y = \sin 2x$. For eac value of x, the y-value for $\sin x + \sin 2x$ is the sum of the y-values for $\sin x$ an $\sin 2x$.

More generally, a function defined by an equation of form $y = f(x) + g(x)$ may be considered as the sum of the two functions given by the equations $y = f(x)$ and $y = g(x)$. For each value of x, the y-value for $f(x) + g(x)$ is the sum of the y-values for $f(x)$ and $g(x)$.

To graph $y = f(x) + g(x)$, we first graph each of the equations $y = f(x)$ and $y = g(x)$ as dashed curves on the same set of axes. Then, for various values of x, we compute the sum of the y-values for $f(x)$ and $g(x)$ and plot these points. These new points lie on the graph of $y = f(x) + g(x)$. If enough of these points are plotted, the curve $y = f(x) + g(x)$ can then be sketched.

EXAMPLE: We sketch the graph of

$$y = \sin x + \sin 2x$$

for $0 \le x \le 2\pi$.

The graphs of $y = \sin x$ and $y = \sin 2x$ are drawn as dashed curves, as in Figure 6–16. Then some points on the graph of $y = \sin x + \sin 2x$ should be plotted. For this example, we have worked out a table of approximate values of $\sin x + \sin 2x$ for various values of x. See Figure 6–17.

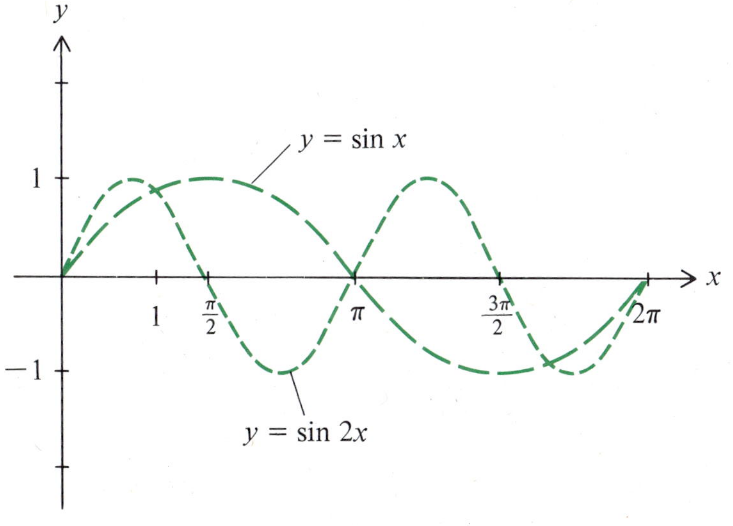

Figure 6–16

The points (x, $\sin x + \sin 2x$) obtained from the table are then plotted, as in Figure 6–18. Finally, the graph of $y = \sin x + \sin 2x$ is sketched through these points. See Figure 6–18 again. ■

It can be shown that the function defined by the equation $y = \sin x + \sin 2x$ is periodic with period 2π. Therefore the portion of the graph in Figure 6–18 is

x	$\sin x$	$\sin 2x$	$\sin x + \sin 2x$
0	0.000	0.000	0.00
$\frac{\pi}{6}$	0.500	0.866	1.37
$\frac{\pi}{4}$	0.707	1.000	1.71
$\frac{\pi}{3}$	0.866	0.866	1.73
$\frac{\pi}{2}$	1.000	0.000	1.00
$\frac{2\pi}{3}$	0.866	−0.866	0.00
$\frac{3\pi}{4}$	0.707	−1.000	−0.29
$\frac{5\pi}{6}$	0.500	−0.866	−0.37
π	0.000	0.000	0.00
$\frac{7\pi}{6}$	−0.500	0.866	0.37
$\frac{5\pi}{4}$	−0.707	1.000	0.29
$\frac{4\pi}{3}$	−0.866	0.866	0.00
$\frac{3\pi}{2}$	−1.000	0.000	−1.00
$\frac{5\pi}{3}$	−0.866	−0.866	−1.73
$\frac{7\pi}{4}$	−0.707	−1.000	−1.71
$\frac{11\pi}{6}$	−0.500	−0.866	−1.37
2π	0.000	0.000	0.00

Figure 6–17

one complete cycle. The graph of $y = \sin x + \sin 2x$, for all x, is as indicated Figure 6–19. This more complicated sort of wave is typical of graphs which resu from the sum of two sinusoidal functions.

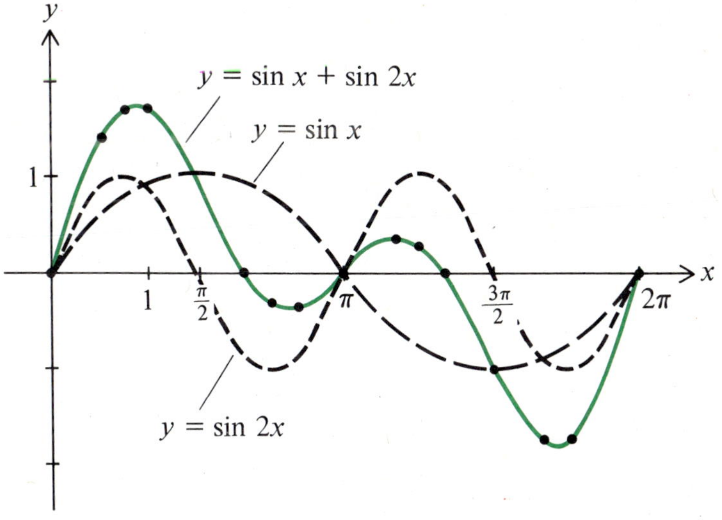

Figure 6–18
$\{(x, y) \mid y = \sin x + \sin 2x, 0 \leq x \leq 2\pi\}$

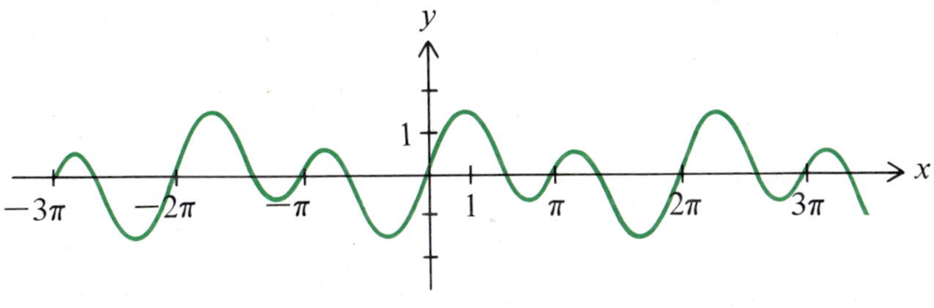

Figure 6–19
$\{(x, y) \mid y = \sin x + \sin 2x\}$

In the preceding example, we plotted points from a table of values. That procedure is not usually necessary. After the curves $y = f(x)$ and $y = g(x)$ have been sketched as dashed curves, it is often possible to "eyeball" enough points on the graph of $y = f(x) + g(x)$ so that the graph may be sketched. For a given value of x, we look at the points (x, y) on the dashed graphs. For each of these points, its height above or below the x-axis is the y-value. We add these heights to get the corresponding point (x, y) on the graph of $y = f(x) + g(x)$.

This works particularly well for those values of x such that $f(x)$ or $g(x)$ is 0. If, for example, $f(x) = 0$, then $f(x) + g(x) = 0 + g(x) = g(x)$. Therefore, for this x, the point (x, y) on the graph of $y = f(x) + g(x)$ is the same as the point (x, y) on the graph of $y = g(x)$.

EXAMPLE: Sketch the graph of

$$y = 3 \sin x + \cos x$$

for $0 \leq x \leq 2\pi$.

The function defined by $y = 3 \sin x + \cos x$ is the sum of the two sinusoid functions given by $y = 3 \sin x$ and $y = \cos x$. The graphs of these two functions are drawn as dashed curves, and then the sum $y = 3 \sin x + \cos x$ may be sketched. See Figure 6–20.

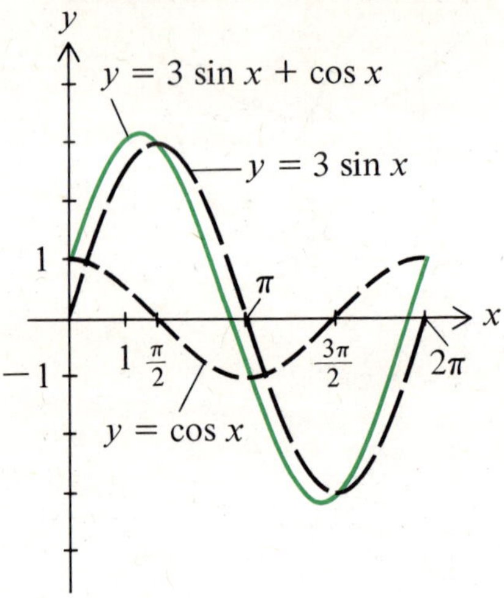

Figure 6–20
$\{(x, y) \mid y = 3 \sin x + \cos x\}$

Note in Figure 6–20 that $3 \sin x = 0$ for $x = 0$, π, and 2π. At those x-values the graph of $y = 3 \sin x + \cos x$ crosses the graph of $y = \cos x$. Similarly, $\cos x =$ for $x = \pi/2$ and $x = 3\pi/2$. At these values of x, the graph of $y = 3 \sin x + \cos$ meets the graph of $y = 3 \sin x$. ■

EXAMPLE: Sketch the graph of

$$y = 2 \sin \left(x - \frac{\pi}{3}\right) - 4 \sin 2x$$

for $-2\pi \leq x \leq 2\pi$.

For $-2\pi \leq x \leq 2\pi$, the graphs of the two sinusoidal functions defined b $y = 2 \sin (x - \pi/3)$ and $y = -4 \sin 2x$ are shown as dashed curves in Figure 6–2 The solid graph in Figure 6–21 is the graph of the sum,

$$y = 2 \sin \left(x - \frac{\pi}{3}\right) - 4 \sin 2x \quad ■$$

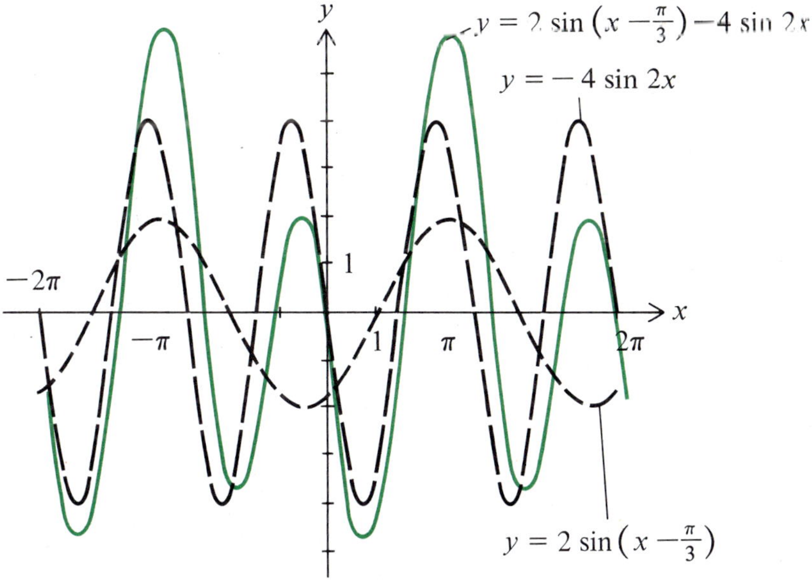

Figure 6–21
$\{(x, y) \mid y = 2\sin(x - \pi/3) - 4\sin 2x,\ -2\pi \le x \le 2\pi\}$

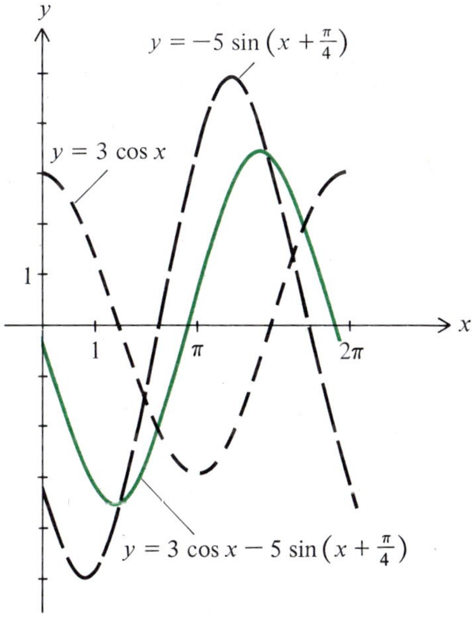

Figure 6–22
$\{(x, y) \mid y = 3\cos x - 5\sin(x + \pi/4)\}$

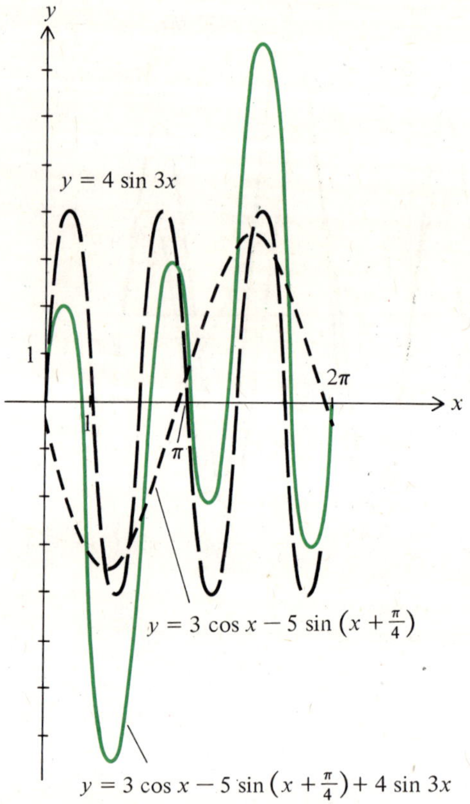

Figure 6–23
$\{(x, y) \mid y = 3 \cos x - 5 \sin (x + \pi/4) + 4 \sin 3x\}$

Still more complicated wave forms can be obtained as graphs of sums of thr (or more) sinusoidal functions.

EXAMPLE: Sketch the graph, for $0 \leq x \leq 2\pi$, of

$$y = 3 \cos x - 5 \sin \left(x + \frac{\pi}{4}\right) + 4 \sin 3x$$

The graphs of $y = 3 \cos x$ and $y = -5 \sin (x + \pi/4)$ are shown a dashed curves in Figure 6–22, together with the graph of their sum $y =$ $3 \cos x - 5 \sin (x + \pi/4)$. This last graph, and that of $y = 4 \sin 3x$, are shown a dashed curves in Figure 6–23. The solid graph in Figure 6–23 is the grap of the sum of these last two functions, and thus is the graph of

$$y = 3\cos x - 5\sin\left(x + \frac{\pi}{4}\right) + 4\sin 3x$$

as required. ■

6–3 *Exercises*

Sketch the graph for $0 \le x \le 2\pi$.

1. $y = \sin x + \cos x$
2. $y = \sin x - \cos x$
3. $y = 2\sin x/2 + 3\cos x$
4. $y = 2\sin x/4 - 3\cos x$
5. $y = \sin x - \sin 2x$
6. $y = \sin 2x - \sin x$
7. $y = \sin 2x + \sin 3x$
8. $y = \sin x + \sin 3x$

Sketch the graph for $-2\pi \le x \le 2\pi$.

9. $y = 3\sin(x - \pi/4) + \sin 2x$
10. $y = 3\sin(x - \pi/4) + \cos 2x$
11. $y = -\frac{1}{2}\sin(x + \pi/6) + 4\sin 3x$
12. $y = -\frac{1}{2}\sin(x + \pi/6) + 4\cos 3x$
13. $y = \sin x + \sin 2x + \sin 3x$
14. $y = \sin x + \sin x/2 + \sin x/4$

6-4 Applications

A wide variety of periodic phenomena in the natural sciences can be described by trigonometric functions. A few of these applications are now illustrated.

Uniform circular motion: A wheel of radius r rotates at a constant rate. We would like to write equations to describe the motion of a point P on the circumference of the wheel.

The wheel is placed in the coordinate plane with its center at the origin. We assume that at time $t = 0$ the point P starts on the x-axis at $(r, 0)$, as in Figure 6–24.

The usual practice is to measure the rate of rotation in radians per unit time. This rate is called the *angular velocity*; its symbol is ω, the Greek letter omega.

Let us now consider the position of the wheel at time t (Figure 6–25). From the starting time, t units of time have elapsed. The angular velocity is ω radians per unit time. Therefore the wheel has rotated through ωt radians.

Let (x, y) be the coordinates of P. From Figure 6–25, we have

$$\frac{x}{r} = \cos \omega t$$

$$\frac{y}{r} = \sin \omega t$$

Therefore

$x = r \cos \omega t$
$y = r \sin \omega t$

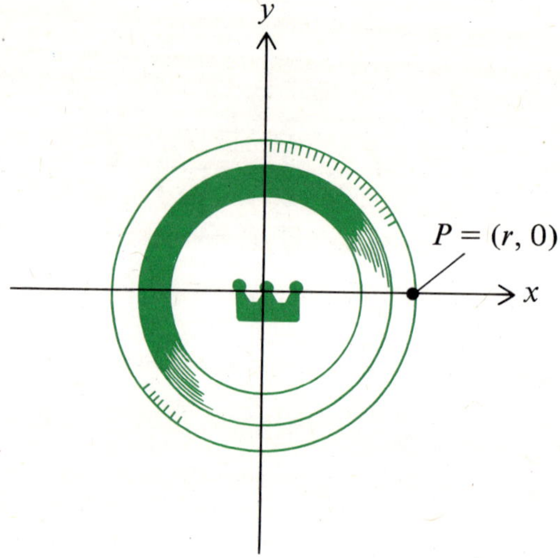

Figure 6–24
At time $t = 0$, P is at $(r, 0)$

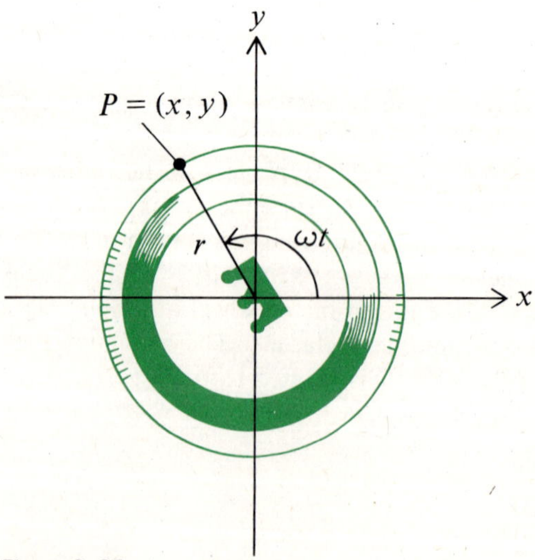

Figure 6–25
At time t, P is at (x, y)

These two equations together describe the motion of P. At time t, the location of P is given by $P = (x, y)$, where $x = r \cos \omega t$ and $y = r \sin \omega t$.

EXAMPLE: The city of Singapore is located almost exactly on the equator. Coordinate axes are drawn in the plane of the equator so that the center of the earth is at the origin and, at noon, Singapore is on the positive x-axis. The radius of the earth is approximately 3963 miles. What are the coordinates of Singapore at 2 a.m.?

We measure time in hours, with noon corresponding to $t = 0$, so that, at $t = 0$ Singapore is at (3963, 0).

The rate of the earth's rotation is 2π radians every 24 hours. Therefore

$$\omega = \frac{2\pi}{24} = \frac{\pi}{12} \text{ radians per hour}$$

Thus the coordinates of Singapore at time t are (x, y), where

$$x = 3963 \cos \left(\frac{\pi}{12}\right) t$$

$$y = 3963 \sin \left(\frac{\pi}{12}\right) t$$

At 2 a.m., $t = 14$, and so at 2 a.m.

$$x = 3963 \cos \frac{14\pi}{12} = 3963 \cos \frac{7\pi}{6}$$

$$y = 3963 \sin \frac{14\pi}{12} = 3963 \sin \frac{7\pi}{6}$$

Substituting $\cos \frac{7\pi}{6} = -\frac{\sqrt{3}}{2}$ and $\sin \frac{7\pi}{6} = -\frac{1}{2}$, we have approximately

$$x = -3963 \cdot \left(\frac{\sqrt{3}}{2}\right) = -3432$$

$$y = -3963 \cdot \tfrac{1}{2} = -1982$$

and so at 2 a.m. the location of Singapore is $(-3432, -1982)$. ■

Simple harmonic motion: Suppose the position y of an object, measured along an axis, is given by an equation of the form

$$y = a \sin b(t - c)$$

We then say that the motion of the object along the axis is *simple harmonic motion*. As in Section 6-2, we refer to $|a|$ as the amplitude, c as the phase shift, and $2\pi/b$ as the period.

Because t represents time, we interpret $2\pi/b$ as the time required for one complete cycle. It follows that there are $b/2\pi$ complete cycles per unit time. Th number of complete cycles per unit time is the *frequency*, which we shall deno by f. Thus

$$f = \frac{b}{2\pi} \quad \text{or, equivalently, } b = 2\pi f$$

and the equation of motion may be rewritten as

$$y = a \sin 2\pi f(t - c)$$

EXAMPLE: A weight hangs down from the end of a spring, see Figure 6–26. Th spring is stretched, and then it is released at $t = 0$. It is a fact of physics that th height y of the weight, measured from its position at rest, is given approximate by an equation

$$y = a \sin \left(bt - \frac{\pi}{2}\right)$$

This is an equation of simple harmonic motion. The amplitude indicates the max mum length the spring is extended or compressed during the motion. The constar b depends on physical properties of the spring.

Rewriting the equation as

$$y = a \sin b\left(t - \frac{\pi}{2b}\right)$$

we see that the phase shift is $\pi/2b$. The period is $2\pi/b$, and for the frequency w have $f = b/2\pi$. ■

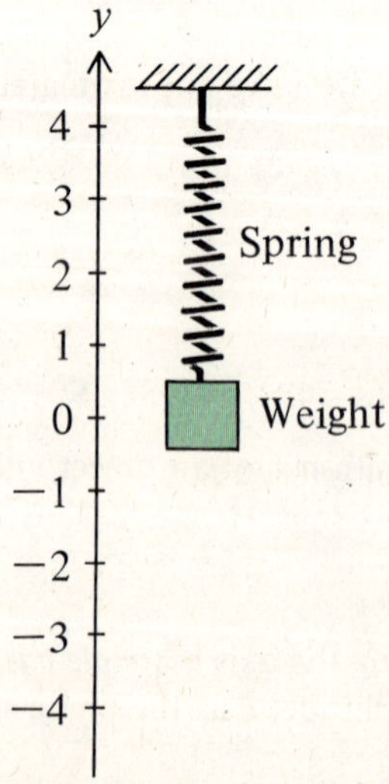

Figure 6–26
The spring at rest

EXAMPLE: We return to the point P on the circumference of the rotating wheel (see the earlier discussion of uniform circular motion). The height y was found to be

$$y = r \sin \omega t$$

Thus, looking only at the height of P, we have simple harmonic motion. The frequency is $\omega/2\pi$. ■

Sound waves: As a sound wave reaches the ear, the pressure of the air pulls and pushes the eardrum back and forth. The inward pressure p on the eardrum may be expressed as a sinusoidal function of time:

$$p = a \sin bt$$

(Negative values of p correspond to outward pressure on the eardrum.) The amplitude depends on the loudness of the sound.

As we saw in the discussion of simple harmonic motion, the frequency f of the vibration is $f = b/2\pi$, and so $b = 2\pi f$. Thus the equation describing a sound wave of frequency f is

$$p = a \sin 2\pi ft$$

EXAMPLE: A piano is tuned so that middle C gives a frequency of 256 cycles per second. What is the equation for the sound wave produced?

Because $f = 256$, the equation is

$$p = a \sin 512\pi t$$

where a depends on the loudness. ■

Overtones: For a tone with frequency f, the tones that have for frequencies the integral multiples of f (such as $2f$, $3f$, and so on) are the *overtones.*

Actually the sound produced by one note on the piano is not a pure tone—that is, the equation is not simply

$$p = a \sin 2\pi ft$$

for just one frequency f. Overtones of the basic frequency will also be produced. The equation for p is more accurately a sum of terms $a_k \sin 2\pi kft$ for $k = 1, 2, 3,$ and so on. The relative size of the amplitudes a_k for the overtones determine the quality of the tone. It is because of a different mix of overtones that the same note has a different sound when played on a different instrument.

EXAMPLE: When middle C is played on the Trigophone, the first three overtones are produced at half the amplitude of the basic frequency. Assuming the other overtones are negligible, find the equation for the sound wave.

The frequencies of the first three overtones are $2 \cdot 256 = 512$, $3 \cdot 256 = 768$, and $4 \cdot 256 = 1024$. Letting a be the amplitude for the basic frequency, we have

$$p = a \sin 512\pi t + \frac{a}{2} \sin 1024\pi t + \frac{a}{2} \sin 1536\pi t + \frac{a}{2} \sin 2048\pi t$$

This equation defines p as a sum of sinusoidal functions. The graph might b sketched by the method of Section 6-3. However, you should be cautioned th the work involved to do this would be most laborious. ■

Moving waves: Imagine waves moving across the surface of the water. Th distance from the crest of one wave to the crest of the next is the *wave length*; i symbol is λ, the Greek letter lambda. See Figure 6–27.

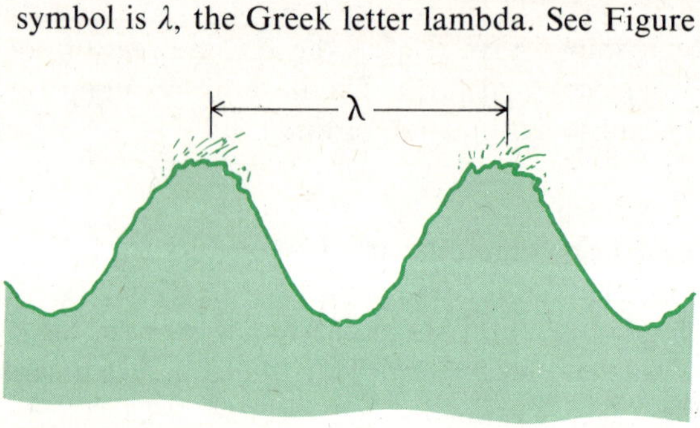

Figure 6–27
The wave length is λ

Consider now a buoy riding the waves. The buoy moves up and down. Fo the purposes of this discussion, the motion is assumed to be simple harmon motion.

Let f be the frequency of this up-and-down vibration. Evidently f is the numb of waves passing per unit time, because each wave produces one complete cycle o the buoy's motion.

We now ask: how fast are the waves moving across the surface of the water

To answer this, we look at matters from the point of view of the buoy. I 1 unit of time, the number of waves passing is f. The length of each wave is . Therefore the waves have moved $f\lambda$ units of distance in 1 unit of time. In othe words, the velocity v of the moving waves is given by

$$v = f\lambda$$

This is an important formula. It is valid not just for waves moving throug the water, but for sound waves, light waves, radio waves, and so on.

EXAMPLE: The speed of sound through the air is approximately 1100 feet pe second. What is the wave length for a sound wave pitched at middle C?

We have

$$v = f\lambda$$

where $v = 1100$ feet per second and $f = 256$ cycles per second. Thus

$$1100 = 256\lambda$$
$$\lambda = \tfrac{1100}{256} \text{ feet}$$

This works out to a wave length of about 4 feet, 4 inches. You might check this by measuring the length of a middle C string on a piano. ■

6–4 Exercises

1. The equations of motion for a point on the circumference of a rotating wheel are

$$x = 5 \cos 4\pi t$$
$$y = 5 \sin 4\pi t$$

where t represents time, measured in seconds.

(a) What is the angular velocity of the rotation?

(b) How much time is required for one complete revolution?

2. A point P is on the circumference of a rotating wheel of radius r.

(a) How far does P travel around the circumference in 2π radians of rotation?

(b) Show that the distance s through which P travels in θ radians of rotation is given by $s = r\theta$.

3. Elizabeth rides a 10-speed bicycle with wheels 27 inches in diameter. What is the angular velocity of the rotation of the wheels when she is going 10 miles per hour?

4. The radius of the flywheel in Frank's sportscar is 25 centimeters. The engine speed registered on the tachometer is 3600 revolutions per minute.

(a) What are the equations of motion for a point on the rim of the flywheel?

(b) How fast is that point traveling along the circumference?

5. A weight hanging on the end of a spring is pulled down 5 centimeters below its position at rest, then released. As the weight moves upward, the first time it passes the rest position is at $t = 3$. Find the equation of motion.

6. The up-and-down motion of a cork bobbing in the ocean approximates simple harmonic motion. The crest of each wave is 30 feet above the trough, and the waves are passing at the rate of one every 6 seconds. Find an equation for the height y of the cork above sea level at time t.

7. Most symphony orchestras are tuned to *concert pitch*, with A above middle C corresponding to 440 cycles per second.

(a) Give the equation describing a pure tone of unit amplitude at this pitch.

(b) What is the frequency of the second overtone?

8. The interval separating sounds with frequency f and $2f$ is known in music as an *octave*.

(a) What is the frequency for a sound one octave below middle C?

(b) Which overtone of a given note is two octaves above it?

9. A musician plays two notes simultaneously with equal volume. The two frequencies are in the ratio 2 : 3. Make a rough sketch of the graph of the equation describing the sound produced.

10. **(a)** What is the wave length of a sound wave pitched at Concert A (440 cycles per second)?

(b) How are the wave lengths for a given pitch and its overtones related?

11. The frequency of radio waves is usually expressed in kiloHertz (1 kiloHertz 1000 cycles per second) for AM waves and in megaHertz (1 megaHertz = 1,000,0 cycles per second) for FM. The speed of radio waves is approximately 186,3 miles per second.

(a) Find the wave length, to the nearest foot, for the signal received at 10 on the AM dial.

(b) What is the wave length, to the nearest foot, for the signal at 100 on t FM dial?

12. The Perfecto Tuning Fork Company is testing a sample of its A-440 Conce Tuning Fork. For the test, both the sample and a perfect fork, known to vibra at 440 cycles per second, are struck equally hard at the same time. Let f be t frequency for the fork being tested, and assume that each fork produces a wave wi unit amplitude.

(a) Find the equation for the sound produced.

(b) Suppose $f = 438$ (so that the fork is defective). Verify that the equation $p = \sin 880\pi t + \sin 876\pi t$.

(c) Use the identity from Exercise 17 of Section 4-3 to rewrite the equati in (b) as $p = 2 \sin 878\pi t \cos 2\pi t$.

(d) Try to figure out what the observer will hear by making a rough sketch the graph of p as a function of t.

(e) Ask your physics teacher for a demonstration of beats.

Review exercises (Chapter 6)

Explain and/or illustrate with an example.

1. asymptote

2. sinusoidal function

3. sine wave

4. amplitude

5. period

6. phase shift

7. simple harmonic motion

8. frequency

9. wave length

Sketch the graph.

10. $y = \tan x$

11. $y = \csc x$

12. $y = 2 \sin 6x$

13. $y = 2 \cos 6x$

14. $y = \sin \frac{1}{4}(x - \pi/4)$

◇**15.** $y = 4 \sin 2x - \cos x$

◇**16.** $y = \sin x + \cos x + \sin 2x$

◇**17.** $y = \sin x + \tan x$

◇**18.** In an alternating current circuit, we are accustomed to seeing the voltage given as a constant. More accurately, however, the instantaneous voltage E is function of time, described by the equation $E = V\sqrt{2} \sin bt$. In most homes the United States, $V = 110$ volts and the frequency is 60 cycles per second.

(a) Find the equation for E.

(b) Find the peak value of the instantaneous voltage.

Complex numbers and trigonometry

'rom the point of view of algebra, quadratic equations are the reason for studying complex numbers. Because negative numbers do not have real number square roots, we have to invent a number system larger than the real numbers before we can algebraically solve such a simple equation as $x^2 + 1 = 0$.

Here, though, we use the approach of trigonometry to study the rich geometric structure of the complex number system. Trigonometry provides an amazing way to simplify multiplication, division, exponentiation, and extraction of roots in the complex number system.

-1 Complex numbers

Because there is no real number x such that $x^2 = -1$, it is not possible for us to define $\sqrt{-1}$ as a real number. However, it is possible to define $\sqrt{-1}$ in a larger number system, the set of complex numbers. We denote this set by C. The complex number system C contains the real number system, but C also contains some new numbers, called *imaginary numbers*. We will find that $\sqrt{-1}$ makes good sense as an imaginary number.

It is convenient to define the set C in terms of real numbers and a symbol which will turn out to be the square root of -1. The definition is as follows:

By a complex number we mean any expression of the form $a + bi$, in which a and denote real numbers.

EXAMPLE: $2 + 7i$ is a complex number. ■

EXAMPLE: $-\sqrt{2} + 2\pi i$ is a complex number. ■

EXAMPLE: $3 + 0i$ is a complex number. ■

EXAMPLE: $0 - 15i$ is a complex number. ■

The two complex numbers $a = bi$ and $c + di$ are equal if and only if bot $a = c$ and $b = d$.

EXAMPLE: $2 + 5i = \frac{6}{3} + 5i$ ■

EXAMPLE: $-2 + 7i \neq -1 + 7i$, because $-2 \neq -1$ ■

The real number a is called the *real part* of the complex number $a + bi$, and b called the *imaginary part*.

EXAMPLE: 4 is the real part of $4 - 6i$. ■

EXAMPLE: $-3\frac{1}{2}$ is the imaginary part of $2\frac{1}{4} - 3\frac{1}{2}i$. ■

Given a real number a, we may associate with it the complex number $a + 0i$. I this way we see that there is an exact correspondence between the set R of re numbers and the subset of C consisting of the complex numbers with the imagina part equal to zero. Although there is a fine logical distinction between the re number a and the complex number $a + 0i$, it is convenient to ignore this subtlet Therefore, we will identify the real number a with the complex number $a + 0i$. I this way we may think of the set of real numbers as a subset of the set of comple numbers; in symbols, $R \subseteq C$. Therefore, in particular, $0 = 0 + 0i$ and $1 = 1 + 0i$ a complex numbers.

Those complex numbers $a + bi$ in which $b \neq 0$ are not real numbers. The numbers are called *imaginary numbers*.

EXAMPLE: $-3 + 7.48i$ is an imaginary number. ■

EXAMPLE: $0 + 4i$ is an imaginary number. ■

EXAMPLE: $-6\sqrt{2}$ is a real number and therefore a complex number, but it is n an imaginary number. ■

The term "imaginary number" is perhaps misleading, but it is always use Imaginary numbers are neither more nor less concrete than real numbers. Each ty of number is an abstract concept. The real number system does have a simple an appealing interpretation, using the number line, but this does not make the syste more "real" or less "imaginary."

If a complex number is of the form $0 + bi$, so that its real part is zero, we call it a *pure imaginary number*. It is often written simply as bi.

EXAMPLE: $2i(=0 + 2i)$ is a pure imaginary number. ■

The addition process for complex numbers is very straightforward. We add the real parts and pure imaginary parts separately, according to the following definition:

If $a + bi$ and $c + di$ are complex numbers, then

$$(a + bi) + (c + di) = (a + c) + (b + d)i.$$

EXAMPLE: $$\begin{aligned}(2 + 8i) + (-5 + 4i) &= (2 - 5) + (8 + 4)i \\ &= -3 + 12i\end{aligned}$$ ■

The subtraction process is similar. In symbols,

$$(a + bi) - (c + di) = (a - c) + (b - d)i$$

EXAMPLE: $$\begin{aligned}(3 + 5i) - (2 - 3i) &= (3 - 2) + [5 - (-3)]i \\ &= 1 + 8i\end{aligned}$$ ■

EXAMPLE: $$\begin{aligned}4i - (2 + i) &= (0 + 4i) - (2 + i) \\ &= (0 - 2) + (4 - 1)i \\ &= -2 + 3i\end{aligned}$$ ■

Multiplication of complex numbers is not quite so straightforward. The whole idea was to create a number system in which -1 has a square root. The pure imaginary number i is going to be the square root of -1, and so we want to have $i^2 = -1$. Let us try to multiply two complex numbers together, following the usual laws of algebra, but putting $i^2 = -1$. We first work out a particular example:

EXAMPLE: $$\begin{aligned}(2 + 3i)(4 + i) &= 8 + 14i + 3i^2 \\ &= 8 + 14i - 3 \\ &= 5 + 14i\end{aligned}$$ ■

Let us now use the same method to guess the proper general formula:

$$\begin{aligned}(a + bi)(c + di) &= ac + (ad + bc)i + bdi^2 \\ &= ac + (ad + bc)i - bd \\ &= (ac - bd) + (ad + bc)i\end{aligned}$$

The preceding formula is the only one possible if we are going to have $i^2 = -1$ and otherwise follow the usual laws of algebra. We shall therefore adopt it as the formal definition of multiplication of complex numbers, as follows:

If $a + bi$ and $c + di$ are complex numbers, then

$$(a + bi)(c + di) = (ac - bd) + (ad + bc)i$$

EXAMPLE: $$\begin{aligned}(7 + 6i)(2 - 5i) &= 14 - 23i - 30i^2 \\ &= 44 - 23i\end{aligned}$$ ■

EXAMPLE: $(3i)(4 - 2i) = 12i - 6i^2$
$= 6 + 12i$ ■

EXAMPLE: $i^3 = i^2 i = (-1)i = -i$, and $i^4 = i^3 i = (-i)i = -i^2 = 1$ ■

EXAMPLE: $(\sqrt{2} - i)(\sqrt{2} + i) = 2 - i^2 = 3$ ■

In the complex number system, it is possible to divide by a nonzero comple number. However, if we write $(2 + 6i) \div (1 + i)$ as the fraction $(2 + 6i)/(1 + i$ we cannot recognize this fraction as a complex number of the form $a + bi$. Th problem is like that of rationalizing the denominator of a fraction when the de nominator contains a radical. We solve the problem by multiplying numerato and denominator by the conjugate of the denominator. For complex numbers, th conjugate is defined as follows:

The conjugate of the complex number $a + bi$ is the complex number $a - bi$.

EXAMPLE: The conjugate of $4 + 2i$ is $4 - 2i$ ■

EXAMPLE: The conjugate of $-5i$ is $5i$. ■

EXAMPLE: The conjugate of 3 is 3. ■

Let us now return to the division problem $(2 + 6i)/(1 + i)$. This fraction can b simplified by multiplying the numerator and denominator by the conjugate of th denominator, as follows:

EXAMPLE:
$$\frac{2 + 6i}{1 + i} = \frac{(2 + 6i)(1 - i)}{(1 + i)(1 - i)} = \frac{2 + 4i - 6i^2}{1 - i^2}$$
$$= \frac{8 + 4i}{2}$$
$$= 4 + 2i \quad ■$$

Here is another example of this process:

EXAMPLE:
$$\frac{-1 + 4i}{2 - 3i} = \frac{(-1 + 4i)(2 + 3i)}{(2 - 3i)(2 + 3i)}$$
$$= \frac{-2 + 5i + 12i^2}{4 - 9i^2}$$
$$= \frac{-14 + 5i}{13}$$
$$= -\tfrac{14}{13} + \tfrac{5}{13}i \quad ■$$

Because $i^2 = -1$, we see that i is a square root of -1. However, so is $-$ because $(-i)^2 = i^2 = -1$. Therefore, -1 has the two square roots i and $-i$ in th complex number system. *The symbol $\sqrt{-1}$ is reserved for i, the principal square roo*

The complex number system actually provides us with two square roots for every negative real number. We note that $(i\sqrt{a})^2 = i^2a = -a$ and, similarly, $(-i\sqrt{a})^2 = i^2a = -a$. Thus, if a is a positive real number, then $i\sqrt{a}$ and $-i\sqrt{a}$ are each square roots of $-a$. *The symbol* $\sqrt{-a}$ *denotes the principal square root* $i\sqrt{a}$.

EXAMPLE: $\sqrt{-4} = i\sqrt{4} = 2i$ ■

EXAMPLE: $\sqrt{-7} = i\sqrt{7}$ ■

EXAMPLE: $-\sqrt{-25} = -i\sqrt{25} = -5i$ ■

A very important stipulation must be made here.

Warning: The formula $\sqrt{a}\sqrt{b} = \sqrt{ab}$ is not valid when a and b are negative numbers.

EXAMPLE: To simplify $\sqrt{-4}\sqrt{-9}$, we should first express each radical in terms of i. Thus,

$$\sqrt{-4}\sqrt{-9} = (2i)(3i) = 6i^2 = -6$$

Notice that $\sqrt{-4}\sqrt{-9} \neq \sqrt{(-4)(-9)}$, because $\sqrt{-4}\sqrt{-9} = -6$ while $\sqrt{(-4)(-9)} = \sqrt{36} = 6$. ■

EXAMPLE: $\sqrt{-5}\sqrt{-7} = (i\sqrt{5})(i\sqrt{7}) = i^2\sqrt{35} = -\sqrt{35}$ ■

7-1 *Exercises*

Which of the following terms apply to the given numbers?

(i) real number
(ii) complex number
(iii) imaginary number
(iv) pure imaginary number

1. $3 + 7i$
2. 1
3. $4i$
4. $-\pi$
5. $4 + i\sqrt{10}$
6. $\frac{3}{2}$
7. $2\pi - 5i$
8. $-0.72i$

Find the real part, the imaginary part, and the conjugate.

9. $2 + 5i$
10. $4 - (\sqrt{3})i$
11. $6i$
12. -5
13. 0
14. $-1 + \pi i$
15. $\sqrt{2}$
16. $-i\sqrt{5}$

Perform the indicated operations so that the result is expressed in $a + bi$ form.

17. $(2 + 3i) + (-4 - 7i)$
18. $(2 + i) - (3 + 4i)$
19. $(1 - 5i) + (3 + 6i)$
20. $-6i - (-7 - 3i)$

21. $(1 + 3i)(4 + i)$

22. $(-4 + 6i)(2i)$

23. $3(2 - 8i)$

24. $(7 - \pi i)(4\pi - 3i)$

25. $(\sqrt{2} + i)[\sqrt{3} - (\sqrt{2})i]$

26. $[3 + (\sqrt{5})i] \cdot [3 - (\sqrt{5})i]$

Simplify so that the result is expressed in $a + bi$ form.

27. $\dfrac{1}{2 - i}$

28. $\dfrac{4}{3 + 2i}$

29. $\dfrac{3 + i}{1 + i}$

30. $\dfrac{1}{1 + i\sqrt{2}}$

31. $(2 + 5i) \div (8 - 7i)$

32. $(-4 - 11i) \div (5 + 3i)$

33. $2 \div [4 - (\sqrt{5})i]$

34. $1 \div (a + bi)$

Express in $a + bi$ form.

35. $\sqrt{-9}$

36. $\sqrt{-1}$

37. $\sqrt{-2}$

38. $\sqrt{-13}$

39. $\sqrt[3]{-8}$

40. $\sqrt[3]{-27}$

41. $\sqrt{-1}\sqrt{-16}$

42. $\sqrt{-2}\sqrt{-3}$

43. $i\sqrt{-11}$

44. $i\sqrt{-25}$

45. **(a)** Complete the following table.

$i^0 = 1$
$i^1 = i$
$i^2 = ?$
$i^3 = ?$
$i^4 = ?$
$i^5 = ?$

(b) What can you say in general about i^n, where n is any natural number?

46. **(a)** Find i^{-1}.
(b) What can you say about i^n when n is a negative integer?

47. **(a)** Evaluate $(1 + 4i)(1 - 4i)$.
(b) If a and b are any real numbers, what can be said about the product $a + bi$ with its conjugate?
(c) Is there a formula for factoring $a^2 + b^2$?

48. **(a)** Compute $[\sqrt{2}/2 + (\sqrt{2}/2)i]^2$
(b) Find a square root of the complex number i.
(c) Find another square root of i.

49. Compute $[1/2 + (\sqrt{3}/2)i]^3$ and comment on your result.

50. Compute $(\sqrt{3}/2 + \frac{1}{2}i)^3$, and comment.

7-2 Graphical representation of complex numbers

As we know, a real number can be represented by a point on the number line. This graphical representation adds much to our understanding of real numbers.

The plane, with a rectangular coordinate system labeled as in Figure 7–1, is known as the *complex plane*. As we shall see, each complex number can be represented by a point in the complex plane.

In the complex plane, the horizontal axis is referred to as the *real axis*, and the vertical axis is the *imaginary axis*. To plot the complex number $a + bi$, we mark off the real part a on the real axis and the imaginary part bi on the imaginary axis. Then the complex number $a + bi$ is represented by a point directly above (or below) a, and at a height corresponding to bi.

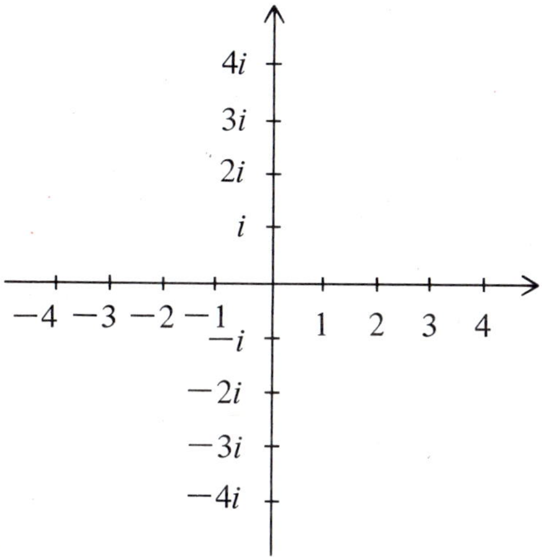

Figure 7–1
The complex plane

EXAMPLE: The complex number $5 + 3i$ is plotted in Figure 7–2. ■

EXAMPLE: The complex numbers $-4 - i$ and $3 - 4i$ are plotted in Figure 7–3. ■

The graph of a real number lies on the real axis, as in the next example.

EXAMPLE: The graph of -6 is shown in Figure 7–4. ■

The graph of a pure imaginary number lies on the imaginary axis.

EXAMPLE: The graph of $3i$ is given in Figure 7–5. ■

For a real number, its absolute value is the distance from the origin to the corresponding point on the number line. For a complex number, a natural extention would be that its absolute value should be the distance from the origin to the corresponding point in the complex plane.

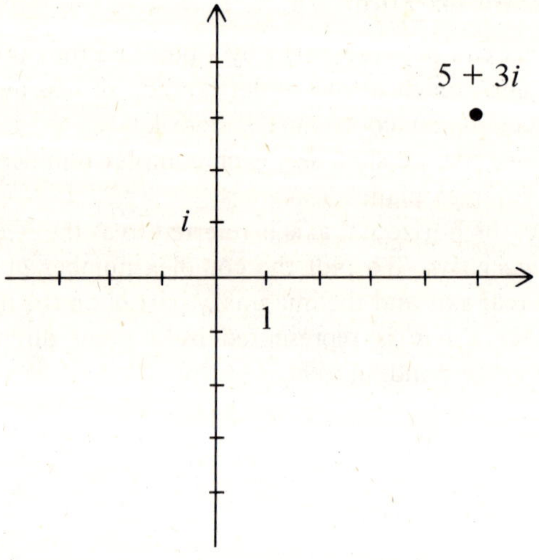

Figure 7–2

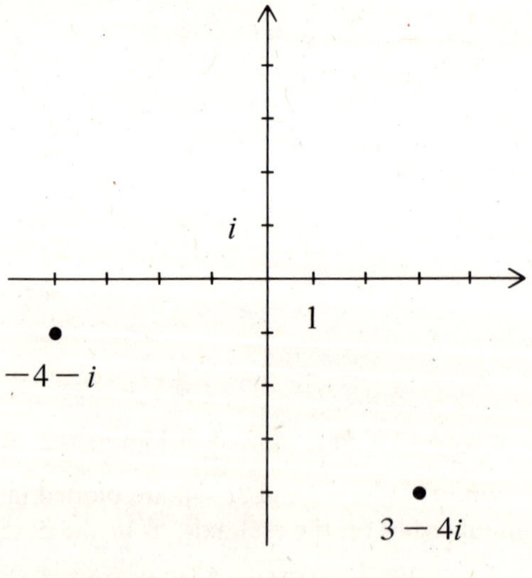

Figure 7–3

By the distance formula (Section 3-1), in the complex plane the distance from the origin to $a + bi$ is $\sqrt{a^2 + b^2}$. We, therefore, make the following definition of absolute value for complex numbers:

If $a + bi$ is a complex number, then
$|a + bi| = \sqrt{a^2 + b^2}$

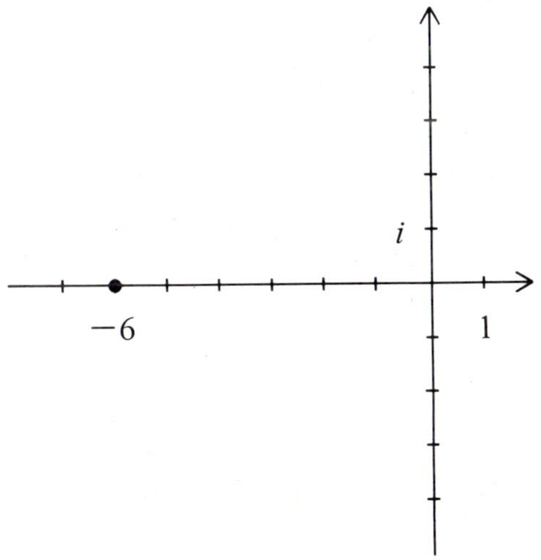

Figure 7–4

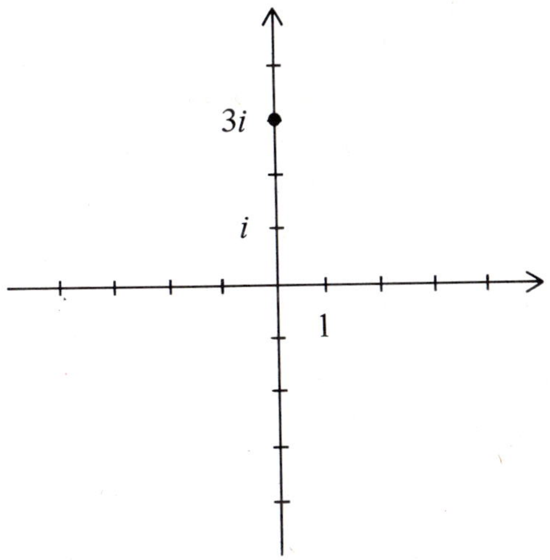

Figure 7–5

EXAMPLE: $|3 + 4i| = \sqrt{3^2 + 4^2} = \sqrt{9 + 16} = \sqrt{25} = 5.$

Note that 5 is the distance from the origin to $3 + 4i$ in the complex plane. See Figure 7–6. ■

EXAMPLE: $|2 - i| = \sqrt{2^2 + (-1)^2} = \sqrt{4 + 1} = \sqrt{5}.$ ■

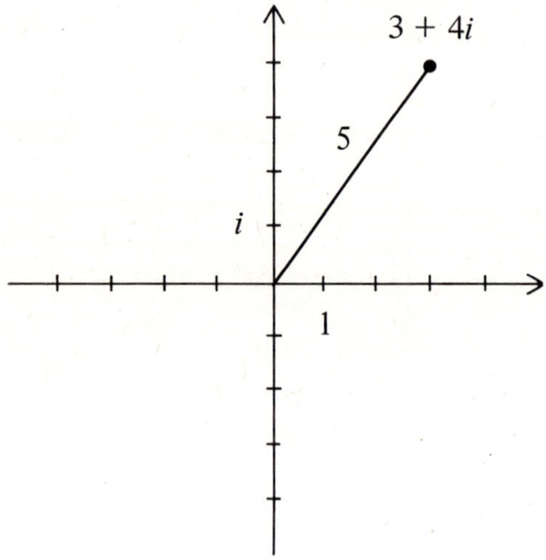

Figure 7–6

EXAMPLE: $|-6| = \sqrt{(-6)^2 + 0^2} = \sqrt{36} = 6$. Note that -6 is 6 units away from the origin (Figure 7–4). Its absolute value is the same whether we think of -6 as a real number or as a complex number. ■

When two complex numbers are multiplied together, the absolute value of the product is the product of the absolute values of the two factors, according to the following formula:

If $a + bi$ and $c + di$ are complex numbers, then

$|(a + bi)(c + di)| = |a + bi|\ |c + di|$

EXAMPLE: Let us verify this formula for $a + bi = 1 - i$ and $c + di = 3 + i$. We find that

$$|1 - i| = \sqrt{1 + 1} = \sqrt{2}$$

and

$$|3 + i| = \sqrt{9 + 1} = \sqrt{10}$$

Thus,

$$|1 - i|\ |3 + i| = \sqrt{2}\sqrt{10} = \sqrt{20}$$

On the other hand,

$$\begin{aligned} |(1 - i)(3 + i)| &= |3 - 2i - i^2| = |4 - 2i| \\ &= \sqrt{16 + 4} = \sqrt{20} \end{aligned}$$

and so

$$|(1-i)(3+i)| = |1-i|\,|3+i|(=\sqrt{20})$$ ■

There is an interesting relationship between the conjugate of a complex number and the absolute value of the complex number. This relationship can be discovered by multiplying the complex number $a + bi$ by its conjugate $a - bi$. We find that

$$(a+bi)(a-bi) = a^2 - b^2i^2 = a^2 + b^2$$

However, we know that

$$|a+bi| = \sqrt{a^2+b^2}$$

Therefore,

$$|a+bi|^2 = a^2 + b^2$$

It follows that $(a+bi)(a-bi) = |a+bi|^2$. In words, *the product of a complex number with its conjugate is the square of its absolute value.*

EXAMPLE: $|3+4i| = 5$

and so

$$(3+4i)(3-4i) = 5^2 = 25$$ ■

7-2 *Exercises*

Plot in the complex plane.

1. $3 + 6i$
2. $5 + 2i$
3. $4 - 9i$
4. $-3 + 2i$
5. $\frac{1}{2} + \frac{1}{2}i$
6. $\frac{1}{2} - \frac{1}{2}i$
7. $2i$
8. $-5i$
9. 4
10. 0
11. $\frac{1}{2} + \frac{1}{2}i\sqrt{3}$
12. $\sqrt{2} - i\sqrt{2}$
13. **(a)** Compare the graphs of $3 + 5i$ and $3 - 5i$.
(b) Discuss in general the relation between the graphs of $a + bi$ and $a - bi$.

Find the absolute value of $a + bi$, and verify that

$$(a+bi)(a-bi) = |a+bi|^2$$

if $a + bi$ is as follows.

14. $3 - 4i$
15. $-i\sqrt{2}$
16. π
17. $-5 + 12i$
18. $1 + i$
19. $-\sqrt{5}$
20. $3i$
21. $\sqrt{3} - i\sqrt{5}$

Verify each of the following.

22. $|(-3+4i)(-1+i)| = |-3+4i|\,|-1+i|$

23. $|(-7-24i)(2i)| = |-7-24i|\,|2i|$

24. $|(1+i\sqrt{3})(1+i\sqrt{3}| = |1+i\sqrt{3}|\,|1+i\sqrt{3}|$

25. $|(2-i)(6+8i)| = |2-i|\,|6+8i|$

26. Use the formula $(a+bi)(a-bi) = |a+bi|^2$ to find a general formula for the reciprocal of $a+bi$.

7-3 Trigonometric representation of complex numbers

Let $a+bi$ be a complex number. Our objective now is to derive an alternative algebraic expression for $a+bi$ using the trigonometric properties of the graph of $a+bi$ in the complex plane.

In Figure 7–7, r is the distance from the origin to $a+bi$. Thus,

$$r = |a+bi| = \sqrt{a^2+b^2} \qquad \text{(from Section 7-2)}$$

Also in Figure 7–7, θ is the angle from the positive real axis to the ray extending from the origin through the graph of $a+bi$. (We are assuming that $a+bi \neq 0$.) The angle θ is called the *argument* of the complex number $a+bi$.

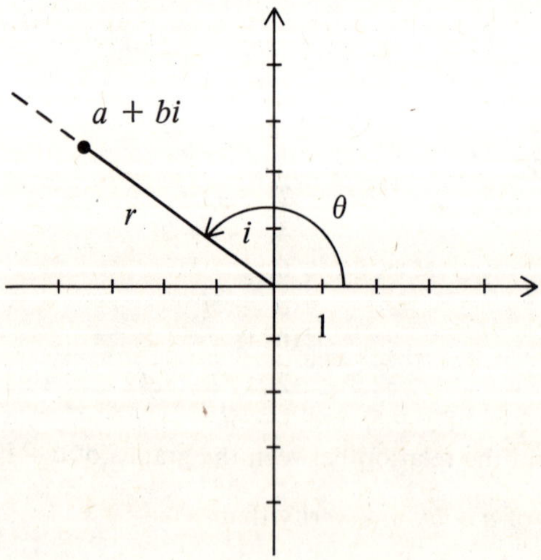

Figure 7–7

EXAMPLE: The complex number $1+i$ is plotted in Figure 7–8. From the formula, $r = \sqrt{1^2+1^2} = \sqrt{2}$. From Figure 7–8, we see that $\theta = 45°$ or $\pi/4$ radians. ■

To be precise, we should allow for addition of any number of complete revolu-

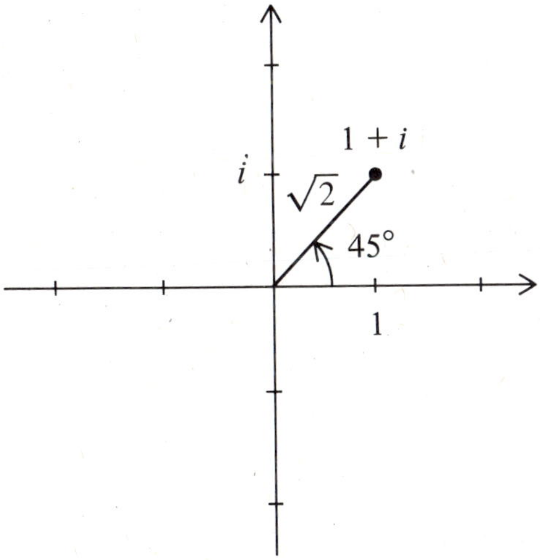

Figure 7–8

tions to θ. Thus, in the preceding example, the possible arguments are $45° + k \cdot 360°$, $k \in J$, or, in radians, $\pi/4 + 2\pi k$, $k \in J$.

In the usual xy-coordinate system on the plane, the x-coordinate of $a + bi$ is a and the y-coordinate is b. Therefore, from Section 3-3, we have

$$\cos \theta = \frac{a}{r}$$

$$\sin \theta = \frac{b}{r}$$

Thus,

$$a = r \cos \theta$$
$$b = r \sin \theta$$

It follows that

$$a + bi = r \cos \theta + (r \sin \theta)i = r(\cos \theta + i \sin \theta)$$

Thus $a + bi$ is expressed in terms of its absolute value r and its argument θ by the important equation:

$$a + bi = r(\cos \theta + i \sin \theta)$$

The form $r(\cos \theta + i \sin \theta)$ is known as the *trigonometric form* of the complex number $a + bi$, while $a + bi$ is called the *rectangular* form.

Note: If $a + bi = 0$, we have $r = 0$, but no angle θ is determined. Because $0 = 0(\cos \theta + i \sin \theta)$ for any angle, we allow any value θ for the argument of the complex number 0.

EXAMPLE: We find the trigonometric form of $1 + i$. As noted above (see Figur 7–8), we may take $r = \sqrt{2}$ and $\theta = 45°$. Thus we have the trigonometric form

$$1 + i = \sqrt{2}\,(\cos 45° + i \sin 45°)$$

or, using radian measure,

$$1 + i = \sqrt{2}\left(\cos \frac{\pi}{4} + i \sin \frac{\pi}{4}\right)$$

The expression $\cos \theta + i \sin \theta$ is commonly abbreviated as cis θ—that is, w make the definition:

$$\text{cis } \theta = \cos \theta + i \sin \theta$$

In this notation,

$$1 + i = \sqrt{2} \text{ cis } 45° = \sqrt{2} \text{ cis } \frac{\pi}{4}$$

EXAMPLE: Express 2 cis 30° in rectangular form.

We have $\cos 30° = \frac{1}{2}\sqrt{3}$ and $\sin 30° = \frac{1}{2}$. Therefore

$$\begin{aligned} 2 \text{ cis } 30° &= 2(\cos 30° + i \sin 30°) \\ &= 2\left(\frac{\sqrt{3}}{2} + \frac{i}{2}\right) \\ &= \sqrt{3} + i \; ■ \end{aligned}$$

EXAMPLE: Find the trigonometric form of $-5i$.

Evidently $r = |-5i| = 5$. From Figure 7–9, we see that $\theta = 3\pi/2 + 2\pi k$, $k \in J$ Using $\theta = 3\pi/2$, we have $-5i = 5 \text{ cis } 3\pi/2$. ■

EXAMPLE: Express 6 cis π in rectangular form.

Because $\cos \pi = -1$ and $\sin \pi = 0$,

$$6 \text{ cis } \pi = 6(\cos \pi + i \sin \pi) = 6 \cdot (-1 + i \cdot 0) = -6 \; ■$$

The general procedure to convert r cis θ to rectangular form is as follows:

1. Evaluate $\cos \theta$ and $\sin \theta$.
2. Compute $a + bi = r(\cos \theta + i \sin \theta)$ by direct calculation.

When we are given the rectangular form $a + bi$, some care is needed in finding the trigonometric form. The following steps are recommended.

1. Sketch the location of $a + bi$ in the complex plane.

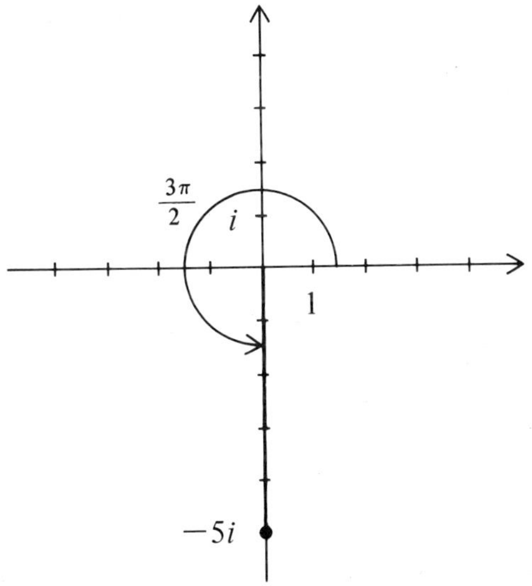

Figure 7–9

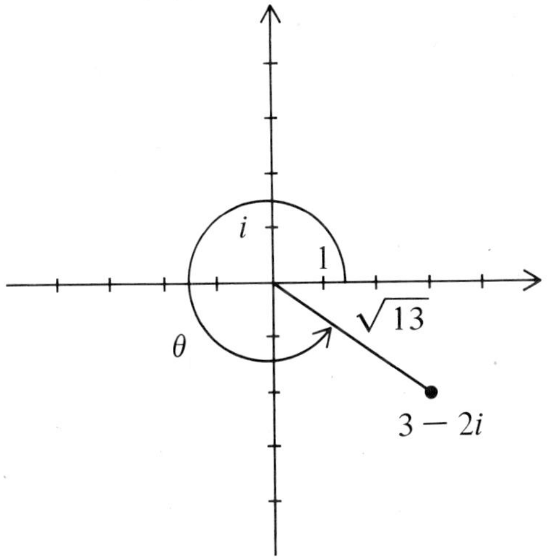

Figure 7–10

2. Evaluate $r = |a + bi| = \sqrt{a^2 + b^2}$.
3. Choose θ in the proper quadrant according to any of the formulas

$$\cos \theta = \frac{a}{r}$$

$$\sin \theta = \frac{b}{r}$$

$$\tan \theta = \frac{b}{a}$$

EXAMPLE: Express $3 - 2i$ in trigonometric form.
This point is shown in Figure 7–10. We have $r = |3 - 2i| = \sqrt{13}$. From Figure 7–10, θ must be a fourth-quadrant angle. We have $\tan \theta = b/a = -2/3 = -0.6667$ From the table of Trigonometric Ratios, we find that the reference angle is approximately 34°, and so we may take $\theta = 360° - 34° = 326°$. Hence the trigonometric form is $\sqrt{13}$ cis 326°. ■

7-3 *Exercises*

Express in rectangular form.

1. $2(\cos 150° + i \sin 150°)$
2. $3(\cos 225° + i \sin 225°)$
3. $\frac{1}{2}$ cis $(-\pi/6)$
4. $\frac{1}{4}$ cis $(-\pi)$
5. $\sqrt{2}$ cis 240°
6. $\sqrt{7}$ cis $\pi/4$
7. $1.8(\cos 23° + i \sin 23°)$
8. $0.4(\cos 131° + i \sin 131°)$

Find r and all possible values of θ for the given complex number.

9. 18
10. 0
11. $-\sqrt{5}$
12. $1 - i$
13. $3 - 5i$
14. $-6 - i$

Express in trigonometric form.

15. -3
16. $-17i$
17. $-3 - 3i$
18. $-4 - (4\sqrt{3})i$
19. $3 - 4i$
20. $-5 + 12i$
21. $1 - 7i$
22. $4 + i$

Perform the indicated operation and express your result in trigonometric form.

23. (3 cis 30°)(2 cis 60°)
24. (2 cis 45°)(5 cis 90°)
25. $(\sqrt{3}$ cis 45°$)^2$
26. (2 cis 60°$)^3$

Let $a + bi$ have trigonometric form r cis θ. Find the trigonometric form for:

27. $a - bi$
28. $-(a + bi)$
29. $-a + bi$

-4 Multiplication of complex numbers

When we work with complex numbers, we may use either rectangular or trigonometric form. Sometimes one is more convenient, sometimes the other.

Usually, to do addition (or subtraction), the rectangular form is easier. The basic formula is

$$(a_1 + b_1 i) + (a_2 + b_2 i) = (a_1 + a_2) + (b_1 + b_2)i$$

In words, we add the real parts and add the imaginary parts. There is no such simple formula for addition of complex numbers in trigonometric form.

However, the situation is different for multiplication. The multiplication formula, in rectangular form, is

$$(a_1 + b_1 i)(a_2 + b_2 i) = (a_1 a_2 - b_1 b_2) + (a_1 b_2 + a_2 b_1)i$$

This is a complicated formula.

There is an easier rule for multiplication of complex numbers in trigonometric form. In words, we multiply the absolute values and add the arguments. More formally, we have the following:

If $r_1 \text{ cis } \theta_1$ and $r_2 \text{ cis } \theta_2$ are complex numbers, then

$$(r_1 \text{ cis } \theta_1)(r_2 \text{ cis } \theta_2) = r_1 r_2 \text{ cis } (\theta_1 + \theta_2)$$

The proof will be based on trigonometric identities. First, however, we present some examples.

EXAMPLE: Compute $(2 \text{ cis } 30°)(3 \text{ cis } 60°)$.
The absolute value of the result is $2 \cdot 3 = 6$. The argument of the result is $30° + 60° = 90°$. Therefore $(2 \text{ cis } 30°)(3 \text{ cis } 60°) = 6 \text{ cis } 90°$.

Note that

$$2 \text{ cis } 30° = 2\left(\frac{\sqrt{3}}{2} + \frac{1}{2}i\right) = \sqrt{3} + i$$

$$3 \text{ cis } 60° = 3\left[\frac{1}{2} + \left(\frac{\sqrt{3}}{2}\right)i\right] = \frac{3}{2} + \left(3\frac{\sqrt{3}}{2}\right)i$$

$$6 \text{ cis } 90° = 6i$$

The result

$$(\sqrt{3} + i)\left[\frac{3}{2} + 3\frac{\sqrt{3}}{2}i\right] = 6i$$

could be verified by computation, but the work is easier in trigonometric form. ■

EXAMPLE: We find (4 cis 27°)(3 cis 11°).
Because $4 \cdot 3 = 12$ and $27° + 11° = 38°$, we have

$$(4 \text{ cis } 27°)(3 \text{ cis } 11°) = 12 \text{ cis } 38°$$ ■

EXAMPLE: Use the trigonometric form to compute

$(1 + i)(-3 - 3i)$.

We have $1 + i = \sqrt{2} \text{ cis } 45°$, and $-3 - 3i = 3\sqrt{2} \text{ cis } 225°$. Therefore,

$$\begin{aligned}(1 + i)(-3 - 3i) &= (\sqrt{2} \text{ cis } 45°)(3\sqrt{2} \text{ cis } 225°)\\ &= 6 \text{ cis } 270°\\ &= -6i\end{aligned}$$ ■

We now give the steps needed to prove

$$(r_1 \text{ cis } \theta_1)(r_2 \text{ cis } \theta_2) = r_1 r_2 \text{ cis } (\theta_1 + \theta_2)$$

We have

$$\begin{aligned}(r_1 \text{ cis } \theta_1)(r_2 \text{ cis } \theta_2) &= r_1(\cos \theta_1 + i \sin \theta_1) \cdot r_2(\cos \theta_2 + i \sin \theta_2)\\ &= r_1 r_2(\cos \theta_1 + i \sin \theta_1)(\cos \theta_2 + i \sin \theta_2)\\ &= r_1 r_2(\cos \theta_1 \cos \theta_2 + i \sin \theta_1 \cos \theta_2 + i \cos \theta_1 \sin \theta_2\\ &\qquad + i^2 \sin \theta_1 \sin \theta_2)\\ &= r_1 r_2[(\cos \theta_1 \cos \theta_2 - \sin \theta_1 \sin \theta_2)\\ &\qquad + i(\sin \theta_1 \cos \theta_2 + \cos \theta_1 \sin \theta_2)]\end{aligned}$$

But, by the identities of Section 4-3,

$$\cos \theta_1 \cos \theta_2 - \sin \theta_1 \sin \theta_2 = \cos (\theta_1 + \theta_2)$$

and

$$\sin \theta_1 \cos \theta_2 + \cos \theta_1 \sin \theta_2 = \sin (\theta_1 + \theta_2)$$

Therefore,

$$\begin{aligned}(r_1 \text{ cis } \theta_1)(r_2 \text{ cis } \theta_2) &= r_1 r_2[\cos(\theta_1 + \theta_2) + i \sin(\theta_1 + \theta_2)]\\ &= r_1 r_2 \text{ cis}(\theta_1 + \theta_2)\end{aligned}$$

Here are some further examples.

EXAMPLE: Find $(-\sqrt{6} - i\sqrt{6})^2$.

We have

$$-\sqrt{6} - i\sqrt{6} = \sqrt{12} \text{ cis } \frac{5\pi}{4}.$$

Therefore

$$(-\sqrt{6} - i\sqrt{6})^2 = \left(\sqrt{12} \text{ cis } \frac{5\pi}{4}\right)\left(\sqrt{12} \text{ cis } \frac{5\pi}{4}\right)$$

$$= 12 \text{ cis } \frac{5\pi}{2}$$

$$= 12 \text{ cis } \frac{\pi}{2}$$

$$= 12i$$ ■

EXAMPLE: We use trigonometric form to verify that $(-8)^2 = 64$.

We have $-8 = 8 \text{ cis } \pi$. Therefore

$$\begin{aligned}(-8)^2 &= (8 \text{ cis } \pi)(8 \text{ cis } \pi)\\ &= 64 \text{ cis } 2\pi\\ &= 64 \text{ cis } 0\\ &= 64\end{aligned}$$ ■

-4 *Exercises*

Express the result of the indicated multiplication in trigonometric form.

1. $(5 \text{ cis } 20°)(4 \text{ cis } 85°)$
2. $(11 \text{ cis } 200°)(4 \text{ cis } 300°)$
3. $(\sqrt{5} \text{ cis } \pi/8)(3 \text{ cis } \pi/4)$
4. $[2 \text{ cis } (-3\pi/8)](5 \text{ cis } 3\pi/4)$

Use trigonometric form to calculate the indicated products.

5. $(2 + 2i)(-7i)$
6. $3i(1 + i\sqrt{3})$
7. $(\sqrt{3} - i)(2 - 2i\sqrt{3})$
8. $(-7 + 7i)(\sqrt{5} + i\sqrt{5})$
9. $(-2i)^2$
10. $(-3 - 3i)^2$
11. $(\sqrt{3} - i)^2$
12. $(\sqrt{3} - i)^3$

Assume that $a + bi$ has trigonometric form $r \text{ cis } \theta$, and hence $a - bi$ has trigonometric form $r \text{ cis } (-\theta)$.

13. Find the trigonometric form for $(a + bi)^2$.
14. Find the trigonometric form for $(a + bi)^3$.
15. Guess the trigonometric form for $(a + bi)^n$, $n \in N$.
16. Find the trigonometric form for $(a + bi)(a - bi)$.
17. Use the result of Exercise 16 to find the rectangular form for $(a + bi)(a - bi)$.

The formula, in trigonometric form, for division by a nonzero complex number, is:

$$(r_1 \text{ cis } \theta_1) \div (r_2 \text{ cis } \theta_2) = \left(\frac{r_1}{r_2}\right) \text{ cis } (\theta_1 - \theta_2)$$

Use this formula to compute:

18. $(5 \text{ cis } 20°) \div (4 \text{ cis } 85°)$
19. $(11 \text{ cis } 200°) \div (4 \text{ cis } 300°)$
20. $(\sqrt{5} \text{ cis } \pi/8) \div (3 \text{ cis } \pi/4)$
21. $[2 \text{ cis } (-3\pi/8)] \div (5 \text{ cis } 7\pi/4)$
22. Prove the above formula for division in trigonometric form.

7-5 Powers and roots of complex numbers

Consider the complex number $r \text{ cis } \theta = r(\cos \theta + i \sin \theta)$ given in trigonometr form. Let us use the trigonometric form to compute the square of this numbe First, we have:

$$(r \text{ cis } \theta)^2 = (r \text{ cis } \theta)(r \text{ cis } \theta)$$

By the formula for multiplication of complex numbers in trigonometric form,

$$(r \text{ cis } \theta)(r \text{ cis } \theta) = r \cdot r \text{ cis } (\theta + \theta)$$

Evidently

$$r \cdot r \text{ cis } (\theta + \theta) = r^2 \text{ cis } 2\theta$$

and therefore

$$(r \text{ cis } \theta)^2 = r^2 \text{ cis } 2\theta$$

Similarly, the cube of $r \text{ cis } \theta$ is computed as follows:

$$\begin{aligned}(r \text{ cis } \theta)^3 &= (r \text{ cis } \theta)^2(r \text{ cis } \theta)\\ &= (r^2 \text{ cis } 2\theta)(r \text{ cis } \theta)\\ &= r^2 \cdot r \text{ cis } (2\theta + \theta)\\ &= r^3 \text{ cis } 3\theta\end{aligned}$$

By the same method, we could compute $(r \text{ cis } \theta)^4$. The result is

$$(r \text{ cis } \theta)^4 = r^4 \text{ cis } 4\theta$$

The formulas

$$\begin{aligned}(r \text{ cis } \theta)^2 &= r^2 \text{ cis } 2\theta\\ (r \text{ cis } \theta)^3 &= r^3 \text{ cis } 3\theta\\ (r \text{ cis } \theta)^4 &= r^4 \text{ cis } 4\theta\end{aligned}$$

suggest a general rule. This rule, known as *De Moivre's theorem*, may be stated a as follows.

If n is a natural number, then

$$(r \text{ cis } \theta)^n = r^n \text{ cis } n\theta$$

EXAMPLE: $(3 \text{ cis } 10°)^2 = 9 \text{ cis } 20°$. ■

EXAMPLE: $(2 \text{ cis } \pi/2)^4 = 2^4 \text{ cis } 4\pi/2 = 16 \text{ cis } 2\pi = 16.$ ■

The calculations to raise a complex number to a power are much easier in trigonometric form, using De Moivre's theorem, than in rectangular form.

EXAMPLE: Compute $(1 - i)^4$.
In trigonometric form, $1 - i = \sqrt{2} \text{ cis } (-45°)$. Therefore

$$(1 - i)^4 = [\sqrt{2} \text{ cis } (-45°)]^4 = (\sqrt{2})^4 \text{ cis } (-180°) = 4 \text{ cis } 180° = 4.$$ ■

EXAMPLE: Use trigonometric form to find i^7.

We have $i = 1 \text{ cis } \pi/2$. Therefore $i^7 = 1^7 \text{cis } 7\pi/2 = 1 \text{ cis } 3\pi/2 = -i.$ ■

EXAMPLE: Compute $(\sqrt{3} + i)^5$.

We have $\sqrt{3} + i = 2 \text{ cis } 30°$. Therefore,

$$\begin{aligned}(\sqrt{3} + i)^5 &= (2 \text{ cis } 30°)^5 \\ &= 2^5 \text{ cis } 150° \\ &= 32\left(-\frac{\sqrt{3}}{2} + \frac{1}{2}i\right) \\ &= -16\sqrt{3} + 16i\end{aligned}$$ ■

Let us now look at the equation

$$(r \text{ cis } \theta)^n = r^n \text{ cis } n\theta$$

from another point of view. It says that an nth root of $r^n \text{ cis } n\theta$ is $r \text{ cis } \theta$.

Let us make the substitutions $s = r^n$, so that $r = \sqrt[n]{s}$, and $\varphi = n\theta$, so that $\theta = \varphi/n$.[1] The equation

$$(r \text{ cis } \theta)^n = r^n \text{ cis } n\theta$$

becomes

$$\left(\sqrt[n]{s} \text{ cis } \frac{\varphi}{n}\right)^n = s \text{ cis } \varphi$$

Thus an nth root of $s \text{ cis } \varphi$ is $\sqrt[n]{s} \text{ cis } \varphi/n$, or, equivalently, we have:

An nth root of $r \text{ cis } \theta$ is $\sqrt[n]{r} \text{ cis } \theta/n$.

EXAMPLE: A square root of $9 \text{ cis } \pi$ is $\sqrt{9} \text{ cis } \pi/2 = 3 \text{ cis } \pi/2$. And indeed, $9 \text{ cis } \pi = -9$, $3 \text{ cis } \pi/2 = 3i$, and $(3i)^2 = -9$. ■

However, this example raises an important question. The method gave us one square root of -9, namely $3i$. We know that -9 has another square root, namely $-3i$. How could this other square root have been obtained?

The answer lies in the ambiguity of the argument chosen for -9. Instead π, let us use the argument $\pi + 2\pi = 3\pi$. Thus we write

$$-9 = 9 \text{ cis } 3\pi$$

and so a square root for -9 is $\sqrt{9} \text{ cis } 3\pi/2 = 3 \text{ cis } 3\pi/2 = -3i$.

To get both square roots, we use two consecutive values for the argument.

EXAMPLE: Find both square roots of $4i$.

We use two consecutive values for the argument of $4i$ and write

$$4i = 4 \text{ cis } \frac{\pi}{2}$$

$$4i = 4 \text{ cis } \left(\frac{\pi}{2} + 2\pi\right) = 4 \text{ cis } \frac{5\pi}{2}$$

Hence the square roots of $4i$ are:

$$\sqrt{4} \text{ cis } \frac{1}{2} \cdot \frac{\pi}{2} = \sqrt{4} \text{ cis } \frac{\pi}{4} = 2\left[\left(\frac{\sqrt{2}}{2}\right) + \left(\frac{\sqrt{2}}{2}\right) i\right] = \sqrt{2} + (\sqrt{2})\, i$$

$$\sqrt{4} \text{ cis } \frac{1}{2} \cdot \frac{5\pi}{2} = \sqrt{4} \text{ cis } \frac{5\pi}{4} = 2\left[-\left(\frac{\sqrt{2}}{2}\right) - \left(\frac{\sqrt{2}}{2}\right) i\right] = -\sqrt{2} - (\sqrt{2})\, i \ ■$$

The general situation is this. If n is a natural number, then an nth root of r ci is given by $\sqrt[n]{r} \text{ cis } \theta/n$. However, if $r \text{ cis } \theta \neq 0$, then this number has n differe nth roots. They can be obtained from n consecutive values θ, $\theta + 2\pi$, $\theta + 2 \cdot 2\pi$, . $\theta + (n-1)2\pi$ for the argument of $r \text{ cis } \theta$. These nth roots are $\sqrt[n]{r} \text{ cis } \theta/n$, $\sqrt[n]{r}$ $(\theta + 2\pi)/n$, $\sqrt[n]{r} \text{ cis } (\theta + 2 \cdot 2\pi)/n, \ldots, \sqrt[n]{r} \text{ cis } [\theta + (n-1) \cdot 2\pi]/n$. Thus, stated f mally, we have De Moivre's theorem for roots.

Let n be a natural number and $r \text{ cis } \theta$ be a complex number with $r > 0$. Then r ci has n different nth roots. They are

$$\sqrt[n]{r} \text{ cis } \frac{\theta}{n}$$

$$\sqrt[n]{r} \text{ cis } \frac{\theta + 2\pi}{n}$$

$$\sqrt[n]{r} \text{ cis } \frac{\theta + 2 \cdot 2\pi}{n}$$

$$\vdots$$

$$\sqrt[n]{r} \text{ cis } \frac{\theta + (n-1)2\pi}{n}$$

[1] The symbol φ is the Greek letter phi.

EXAMPLE: Find all cube roots of -1.
Because we are finding cube roots, we use three consecutive values for the argument of -1. For this example, we will work in degrees. Thus we write:

$-1 = 1 \text{ cis } 180°$
$-1 = 1 \text{ cis } 540°$
$-1 = 1 \text{ cis } 900°$

Hence the cube roots of -1 are:

$$\sqrt[3]{1} \text{ cis } \frac{180°}{3} = 1 \text{ cis } 60° = \frac{1}{2} + \left(\frac{\sqrt{3}}{2}\right) i$$

$$\sqrt[3]{1} \text{ cis } \frac{540°}{3} = 1 \text{ cis } 180° = -1$$

$$\sqrt[3]{1} \text{ cis } \frac{900°}{3} = 1 \text{ cis } 300° = \frac{1}{2} - \left(\frac{\sqrt{3}}{2}\right) i \ ■$$

EXAMPLE: Express the fourth roots of 16 cis 88° in trigonometric form.
We use the four forms:

$16 \text{ cis } 88° = 16 \text{ cis } 88°$
$16 \text{ cis } 88° = 16 \text{ cis } 448°$
$16 \text{ cis } 88° = 16 \text{ cis } 808°$
$16 \text{ cis } 88° = 16 \text{ cis } 1168°$

Therefore the fourth roots of 16 cis 88° are:

$$\sqrt[4]{16} \text{ cis } \frac{88°}{4} = 2 \text{ cis } 22°$$

$$\sqrt[4]{16} \text{ cis } \frac{448°}{4} = 2 \text{ cis } 112°$$

$$\sqrt[4]{16} \text{ cis } \frac{808°}{4} = 2 \text{ cis } 202°$$

$$\sqrt[4]{16} \text{ cis } \frac{1168°}{4} = 2 \text{ cis } 292° \ ■$$

To convert the results of the preceding example to rectangular, decimal form, we would use the table of values of the trigonometric ratios.

-5 *Exercises*

Compute the indicated powers.

1. $(3 \text{ cis } 24°)^2$
2. $(3 \text{ cis } 24°)^3$
3. $(2 \text{ cis } \pi/4)^4$
4. $(2 \text{ cis } \pi/4)^3$

5. $(1 + i)^3$

6. $(1 + i)^4$

7. $(\sqrt{3} - i)^3$

8. $(\sqrt{3} - i)^2$

Express in trigonometric form.

9. All square roots of 16 cis 18°

10. All square roots of 9 cis 40°

11. All cube roots of 8 cis 159°

12. All fourth roots of cis 160°

13. All fifth roots of 1

14. All fifth roots of -1

Express in rectangular form.

15. All square roots of $-16i$

16. All square roots of $1 + (\sqrt{3})i$

17. All cube roots of 27

18. All fourth roots of 1

19. All sixth roots of 1

20. All sixth roots of -1

21. **(a)** For each of Exercises 9, 10, 15, and 16, plot the two results as points in t complex plane.

(b) Make a general statement about the location of the two square roots of complex number in the complex plane.

22. **(a)** For each of Exercises 11 and 17, plot the three results as points in t complex plane.

(b) Make a general statement about the location of the three cube roots of complex number in the complex plane.

23. Use your results from Exercises 12 through 14 and Exercises 18 through 20 try to make a general statement about the location of the nth roots of a compl number in the complex plane.

24. Is the formula of De Moivre's theorem, $(r \text{ cis } \theta)^n = r^n \text{ cis } n\theta$, valid if $n = ($ Explain.

25. Is the formula of De Moivre's theorem valid for $n = -1$? Why (not)?

26. Is the formula of De Moivre's theorem valid when n is any negative intege What is your reasoning?

Review exercises (Chapter 7)

Explain and/or illustrate with an example.

1. complex number

2. imaginary number

3. conjugate of a complex number

4. absolute value of a complex number

5. argument of a complex number

6. complex plane

7. imaginary axis
8. trigonometric form
9. De Moivre's theorem

Plot in the complex plane.

10. $-7 + 4i$
11. $4 \text{ cis } 2\pi/3$
12. Express $-3 + 3i$ in trigonometric form.
13. Express cis $300°$ in rectangular form.

Evaluate or simplify.

14. $(2 + 5i)(3 - 11i)$
15. $(2 - 7i) \div (5 + 4i)$
16. $|(3 - 4i)(7 + 24i)|$
17. $(2 \text{ cis } 51°)[5 \text{ cis } (-113°)]$
18. $(4 \text{ cis } \pi/7)[4 \text{ cis } (-\pi/7)]$
19. i^{10}
20. $[2 - (2\sqrt{3})i]^3$
21. Find all cube roots of 8.
22. Plot all fourth roots of $16i$ in the complex plane.

8 Inverse trigonometric functions

Often it is necessary to solve an equation $y = f(x)$ for x as a function of y. The function we a looking for is called the inverse function of f.

In this chapter we define inverse functions of the various trigonometric fun tions. An application to graphing is given in Section 8-4. A different type of applic tion appears in Appendix A, where logarithmic functions are presented as t inverse functions of exponential functions.

8-1 Inverse functions

Let us recall the graph of the function defined by $y = 2x + 1$ (see Figure 8–1). W know that the graph in Figure 8–1 is the graph of a function $y = f(x)$ becau $y = 2x + 1$ expresses y as a function of x. The domain of f is the set of real number as is the range.

This function $f(x) = 2x + 1$ has an important additional property: For each in the range of f, there is only one number x in the domain of f such that $y = f(x$ Geometrically, the corresponding property is that each horizontal line intersec

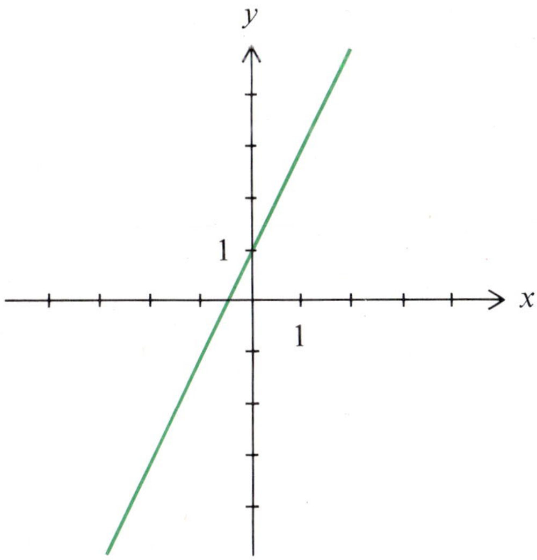

Figure 8–1
$\{(x, y) | y = 2x + 1\}$

the graph of f in at most one point. Such a function is called a *one-to-one* function, according to the following definition:

The function f is a one-to-one function if for each y in the range of f there is only one number x in the domain of f such that $y = f(x)$.

In terms of ordered pairs, the function f is one-to-one if no two different ordered pairs of f have the same second component.

EXAMPLE: The function f defined by $y = 2x + 1$ is a one-to-one function, because for each $y \in R$ there is only one x such that $y = 2x + 1$. (See Figure 8–1). ■

EXAMPLE: The function f defined by $y = x^2$ is not a one-to-one function. Geometrically (see Figure 8–2) there is at least one (and in fact many) horizontal line which meets the graph in more than one point. Or, from the point of view of ordered pairs, both $(3, 9) \in f$ and $(-3, 9) \in f$. Thus there are two different ordered pairs of f which have the same second component, and so f is not a one-to-one function.

EXAMPLE: The function f with $y = x^2$ but with domain $\{x | x \geq 0\}$ is a one-to-one function. Its graph is shown in Figure 8–3.

EXAMPLE: The function f given by $f(x) = 3x$ is a one-to-one function. The graph of f is shown in Figure 8–4.

Let us now look more closely at the set of ordered pairs $f = \{(x, y) | y = f(x)\}$, where f is any one-to-one function. The new set of ordered pairs obtained by reversing the order of the two components in each ordered pair of f is called the

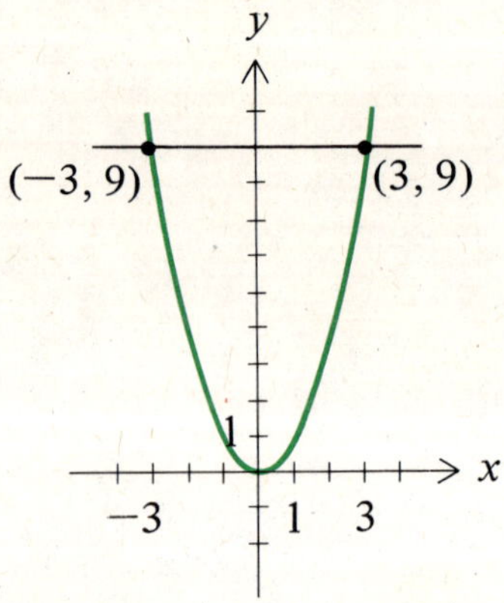

Figure 8–2
$\{(x, y) \mid y = x^2\}$ ■

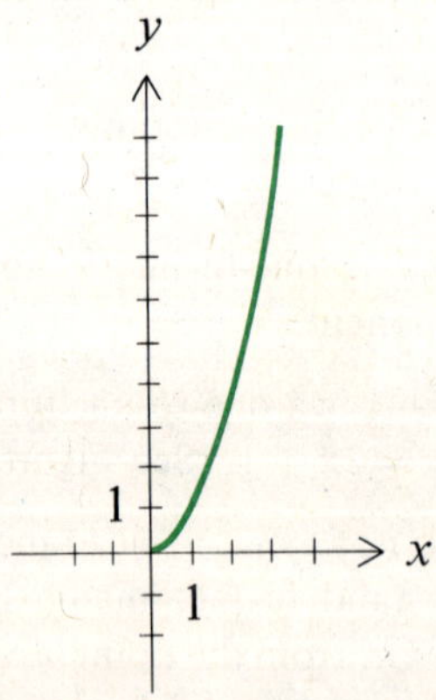

Figure 8–3
$\{(x, y) \mid y = x^2, x \geq 0\}$ ■

inverse of *f*. In symbols, this set is denoted f^{-1} (read "*f* inverse"), and so

$$f^{-1} = \{(y, x) \mid y = f(x)\}$$

Because *f* is one-to-one, no two ordered pairs of *f* have the same second co ponent. Therefore, no two ordered pairs of f^{-1} have the same *first* component, a so f^{-1} is a function.

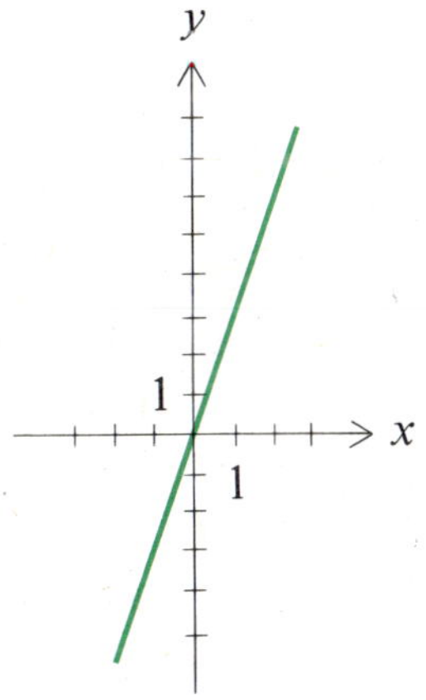

Figure 8–4
$\{(x, y) | y = 3x\}$ ■

It is usually more convenient to use x, rather than y, as the first component of an ordered pair. For that reason, we now exchange the letters x and y in our formula for f^{-1}, and so the definition of f^{-1} is the following:

If $f = \{(x, y) | y = f(x)\}$ is a one-to-one function, then f^{-1} is the function defined by $f^{-1} = \{(x, y) | x = f(y)\}$.

Because $(x, y) \in f^{-1}$ means $y = f^{-1}(x)$, it follows that $y = f^{-1}(x)$ if and only if $x = f(y)$.

Note: In the symbol f^{-1}, the -1 is not an exponent. It simply indicates that we are referring to the inverse of f.

Because the first components of ordered pairs of f are the second components of ordered pairs of f^{-1} and vice versa, we see that *the domain of f is the range of f^{-1}* and *the range of f is the domain of f^{-1}*.

EXAMPLE: Let f be the function defined by $f(x) = 3x$. Then f is a one-to-one function; its domain and range are R. Therefore, f^{-1} is a function with R as its domain and range. We have $f^{-1} = \{(x, y) | x = 3y\}$, and $y = f^{-1}(x)$ if and only if $x = 3y$. As a specific illustration, because $6 = f(2)$, then $2 = f^{-1}(6)$. Similarly, because $3 = f(1)$, we have $1 = f^{-1}(3)$. ■

EXAMPLE: Let f be the function defined by $f(x) = 2x + 1$. Then f is a one-to-one function; its domain and range are R. Therefore, $f^{-1} = \{(x, y) | x = 2y + 1\}$ is a function with domain and range R. Whenever $x = 2y + 1$, then $y = f^{-1}(x)$. Specifically, because $7 = f(3) = 2 \cdot 3 + 1$, we have $3 = f^{-1}(7)$. Because $1 = f(0) = 2 \cdot 0 + 1$, we have $0 = f^{-1}(1)$. Because $-5 = f(-3)$, we have $-3 = f^{-1}(-5)$. ■

When a one-to-one function f is described by an equation $y = f(x)$, we can often calculate the inverse function by the following two-step process. The justific tion for this process is that $x = f(y)$ if and only if $y = f^{-1}(x)$.

1. In the equation $y = f(x)$, exchange x and y to obtain $x = f(y)$.
2. Solve the resulting equation for y to obtain $y = f^{-1}(x)$.

EXAMPLE: Find f^{-1} if f is the function given by $f(x) = 3x$.
We take the original equation $y = 3x$ and exchange y and x, obtaining $x = 3y$. V then solve for y. The result is $y = x/3$, and so $f^{-1}(x) = x/3$. ■

EXAMPLE: If f is defined by $y = 2x + 1$, find $f^{-1}(x)$. Exchanging y and x, we ha $x = 2y + 1$. If we solve for y, we get $y = \frac{1}{2}x - \frac{1}{2}$, and so $f^{-1}(x) = \frac{1}{2}x - \frac{1}{2}$. ■

There is a close connection between the graph of any one-to-one function f a its inverse function f^{-1}. In fact, the graph of f^{-1} is obtained from the graph o by reflecting the graph of f in the line $y = x$. Figure 8–5 shows the graph of y $f(x) = 3x$, together with the dashed line $y = x$. Imagine folding the paper on t dashed line, bringing the two parts together, and then tracing the graph onto t other half of the paper. The result is the graph of $y = f^{-1}(x) = x/3$ (see Figu 8–6).

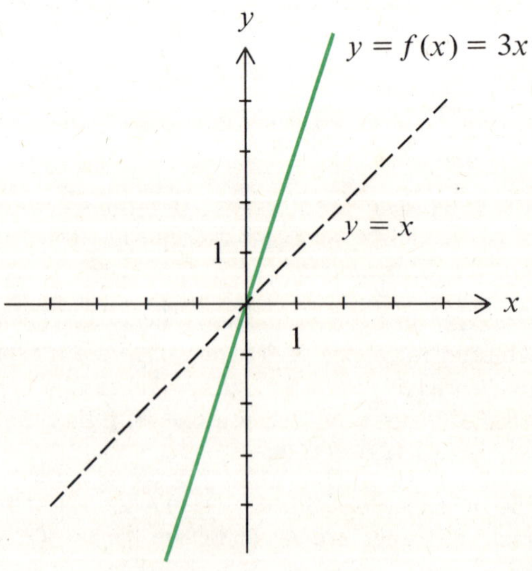

Figure 8–5

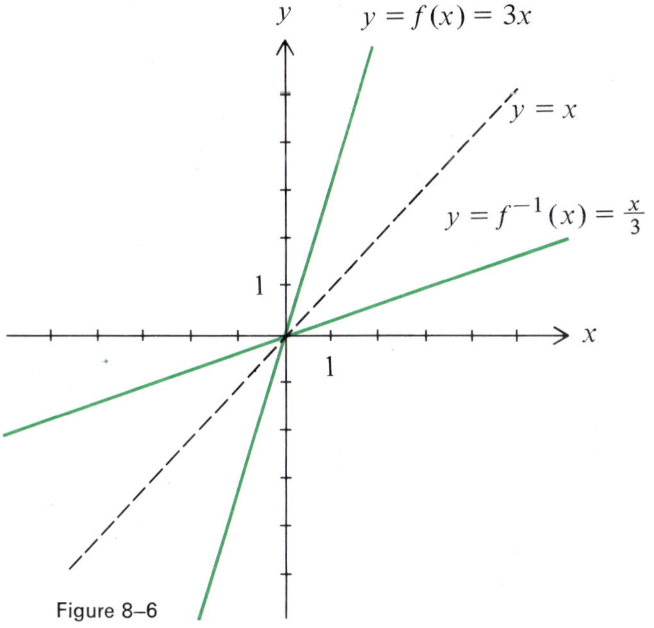

Figure 8–6

EXAMPLE: The graphs of $y = f(x) = 2x + 1$ and $y = f^{-1}(x) = \frac{1}{2}x - \frac{1}{2}$ are shown in Figure 8–7. Note again that the graph of $y = f^{-1}(x)$ is the reflection of the graph of $y = f(x)$ in the line $y = x$. ■

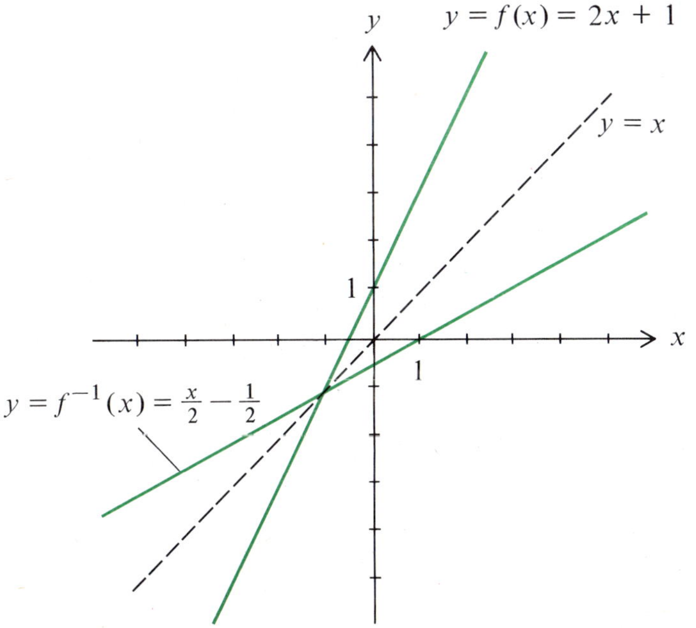

Figure 8–7

If f is not one-to-one, we do not expect to find an inverse function. Obser the difficulty in the following example.

EXAMPLE: If $f(x) = x^2$, find $f^{-1}(x)$.
We replace $y = x^2$ with $x = y^2$ and attempt to solve for y. Because y could be eitl $\sqrt{x}$ or $-\sqrt{x}$, there is no unique solution for y. This reflects the fact that f is not o to-one, and there is no inverse function. ■

8-1 *Exercises*

If the graph of the function f is as shown, is f one-to-one? Does f have an inve function f^{-1}? If so, find the domain and range of f^{-1} and sketch its graph.

1.

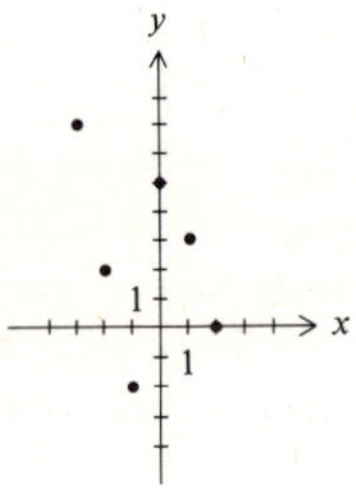

2.

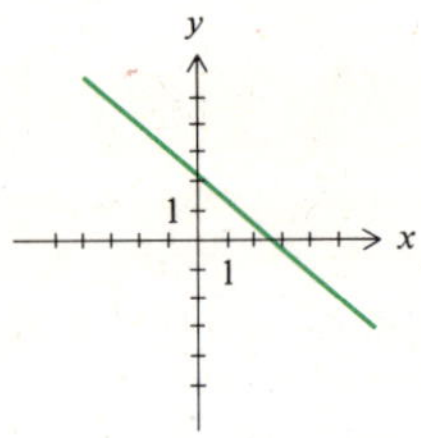

3.

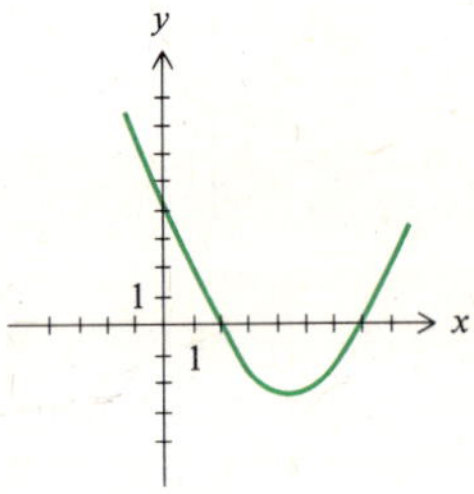

4.

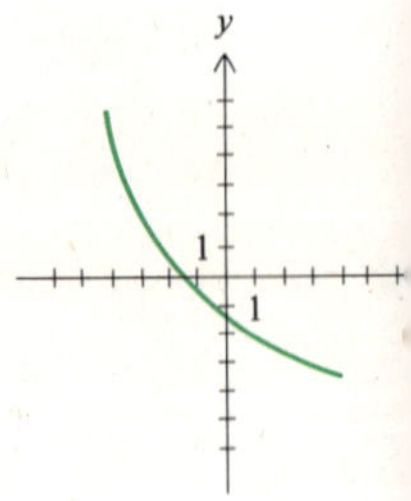

5.

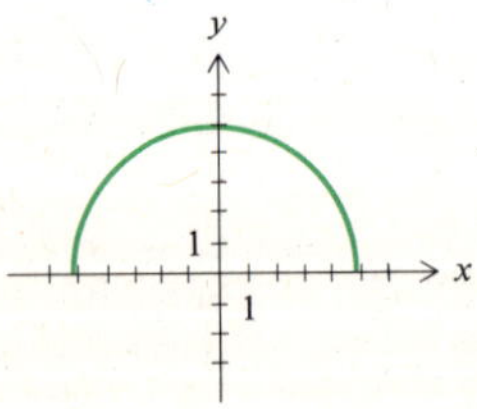

6.

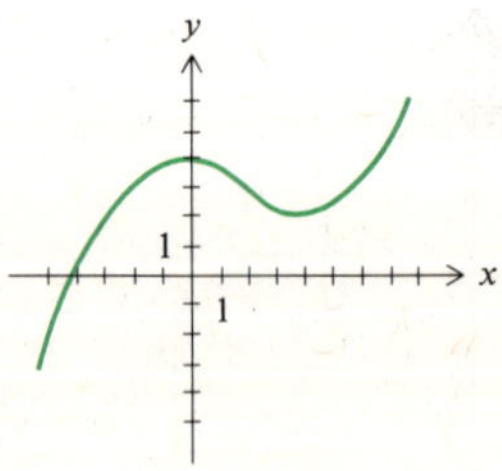

7.

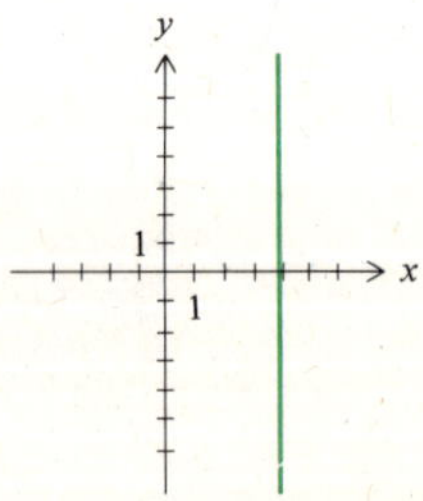

8.

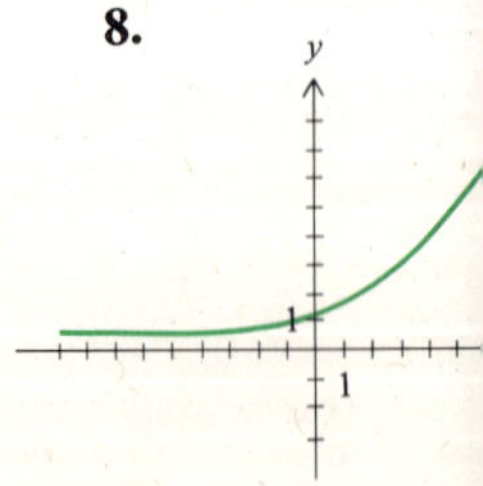

9. If f^{-1} is the inverse function to f, and if $b = f(a)$, what can be said abo $f^{-1}(b)$?

If $f(x) = \frac{1}{2}x$, find the following.
10. $f^{-1}(6)$
11. $f^{-1}(-8)$
12. $f^{-1}(0)$

If $f(x) = x^3$, find the following.
13. $f^{-1}(8)$
14. $f^{-1}(1)$
15. $f^{-1}(0)$

If $f(x) = x + 1$, find the following.
16. $f^{-1}(-2)$
17. $f^{-1}(13)$
18. $f^{-1}(0)$

If $f(x) = 4x - 1$, find the following.
19. $f^{-1}(3)$
20. $f^{-1}(11)$
21. $f^{-1}(0)$

Find $f^{-1}(x)$, where f^{-1} is the inverse function to the given function f defined below and find the domain and range of f^{-1}. It is assumed that the domain of f is R.

22. $f(x) = \frac{1}{4}x$

23. $f(x) = 3x - 8$

24. $f(x) = mx + b$, where $m \neq 0$

25. $f(x) = x^3$

Find $f^{-1}(x)$, where f^{-1} is the inverse function to the given function f, and find the domain and range of f^{-1}. The domain of f is as indicated.

26. $f(x) = x^2$, where $x \geq 0$

27. $f(x) = x^4$, where $x \geq 0$

28. $f(x) = |x|$, where $x < 0$

29. $f(x) = 2\sqrt[3]{x}$, where x is an element of R

30. $f(x) = 1 + \frac{1}{2}\sqrt{x}$, where $x > 0$

Suppose f is a one-to-one function.

31. For x in the domain of f, it is true that $f^{-1}(f(x)) = x$. Explain.

32. For x in the range of f, it is true that $f(f^{-1}(x)) = x$. Explain.

-2 Inverse functions for sine and cosine

The function defined by $y = f(x) = \sin x$ is not a one-to-one function. As we see from the graph of $y = \sin x$ (Figure 8–8), a horizontal line (such as the x-axis) may meet the graph in many points, not at just one point. Or, to work analytically, we may note that $\sin 0 = 0$ and $\sin \pi = 0$. Therefore there are two numbers, 0 and π, in the domain of f that correspond to the same element, that is, 0, in the range. Either way, our conclusion is that $f = \{(x, y) | y = \sin x\}$ is not a one-to-one function.

According to Section 8-1, only the one-to-one functions have inverse functions. Therefore we cannot define an inverse function for $f = \{(x, \sin x), x \in R\}$.

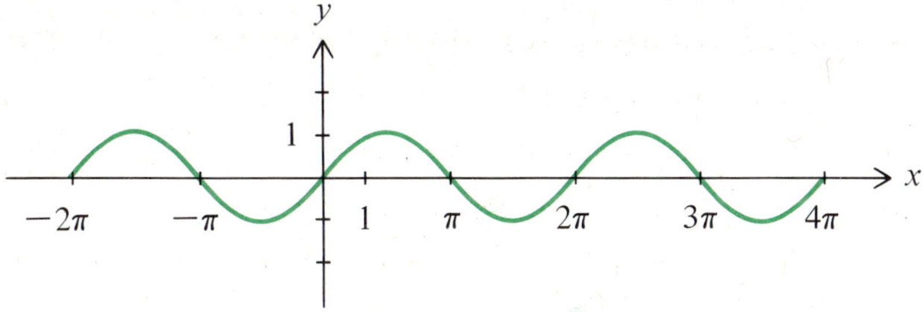

Figure 8–8
$\{(x, y) | y = \sin x\}$

Let us now examine the portion of the sine wave indicated in Figure 8–9. This curve is obtained by restricting the domain of the sine function to $\{x | -\pi/2 \leq x \leq \pi/2\}$. It is the graph of the function g defined by $y = g(x) = \sin x$,

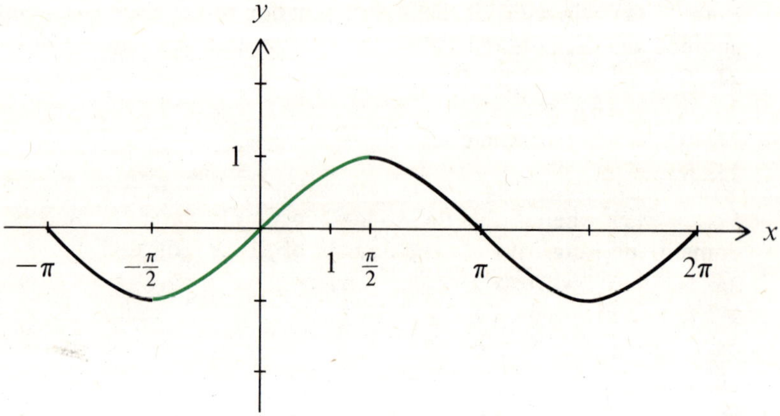

Figure 8–9
$\{(x, y) \mid y = \sin x, \ -\pi/2 \leq x \leq \pi/2\}$

$-\pi/2 \leq x \leq \pi/2$. In set language, $g = \{(x, \sin x) \mid -\pi/2 \leq x \leq \pi/2\}$. From Figu 8–9, it is evident that g is a one-to-one function. The domain of g is $\{x \mid -\pi/2 \leq x \leq \pi/2\}$, and its range is $\{y \mid -1 \leq y \leq 1\} = \{y \mid |y| \leq 1\}$.

Because g is a one-to-one function, it has an inverse function g^{-1}. Our notati for g^{-1} is Arcsin.

Starting with $g = \{(x, y) \mid y = \sin x$ and $-\pi/2 \leq x \leq \pi/2\}$, we exchange t roles of x and y to obtain the following description of Arcsin.

$\text{Arcsin} = \{(x, y) \mid x = \sin y \text{ and } -\pi/2 \leq y \leq \pi/2\}$

It follows that $y = \text{Arcsin } x$ if and only if both $x = \sin y$ and $-\pi/2 \leq y \leq \pi/2$. T domain of Arcsin is $\{x \mid |x| \leq 1\}$, and so Arcsin x is defined only for numbers between -1 and 1.

For such x, we think of Arcsin x as the number y, between $-\pi/2$ and $\pi/$ whose sine is x. The range of Arcsin is $\{y \mid -\pi/2 \leq y \leq \pi/2\} = \{y \mid |y| \leq \pi/2\}$.

EXAMPLE: Arcsin $1 = \pi/2$, because $\sin \pi/2 = 1$ and $\pi/2 \in \{y \mid -\pi/2 \leq y \leq \pi/2\}$.

EXAMPLE: Arcsin $1 \neq 5\pi/2$, even though $\sin 5\pi/2 = 1$, because $5\pi/2 \notin \{y \mid -\pi/2 \leq y \leq \pi/2\}$. ■

EXAMPLE: Arcsin $-\frac{1}{2} = -\pi/6$, because $\sin -\pi/6 = -\frac{1}{2}$ and $-\pi/2 \leq -\pi/6 \leq \pi/2$. ■

EXAMPLE: Arcsin $-\frac{1}{2} \neq 7\pi/6$, even though $\sin 7\pi/6 = -\frac{1}{2}$. ■

EXAMPLE: Arcsin 2 is not defined, because $2 \notin \{x \mid -1 \leq x \leq 1\}$. ■

It is easy to sketch the graph of Arcsin. We know from Section 8-1 that th graph of Arcsin is the reflection of the graph of g in the line $y = x$. This graph shown in Figure 8–10.

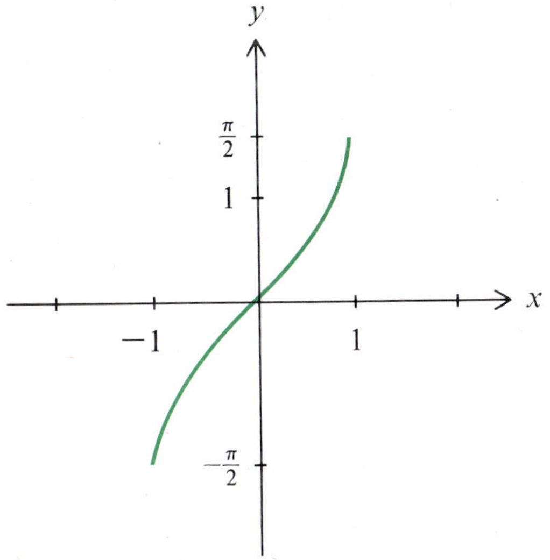

Figure 8–10
$\{(x, y) \mid y = \text{Arcsin } x\}$

Strictly speaking, the value of Arcsin x is a number, not an angle. Nevertheless, we may think of Arcsin x as the radian measure of one particular angle whose sine is x. The measure of this angle may also be expressed in degrees. With this interpretation, the degree measure of Arcsin x will be between $-90°$ and $90°$, because the radian measure is between $-\pi/2$ and $\pi/2$.

EXAMPLE: Arcsin $1 = 90°$. ■

In an earlier example, we said Arcsin $1 = \pi/2$. Mathematically, it is not exactly proper to use the same symbol Arcsin 1 to refer to the number, or radian measure, $\pi/2$, as well as the degree measure, $90°$. However, this practice is common and we shall follow it.

EXAMPLE: Arcsin $-\frac{1}{2} = -30°$. ■

EXAMPLE: Find Arcsin 0.4226.

From the table of Trigonometric Ratios, we have $\sin 25° = 0.4226$, and therefore Arcsin $0.4226 = 25°$. ■

EXAMPLE: Arcsin $-0.4226 = -25°$. ■

EXAMPLE: Arcsin $-0.4226 \neq 335°$, even though $\sin 335° = -0.4226$, because $335°$ is not between $-90°$ and $90°$. ■

We now wish to define an inverse function for the cosine function. The graph of $y = f(x) = \cos x$ is shown in Figure 8–11. As we see from the graph, f is not a one-to-one function.

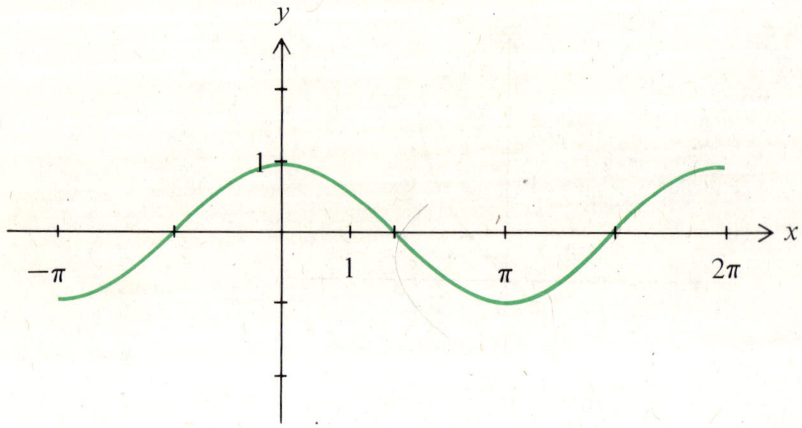

Figure 8–11
$\{(x, y) | y = \cos x\}$.

The portion of the graph corresponding to $\{x | -\pi/2 \le x \le \pi/2\}$ is *not* t graph of a one-to-one function either (see Figure 8–12).

A better choice is the curve obtained by restricting the domain of the cosi function to $\{x | 0 \le x \le \pi\}$ (Figure 8–13). This graph is the graph of the function defined by $y = h(x) = \cos x, 0 \le x \le \pi$. In set language, $h = \{(x, \cos x) | 0 \le x \le$ From Figure 8–13, we see that g is a one-to-one function. The domain of h $\{x | 0 \le x \le \pi\}$, and its range is $\{y | |y| \le 1\}$. The inverse function of h is denot Arccos.

The formal description of h^{-1} = Arccos is as follows:

Arccos $= \{(x, y) | x = \cos y \text{ and } 0 \le y \le \pi\}$.

Therefore $y = \text{Arccos } x$ if and only if both $x = \cos y$ and $0 \le y \le \pi$. The domain Arccos is $\{x | |x| \le 1\}$, and so Arccos x is defined only for numbers x betwe -1 and 1. We think of Arccos x as the number, between 0 and π, whose cosine is The range of Arccos is $\{y | 0 \le y \le \pi\}$.

As with Arcsin, we will measure Arccos x in degrees when it is convenient do so. In degrees, we have $0° \le \text{Arccos } x \le 180°$.

EXAMPLE: Arccos $1 = 0$, or Arccos $1 = 0°$ ■

EXAMPLE: Arccos $1 \ne 2\pi$, because $2\pi \notin \{x | 0 \le x \le \pi\}$. ■

EXAMPLE: Arccos $-\frac{1}{2} = 2\pi/3$ or Arccos $-\frac{1}{2} = 120°$. ■

EXAMPLE: Arccos $0.7771 = 39°$ (from the table of Trigonometric Ratios). ■

EXAMPLE: Arccos $-0.7771 = 141°$. ■

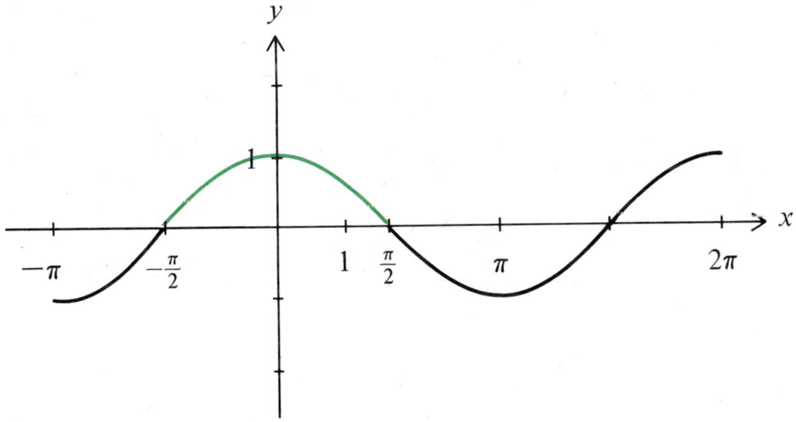

Figure 8–12
$\{(x, y) \mid y = \cos x, \ -\pi/2 \leq x \leq \pi/2\}$

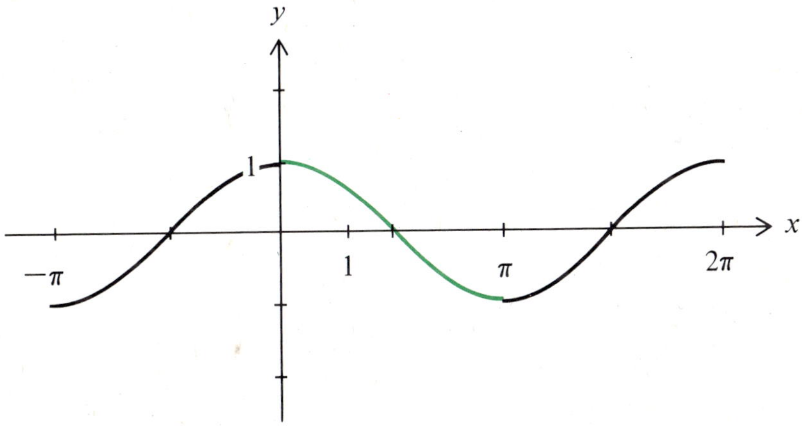

Figure 8–13
$\{(x, y) \mid y = \cos x, \ 0 \leq x \leq \pi\}$

The graph of Arccos is the reflection of the graph of h in the line $y = x$. This graph is shown in Figure 8–14.

Note: The symbols Arcsin and Arccos are not completely standard. Some authors use $\sin^{-1}$ and $\cos^{-1}$, although, technically speaking, the functions $\sin = \{(x, y) \mid y = \sin x\}$ and $\cos = \{(x, y) \mid y = \cos x\}$ do not have inverse functions.

A distinction is often made between Arcsin and arcsin (and between Arccos and arccos). Then $\arcsin x = \{y \mid x = \sin y\}$, while Arcsin x is the specific y such that $x = \sin y$ and $-\pi/2 \leq y \leq \pi/2$. For example, Arcsin $1 = \pi/2$, but $\arcsin 1 = \{\pi/2 + 2k\pi, \ k \in J\}$.

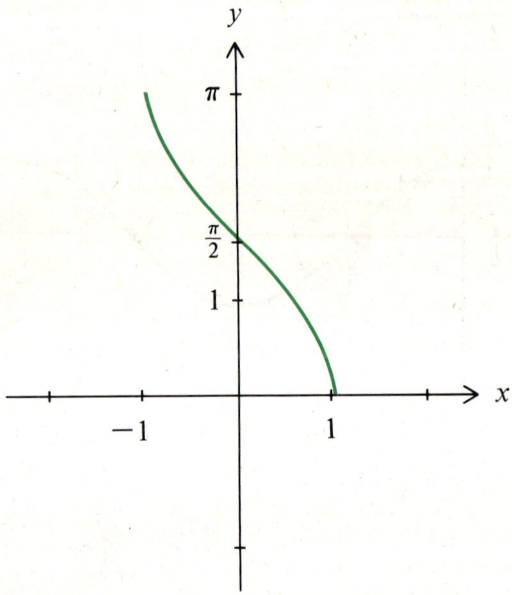

Figure 8–14
$\{(x, y) \mid y = \text{Arccos } x\}$

8-2 *Exercises*

Express in radians.

1. Arcsin $\frac{1}{2}$

2. Arccos $\frac{1}{2}$

3. Arccos -1

4. Arcsin -1

5. Arcsin $\sqrt{2}/2$

6. Arccos $\sqrt{2}/2$

Express in degrees.

7. Arccos $-\sqrt{2}/2$

8. Arcsin $-\dfrac{\sqrt{2}}{2}$

9. Arcsin 0.3584

10. Arccos 0.3584

11. Arccos -0.3584

12. Arcsin -0.3584

Evaluate.

13. sin (Arcsin $\frac{2}{3}$)

14. cos (Arccos $\frac{2}{3}$)

15. cos (Arcsin $\frac{2}{3}$)
16. sin (Arccos $\frac{2}{3}$)
17. sin (Arccos $-\frac{2}{3}$)
18. cos (Arcsin $-\frac{2}{3}$)

Justify the following identities.
19. $\sin(\text{Arcsin } x) = x$, if $-1 \le x \le 1$
20. $\cos(\text{Arccos } x) = x$, if $-1 \le x \le 1$

Find formulas for the following, where $-1 \le x \le 1$.
21. cos (Arcsin x)
22. sin (Arccos x)
23. Arcsin x + Arccos x

Evaluate.
24. Arcsin (sin $3\pi/2$)
25. Arccos (cos $3\pi/2$)
26. For what values of x is it true that Arcsin (sin x) = x?
27. For what values of x is it true that Arccos (cos x) = x?

-3 Other inverse trigonometric functions

The pattern used to define inverse functions for the tangent, cotangent, secant, and cosecant functions is the same as that used to define Arcsin and Arccos. In each case, it is necessary to restrict the domain of the trigonometric function to get a one-to-one function.

Let us consider the portion of the graph of the tangent function indicated in Figure 8–15. This is the graph of a function $\{(x, y) \mid y = \tan x,\ -\pi/2 < x < \pi/2\}$.

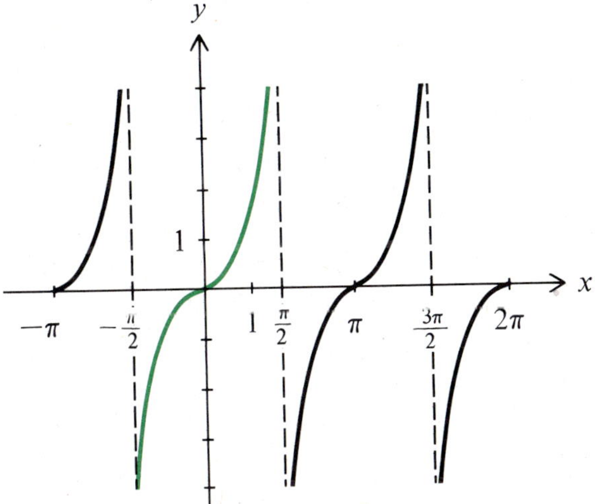

Figure 8–15
$\{(x, y) \mid y = \tan x,\ -\pi/2 < x < \pi/2\}$

The inverse function for this function is Arctan.

Thus,

$$\text{Arctan} = \{(x, y) \mid x = \tan y \quad \text{and} \quad -\pi/2 < y < \pi/2\}.$$

The domain of Arctan is R, and its range is $\{y \mid |y| < \pi/2\}$. The graph of Arctan as shown in Figure 8–16.

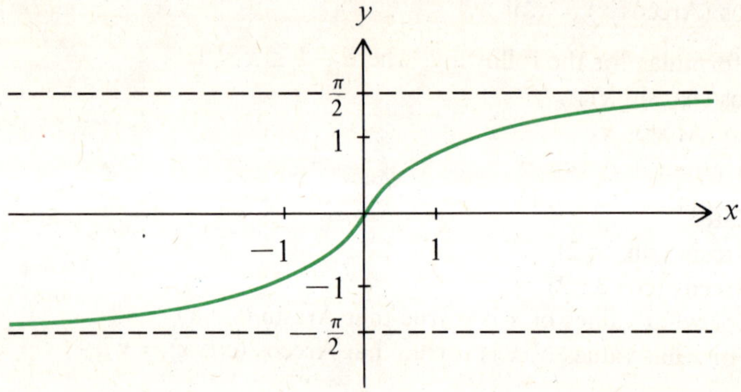

Figure 8–16
$\{(x, y) \mid y = \text{Arctan } x\}$

EXAMPLE: Arctan $1 = \pi/4$ (or 45°). ■

EXAMPLE: Arctan $-0.1405 = -8°$. ■

To define Arccot, we use the portion of the graph of the cotangent functio shown in Figure 8–17. The domain has been restricted to $\{x \mid 0 < x < \pi\}$. Therefo

$$\text{Arccot} = \{(x, y) \mid x = \cot y \quad \text{and} \quad 0 < y < \pi\}.$$

The domain of Arccot is R, and its range is $\{y \mid 0 < y < \pi\}$. The graph of Arccot pictured in Figure 8–18.

EXAMPLE: Arccot $-1 = 3\pi/4$ (or 135°). ■

EXAMPLE: Arccot $-0.1405 = 98°$. ■

To define Arcsec, we use the part of the graph of the secant function suggeste by Figure 8–19. Taking the inverse function, we have:

$$\text{Arcsec} = \{(x, y) \mid x = \sec y \quad \text{and} \quad 0 \le y \le \pi\}$$

The domain of Arcsec is $\{x \mid |x| \ge 1\}$, and its range is $\{y \mid 0 \le y \le \pi, y \ne \pi/2\}$. Th graph of Arcsec is given in Figure 8–20.

EXAMPLE: Arcsec $-1 = \pi$ (or 180°). ■

EXAMPLE: Arcsec $2 = \pi/3$ (or 60°). ■

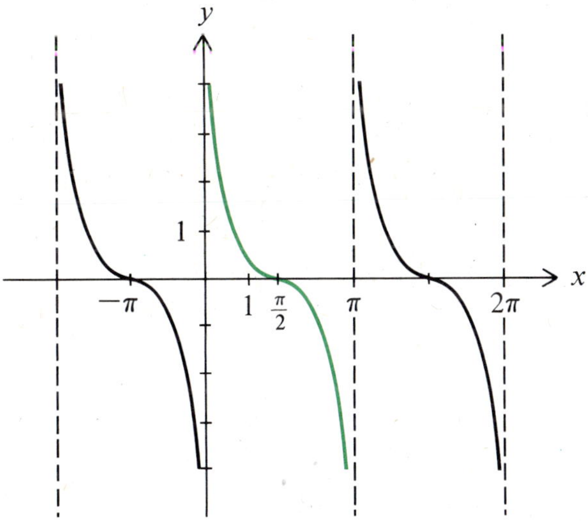

Figure 8–17
$\{(x, y) \mid y = \cot x,\ 0 < x < \pi\}$

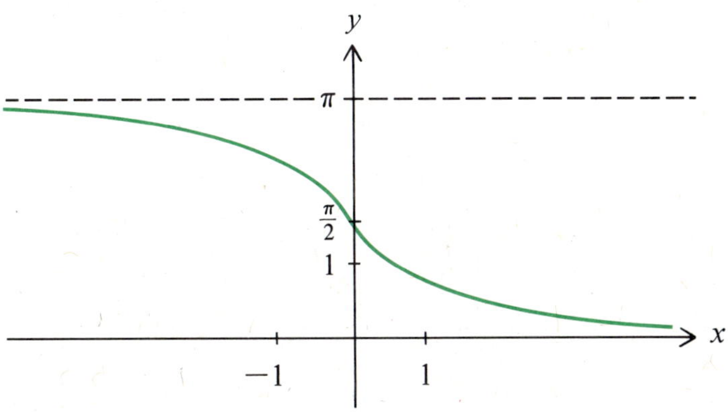

Figure 8–18
$\{(x, y) \mid y = \text{Arccot } x\}$

Finally, to define Arccsc we examine the portion of the graph of the cosecant function shown in Figure 8–21. The inverse function gives us:

$$\text{Arccsc} = \left\{(x, y) \mid x = \csc y \quad \text{and} \quad -\frac{\pi}{2} \leq y \leq \frac{\pi}{2}\right\}$$

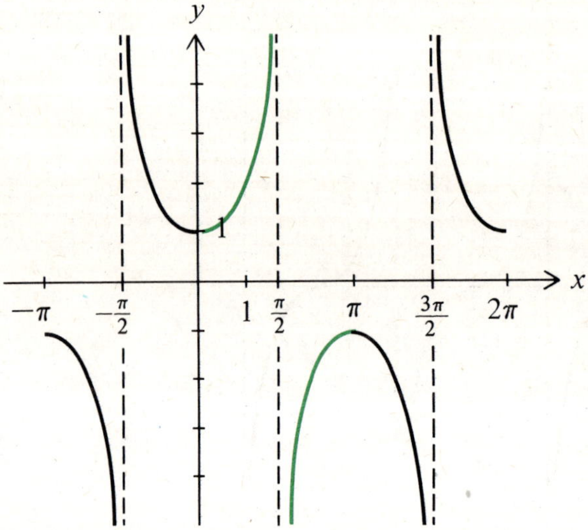

Figure 8–19
$\{(x, y) \mid y = \sec x,\ 0 \leq x \leq \pi,\ x \neq \pi/2\}$

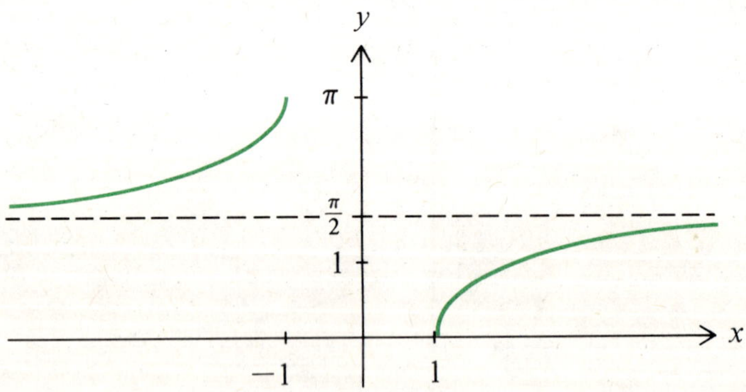

Figure 8–20
$\{(x, y) \mid y = \text{Arcsec } x\}$

The domain of Arccsc is $\{x \mid |x| \geq 1\}$, and its range is $\{y \mid -\pi/2 \leq y \leq \pi/2,\ y \neq$
The graph of Arccsc is as illustrated in Figure 8–22.

EXAMPLE: Arccsc $-1 = -\pi/2$ (or $-90°$). ■

EXAMPLE: Arccsc $2 = \pi/6$ (or $30°$). ■

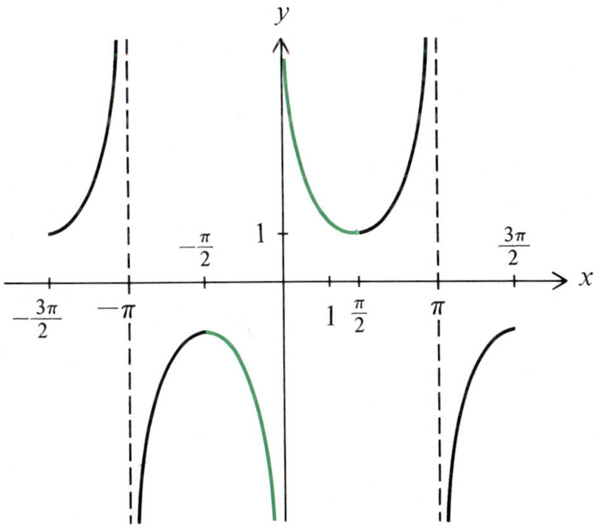

Figure 8–21
$\{(x, y) \mid y = \csc x, -\pi/2 \le x \le \pi/2, x \ne 0\}$

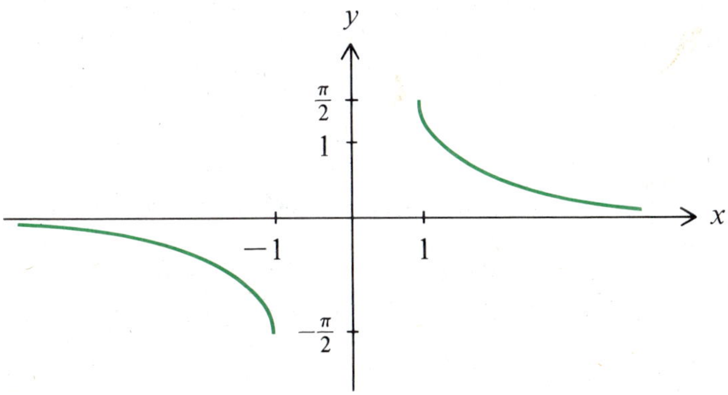

Figure 8–22
$(\{x, y) \mid y = \text{Arccsc } x\}$

-3 *Exercises*

Evaluate.

1. Arctan $\sqrt{3}$

2. Arccot $\sqrt{3}$

3. Arccot $-\sqrt{3}$

4. Arctan $-\sqrt{3}$

5. Arctan 0.2493

6. Arccot -0.2493

7. Arcsec 1

8. Arccsc 1

9. Arccsc $-\sqrt{2}$

10. Arcsec $-\sqrt{2}$

Justify the identities.

11. $\tan(\text{Arctan } x) = x$

12. $\cot(\text{Arccot } x) = x$

13. $\text{Arccot } 1/x = \text{Arctan } x, \; x \neq 0$

14. $\text{Arcsec } 1/x = \text{Arcsin } x, \; |x| \leq 1$

15. $\text{Arccsc } 1/x = \text{Arccos } x, \; |x| \leq 1$

Evaluate.

16. $\text{Arctan } x + \text{Arccot } x$

17. $\text{Arcsec } x + \text{Arccsc } x, \; |x| \geq 1$

Give equations for any asymptotes to the graph.

18. $y = \text{Arctan } x$

19. $y = \text{Arccot } x$

20. $y = \text{Arcsec } x$

21. $y = \text{Arccsc } x$

◇8-4 Linear combinations of sine and cosine

Let us consider functions of the type defined by $y = f(x) = a \sin x + b \cos x$, wher a and b are real numbers. Evidently $g(x) = a \sin x$ defines a sinusoidal functior Because $b \cos x = b \sin(x + \pi/2)$, we see that $h(x) = b \cos x$ is also a sinusoida function. Therefore $f(x) = a \sin x + b \cos x = g(x) + h(x)$ is a sum of sinusoida functions.

In Section 6-3 we developed a method for graphing a sum of sinusoidal func tions. However, that method, which is graphical in nature, is slow and inaccurat Our objective now is to present a much more efficient method for sketching th graph of a function of the type $y = f(x) = a \sin x + b \cos x$.

As a first example, consider the equation

$$y = \frac{\sqrt{3}}{2} \sin x + \frac{1}{2} \cos x$$

This is of form $y = a \sin x + b \cos x$, where $a = \frac{1}{2}\sqrt{3}$ and $b = \frac{1}{2}$. The numbe $a = \frac{1}{2}\sqrt{3}$ and $b = \frac{1}{2}$ satisfy $a^2 + b^2 = 1$, because $(\frac{1}{2}\sqrt{3})^2 + (\frac{1}{2})^2 = \frac{3}{4} + \frac{1}{4} = 1$. This reminiscent of the first Pythagorean identity. For any real number c,

$$(\cos c)^2 + (\sin c)^2 = \cos^2 c + \sin^2 c = 1$$

Therefore, it seems reasonable to look for a real number c such that $a = \cos c$ an $b = \sin c$.

In fact, we recognize that if $c = \text{Arccos } \frac{1}{2}\sqrt{3} = \pi/6$, then $\cos c = \sqrt{3}/2$ an $\sin c = \frac{1}{2}$. We substitute these values in our equation

$$y = \frac{\sqrt{3}}{2} \sin x + \frac{1}{2} \cos x$$

and obtain

$$y = \cos \frac{\pi}{6} \sin x + \sin \frac{\pi}{6} \cos x$$

or, by a basic identity,

$$y = \sin \left(x + \frac{\pi}{6} \right)$$

We know (Section 6-2) that the graph of $y = \sin (x + \pi/6)$ is a sine wave, with amplitude 1, period 2π, and phase shift of $-\pi/6$. One complete cycle is shown in Figure 8–23.

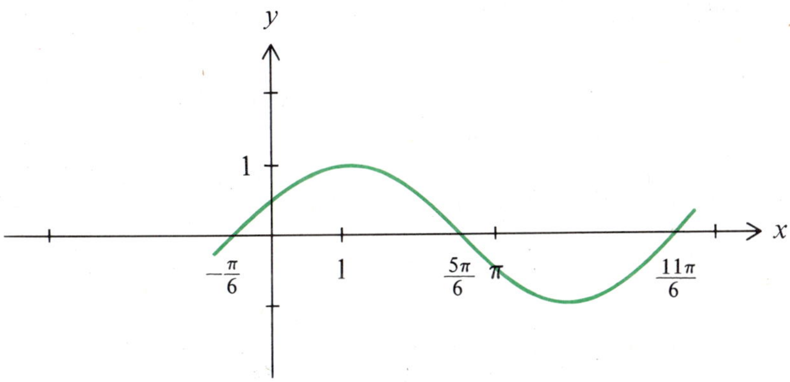

Figure 8–23
$\{(x, y) \mid y = (\sqrt{3}/2) \sin x + (1/2) \cos x, \ -\pi/6 \le x \le 11\pi/6\} =$
$\{(x, y) \mid y = \sin (x + \pi/6), \ -\pi/6 \le x \le 11\pi/6\}$

In the above example, it was crucial to have $a^2 + b^2 = 1$ so that we could find a real number c such that $a = \cos c$ and $b = \sin c$. If we do not have $a^2 + b^2 = 1$, the first step is to rewrite our general equation

$$y - a \sin x + b \cos x$$

as

$$y = \sqrt{a^2 + b^2} \left(\frac{a}{\sqrt{a^2 + b^2}} \sin x + \frac{b}{\sqrt{a^2 + b^2}} \cos x \right)$$

We then work with $a/\sqrt{a^2 + b^2}$ and $b/\sqrt{a^2 + b^2}$, as in the following example.

EXAMPLE: Sketch the graph of $y = 2\sqrt{3} \sin x + 2 \cos x$.

Since $\sqrt{(2\sqrt{3})^2 + 2^2} = 4$, we rewrite this as

$$y = 4\left(\frac{\sqrt{3}}{2} \sin x + \tfrac{1}{2} \cos x\right)$$

We now proceed as above. Let $c = \text{Arccos } \frac{1}{2}\sqrt{3} = \pi/6$, so that $\frac{1}{2}\sqrt{3} = \cos \pi/6$ ar $\frac{1}{2} = \sin \pi/6$. Hence we have

$$y = 4\left(\cos\frac{\pi}{6} \sin x + \sin\frac{\pi}{6} \cos x\right)$$

$$= 4 \sin\left(x + \frac{\pi}{6}\right)$$

The graph, one cycle of which is shown in Figure 8–24, has amplitude 4, period 2 and phase shift of $-\pi/6$. ■

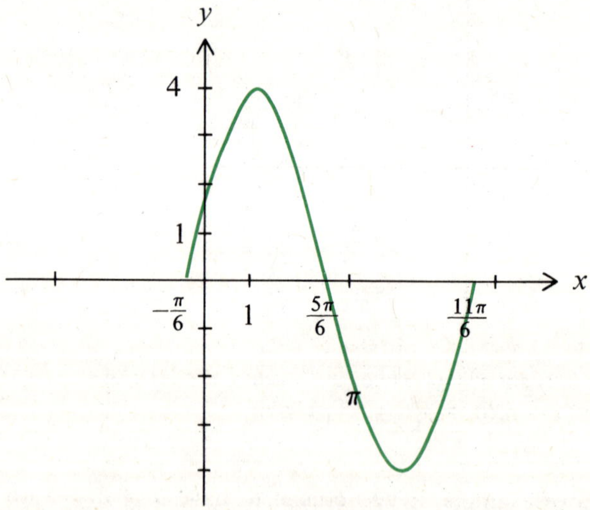

Figure 8–24
$\{(x, y) \mid y = 2\sqrt{3} \sin x + 2 \cos x,\ -\pi/6 \le x \le 11\pi/6\} =$
$\{(x, y) \mid y = 4 \sin (x + \pi/6),\ -\pi/6 \le x \le 11\pi/6\}$

A precaution is necessary when b is negative. Normally we choose c to equal to Arccos $a/\sqrt{a^2 + b^2}$. Therefore c is in Quadrant I or II. In either cas $\sin c$ is positive. Thus, if b is negative, we will have $b/\sqrt{a^2 + b^2} = -\sin c$, n $b/\sqrt{a^2 + b^2} = \sin c$.

EXAMPLE: We graph $y = 2\sqrt{3} \sin x - 2 \cos x$.

First we rewrite this, as above, as

$$y = 4\left(\frac{\sqrt{3}}{2} \sin x - \tfrac{1}{2} \cos x\right)$$

We then let $c = \text{Arccos } \tfrac{1}{2}\sqrt{3} = \pi/6$. Thus $-\tfrac{1}{2} = -\sin \pi/6$. Hence our equation is

$$y = 4\left(\cos\frac{\pi}{6} \sin x - \sin\frac{\pi}{6} \cos x\right)$$
$$= 4 \sin\left(x - \frac{\pi}{6}\right)$$

This graph has phase shift $\pi/6$, not $-\pi/6$. The period is 2π. The graph is sketched in Figure 8–25. ■

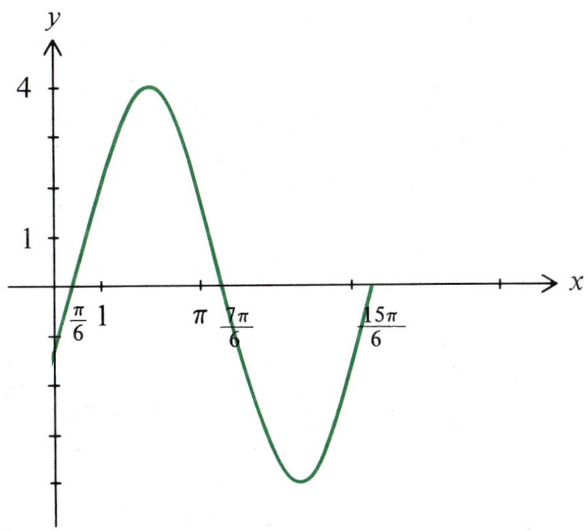

Figure 8–25
$\{(x, y) \mid y = 2\sqrt{3} \sin x - 2 \cos x\} =$
$\{(x, y) \mid y = 4 \sin (x - \pi/6)\}$

Before proceeding to further examples, we review the steps to graph the general equation $y = a \sin x + b \cos x$, assuming a and b are not both zero.

1. Rewrite the equation as

$$y = \sqrt{a^2 + b^2}\left(\frac{a}{\sqrt{a^2 + b^2}} \sin x + \frac{b}{\sqrt{a^2 + b^2}} \cos x\right)$$

2. Let $c = \text{Arccos}\,(a/\sqrt{a^2 + b^2})$, so that

   $a/\sqrt{a^2 + b^2} = \cos c.$
3. Identify $b/\sqrt{a^2 + b^2}$ as either $\sin c$ or $-\sin c$.
4. Rewrite the equation as

   $$y = \sqrt{a^2 + b^2}(\cos c \sin x \pm \sin c \cos x)$$
5. Use a trigonometric identity to rewrite the equation in the form

   $$y = \sqrt{a^2 + b^2} \sin (x \pm c)$$
6. Sketch the graph as a sine wave, with amplitude $\sqrt{a^2 + b^2}$, period 2π, an phase shift of c or $-c$.

EXAMPLE: Sketch the graph of $y = -3 \sin x + 4 \cos x$.

Because $\sqrt{(-3)^2 + 4^2} = 5$, we rewrite this as

$$y = 5(-\tfrac{3}{5} \sin x + \tfrac{4}{5} \cos x)$$

Let $c = \text{Arccos}\, -\tfrac{3}{5}$, so that $-\tfrac{3}{5} = \cos c$. Because $\angle c$ is in Quadrant II, $\sin c$ i positive and so $\tfrac{4}{5} = \sin c$. Hence our equation is

$$\begin{aligned} y &= 5 (\cos c \sin x + \sin c \cos x) \\ &= 5 \sin (x - c) \end{aligned}$$

where $c = \text{Arccos}\, -\tfrac{3}{5}$. From the table of Trigonometric Ratios, an approxima value for $\angle c$ is 127°. Thus the graph of $y = 5 \sin (x - c)$ has period 2π, amplitude and phase shift of approximately 127° or 2.2 radians. This graph is shown i Figure 8–26. ■

EXAMPLE: We sketch the graph of $y = -\sin x - \cos x$.
We rewrite this equation as

$$y = \sqrt{2}\left(-\frac{1}{\sqrt{2}} \sin x - \frac{1}{\sqrt{2}} \cos x\right)$$

Let $c = \text{Arccos}\,(-1/\sqrt{2}) = 3\pi/4$. Then $-1/\sqrt{2} = -\sin c$. Hence our equation i

$$y = 2\left(\cos \frac{3\pi}{4} \sin x - \sin \frac{3\pi}{4} \cos x\right)$$

or

$$y = 2 \sin \left(x - \frac{3\pi}{4}\right)$$

Thus the amplitude is 2, the period is 2π, and the phase shift is $3\pi/4$. See the grap in Figure 8–27. ■

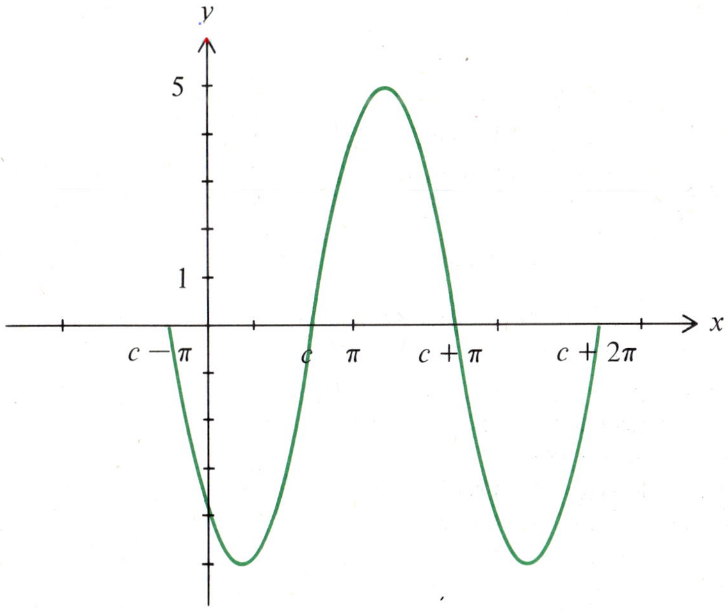

Figure 8–26
$\{(x, y) \mid y = -3 \sin x + 4 \cos x\}$

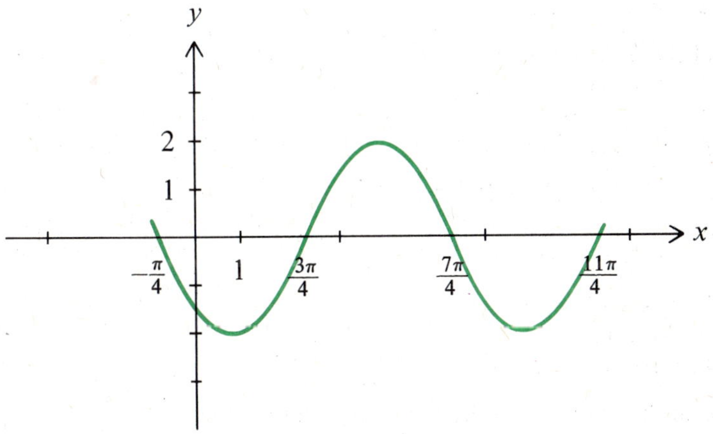

Figure 8–27
$\{(x, y) \mid y = -\sin x - \cos x\}$

◇8-4 *Exercises*

Sketch the graph.

1. $y = \sin x - \cos x$

2. $y = -3 \sin x + 3 \cos x$

3. $y = 4 \sin x + 4\sqrt{3} \cos x$

4. $y = -\sin x - \sqrt{3} \cos x$

5. $y = 4 \sin x - 3 \cos x$

6. $y = -8 \sin x - 6 \cos x$

7. $y = -5 \sin x + 12 \cos x$

8. $y = (\frac{5}{4}) \sin x + 3 \cos x$

9. $y = 2 \sin x + \cos x$

10. $y = 2 \sin x - \cos x$

11. Use the method of this section to analyze the graph of $y = \cos x$.

In the method of this section, we defined c by $c = \text{Arccos}\, a/\sqrt{a^2 + b^2}$. It would possible, instead, to let $c = \text{Arcsin}\, a/\sqrt{a^2 + b^2}$, so that $a/\sqrt{a^2 + b^2} = \sin c$ a $b/\sqrt{a^2 + b^2} = \pm \cos c$. Use the substitution $c = \text{Arcsin}\, a/\sqrt{a^2 + b^2}$ to simplify t following equations of form $y = a \sin x + b \cos x$.

12. $y = \sin x + \cos x$

13. $y = 4 \sin x - 4\sqrt{3} \cos x$

14. $y = -4 \sin x + 3 \cos x$

15. $y = -5 \sin x - 12 \cos x$

16. $y = -2 \sin x + \cos x$

Review exercises (Chapter 8)

Explain and illustrate with an example.

1. one-to-one function

2. inverse function

Find the inverse function, if it exists.

3. $y = f(x) = 4x + 1$

4. $y = f(x) = x + 1$

5. $y = f(x) = x^5 - 1$

6. $y = f(x) = x^4 - 1, x \geq 0$

Evaluate.

7. $\text{Arcsin}\, \frac{1}{2}\sqrt{3}$

8. $\text{Arccos}\, -\frac{1}{2}\sqrt{3}$

9. $\text{Arccot}\, 1$

10. $\text{Arccsc}\, -2$

Sketch the graph and state the domain and range

11. $y = \text{Arcsin}\, x$

12. $y = \text{Arccos}\, x$

13. $y = \text{Arctan}\, x$

14. $y = \text{Arcsec}\, x$

Sketch the graph.

◇**15.** $y = -9 \sin x + 12 \cos x$

◇**16.** $y = 7 \sin x - 24 \cos x$

A

Logarithmic methods

Before calculators and computers came into everyday use, logarithms provided the real short c method for solving many complicated arithmetic problems. One important appl cation was to the computations needed to solve a triangle.

This appendix presents the theory of exponential and logarithmic function followed by an explanation of calculation by logarithms. Exponential and logarit mic functions have numerous uses in higher mathematics and in many other field including physics, chemistry, biology, psychology, and economics.

The table of logarithms, located inside the back cover of the book, is explaine in Sections A-4 and A-5. There is also a table of logarithms of trigonometric rati in Appendix B. This table is discussed in Section A-7.

A-1 Exponential functions

If a is a fixed positive number, then each real number x can be used as an expone to define a real number a^x. To single out a specific example, we let $a = 2$. Then f each real number x, a unique real number 2^x is defined.

This correspondence, which associates to each real number x exactly one real number 2^x, describes a function f, because each x determines a unique value of y. In set language, $f = \{(x, y) \mid y = 2^x\}$. This function f is an example of an *exponential function*.

Let us sketch the graph of the function defined by $f(x) = 2^x$. First we determine a few points on the graph by choosing values of x and calculating $y = 2^x$. These points are then plotted as shown in Figure A–1.

x	$y = 2^x$
-2	$\frac{1}{4}$
-1	$\frac{1}{2}$
0	1
1	2
2	4
3	8

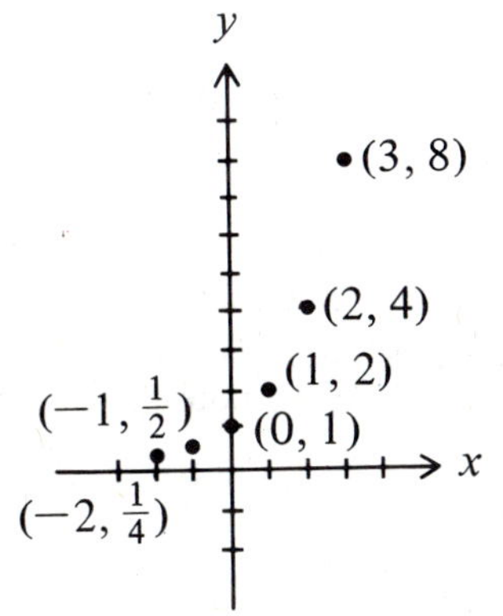

Figure A–1

The points are then connected by a smooth curve. This curve is the graph of the function f defined by $y = 2^x$ (Figure A–2).

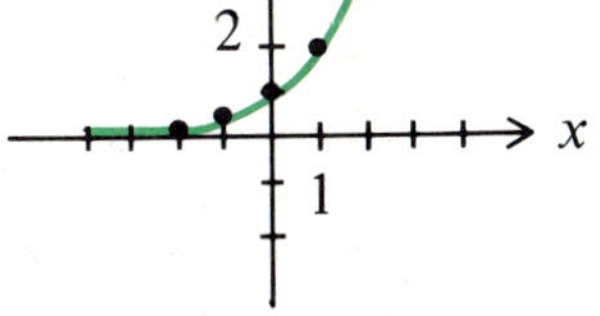

Figure A–2
$\{(x, y) \mid y = 2^x\}$

From the graph in Figure A–2 we may observe the following facts:

1. For each real number x, $2^x > 0$.
2. If $x > 0$, then $2^x > 1$.
3. If $x < 0$, then $0 < 2^x < 1$.
4. $2^0 = 1$.

Now let us replace 2 by any fixed positive number a. Each positive number determines an exponential function f defined by the equation

$$y = a^x$$

This function f is the exponential function with base a. In set language,

$$f = \{(x, y) \mid y = a^x\}$$

When we graph the exponential function, so long as $a > 1$, the graph $f(x) = a^x$ has the same general shape as the graph of $f(x) = 2^x$. However, the grap goes through the point $(1, a)$ rather than $(1, 2)$. Both graphs go through $(0, 1)$. fact, for any $a > 1$, we note the following:

1. For each real number x, $a^x > 0$.
2. If $x > 0$, then $a^x > 1$.
3. If $x < 0$, then $0 < a^x < 1$.
4. $a^0 = 1$.

EXAMPLE: Let us sketch the graph of the function defined by $y = 10^x$. Becau $10 > 1$, the graph has the same general shape as that of $y = 2^x$, but the graph go through the point $(1, 10)$, as shown in Figure A–3.

The graph of the function defined by $y = a^x$ has a somewhat different shape $0 < a < 1$.

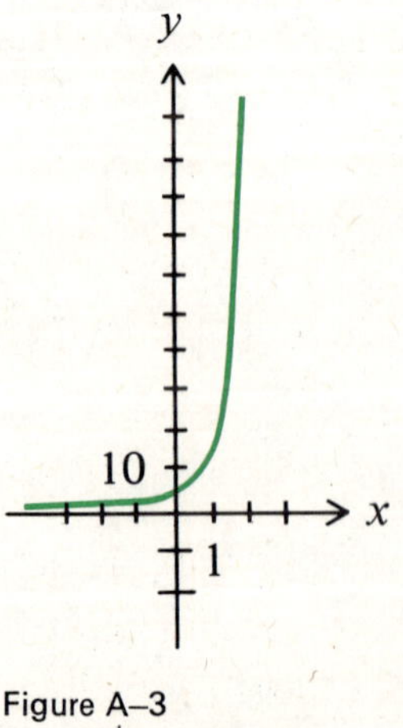

Figure A–3
$\{(x, y) \mid y = 10^x\}$ ■

EXAMPLE We sketch the graph of $y = (\frac{1}{2})^x$. We can plot a few points on the graph first (Figure A–4). Then these points are connected by a smooth curve, as shown in Figure A–5.

x	$y = (\frac{1}{2})^x$
-3	8
-2	4
-1	2
0	1
1	$\frac{1}{2}$
2	$\frac{1}{4}$

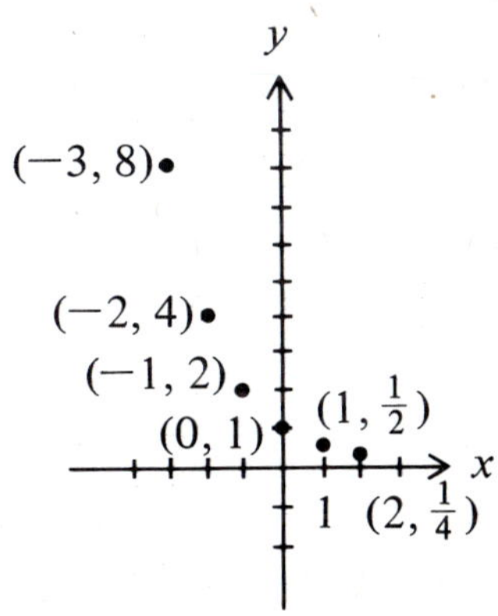

Figure A–4

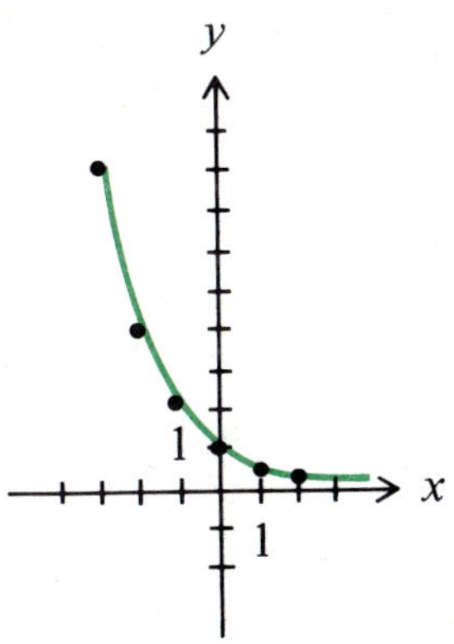

Figure A–5
$\{(x, y) \mid y = (\frac{1}{2})^x\}$ ■

For any a with $0 < a < 1$, the graph of the exponential function with base a has the same general shape as $y = (\frac{1}{2})^x$, graphed in Figure A–5. The general shape is shown in Figure A–6.

From the graph in Figure A–6 we see that if $0 < a < 1$,

1. For each real number x, $a^x > 0$
2. If $x > 0$, then $0 < a^x < 1$
3. If $x < 0$, then $a^x > 1$
4. $a^0 = 1$.

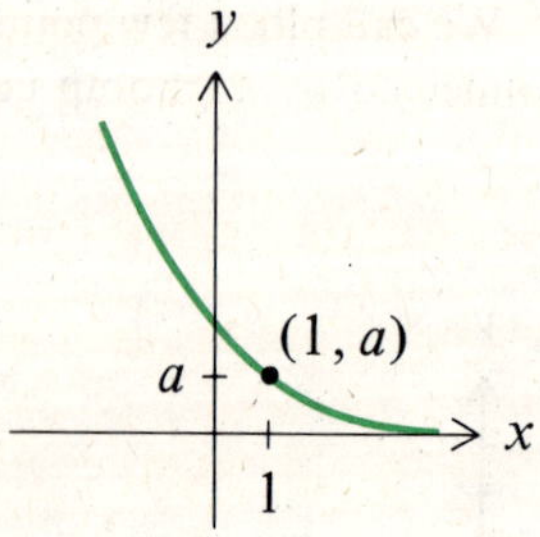

Figure A–6
$\{(x, y) \mid y = a^x, 0 < a < 1\}$

We have not yet considered the exponential function with base 1. Becau $1^x = 1$ for all real numbers x, this function is of no interest to us at this poi On the other hand, in advanced mathematics, the most useful exponential functi is the exponential function with base e, where e is a certain irrational number wi approximate value 2.71828. The graph of the exponential function defined $y = e^x$ is illustrated in Figure A–7.

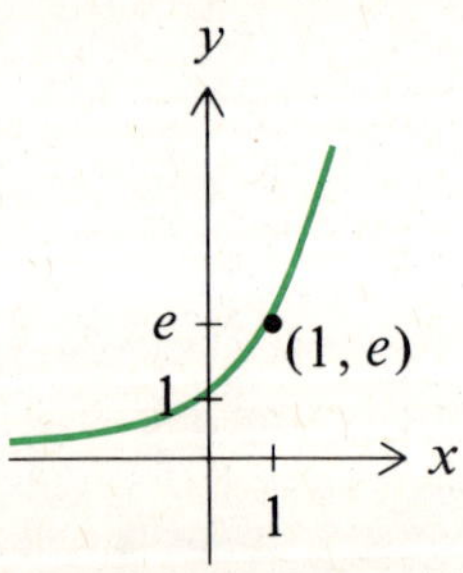

Figure A–7
$\{(x, y) \mid y = e^x\}$

A-1 *Exercises*

Sketch the graph of the exponential functions defined by the following.

1. $y = f(x) = 3^x$
2. $y = f(x) = 5^x$
3. $y = f(x) = (\frac{1}{3})^x$
4. $y = f(x) = (\frac{1}{5})^x$
5. For the exponential function defined in Exercises 1 through 4, find the domai and range.

Let a be a fixed real number with $a > 1$. Supply the appropriate symbols.

6. If x is any real number, then a^x ? 0

7. If x ? 0, then $a^x > 1$.
8. If x ? 0, then $0 < a^x < 1$.
9. a^0 ? 1.
10. Solve Exercises 6 through 9 under the assumption that $0 < a < 1$.

11. Let a be any positive number such that $a \neq 1$. Find the domain and range of the exponential function defined by $f(x) = a^x$.
12. What is the graph of the exponential function defined by $y = a^x$ for $a = 1$?

In Exercises 13 through 15, f is the exponential function with some base a such that $f(x) = a^x$. These problems show how laws of exponents can be used to prove properties of f.
13. Any exponential function f satisfies

$$f(x + y) = f(x) \cdot f(y)$$

Here are the steps—supply the justifications.

$$f(x + y) = a^{x+y}$$
$$a^{x+y} = a^x a^y$$
$$a^x a^y = f(x)f(y)$$
$$f(x + y) = f(x) \cdot f(y)$$

14. If f is an exponential function, then $f(cx) = [f(x)]^c$. Supply the justifications for the following steps.

$$f(cx) = a^{cx}$$
$$a^{cx} = (a^x)^c$$
$$(a^x)^c = [f(x)]^c$$
$$f(cx) = [f(x)]^c$$

15. Find the steps to show that any exponential function f satisfies $f(x - y) = f(x)/f(y)$.
16. If we invest P dollars at an annual rate of r, compounded annually, then the amount A (the worth of the investment after n years) is $A = P(1 + r)^n$, where r is expressed as a decimal. Use this formula to express the amount an investment of \$200 is worth after 7 years if interest is compounded annually at the rate of 6% per year. (Do not attempt to get a numerical answer.)
17. After gold was discovered at Whiskey Gulch, population 100, the population doubled every week. The population P is then given by the formula $P = 100 \times 2^w$, where w is the number of weeks after the strike.

(a) What was the population of Whiskey Gulch five weeks after gold was discovered?

(b) At this rate of growth, how long did it take Whiskey Gulch to become the biggest city in the United States? (The population of New York City is about 8 million.)

18. Harry starts a new job at the magnificent salary of 2¢ per day, but each day is paid twice as much as he was the day before.

(a) How many dollars will he be paid for his work on the nth day?

(b) If he works 30 days, estimate the amount he is to be paid for the last da

A-2 Logarithmic functions

Let us now return to the exponential function with base 2 given by $y = f(x) =$ From the graph (Figure A–8) it is apparent that f is a one-to-one function. domain is R, and its range is the set of positive real numbers.

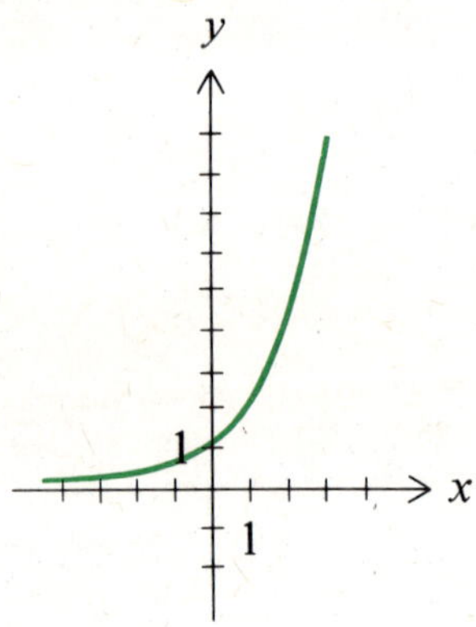

Figure A–8
$\{(x, y) | y = 2^x\}$

Therefore f has an inverse function $f^{-1} = \{(x, y) | x = 2^y\}$. The domain of f is the set of positive real numbers, and its range is R. Whenever $x = f(y) = 2^y$, th we have $y = f^{-1}(x)$.

The inverse function to an exponential function is called a *logarithmic functi* In particular, the inverse function to the exponential function with base 2 is cal the *logarithmic function with base* 2 and is denoted by $\log_2$. From the gene definition of an inverse function, we obtain the following:

Let x be a positive real number. Then $y = \log_2 x$ if and only if $x = 2^y$.

Thus, each time we have an equation

$$x = 2^y$$

we have an equivalent statement

$$y = \log_2 x$$

EXAMPLE: Let us express the equation

$$8 = 2^3$$

by using logarithms.

The corresponding statement, in terms of a logarithmic function, is

$\log_2 8 = 3$ ■

EXAMPLE: Let us express

$\log_2 \frac{1}{2} = -1$

in exponential form.

The corresponding exponential equation is

$2^{-1} = \frac{1}{2}$ ■

EXAMPLE: Let us find $\log_2 1$.

If $y = \log_2 1$, then $2^y = 1$. Because $2^0 = 1$, it must be that $y = 0$. Therefore,

$\log_2 1 = 0$ ■

EXAMPLE: Find $\log_2 32$.

Because

$32 = 2^5$

it follows that

$\log_2 32 = 5$ ■

It is always easy to find $\log_2 2^n$ for any n. The value of $\log_2 2^n$ is simply n. In fact, if $x = 2^n$, then $\log_2 x = n$, and so $\log_2 2^n = n$.

EXAMPLE: $\log_2 64 = 6$, because $2^6 = 64$ ■

EXAMPLE: $\log_2 \frac{1}{16} = -4$, because $2^{-4} = \frac{1}{16}$ ■

We are now going to sketch the graph of this logarithmic function. In the discussion above, we have already found that the points $(8, 3)$, $(\frac{1}{2}, -1)$, $(1, 0)$, $(32, 5)$, $(64, 6)$, and $(\frac{1}{16}, -4)$ lie on the graph of $y = \log_2 x$. These points all lie on the smooth curve shown in Figure A–9. We should make the following observations from the graph:

1. $\log_2 x$ is defined only for $x > 0$, because the domain of the function $\log_2$ is the set of positive real numbers.
2. If $x > 1$, then $\log_2 x$ is positive.
3. If $0 < x < 1$, then $\log_2 x$ is negative.
4. $\log_2 1 = 0$.

As noted in Section 8-1, the graph of f^{-1} is obtained from the graph of f by reflection in the line $y = x$. For the function $y = f(x) = 2^x$ and its inverse $y = f^{-1}(x) = \log_2 x$, this is illustrated in Figure A–10.

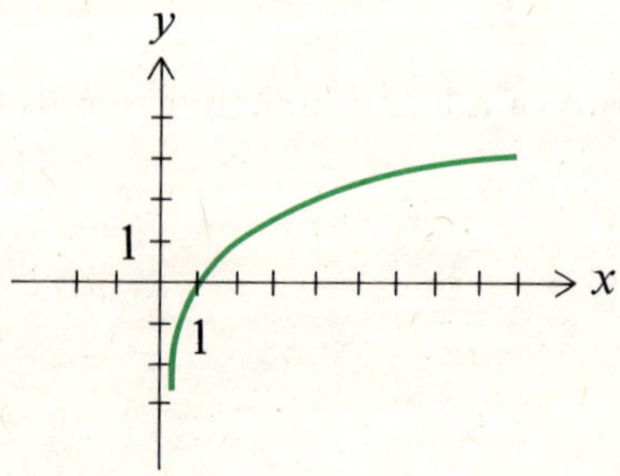

Figure A–9
$\{(x, y) \mid y = \log_2 x\}$

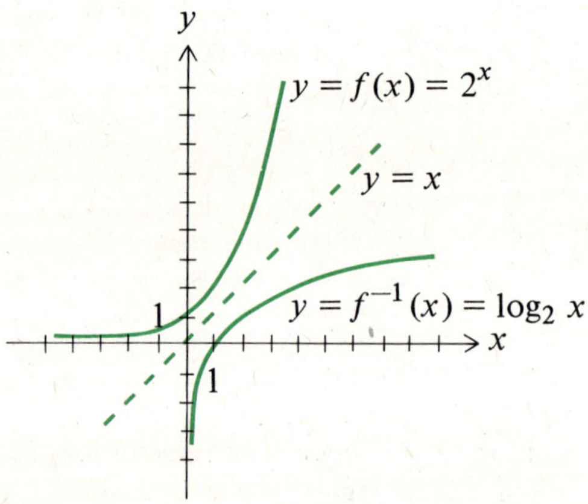

Figure A–10

We are now going to define a logarithmic function using any positive base (except $a = 1$). This function, denoted $\log_a$, is the inverse function to the exponent function defined by $y = a^x$. The domain of $\log_a$ is the set of positive real numbe and the range is R. It is called the *logarithmic function with base a*. From the defi tion of an inverse function we obtain the following:

Let a and x be positive real numbers with $a \neq 1$. Then $y = \log_a x$ if and only $x = a^y$.

Therefore, for each exponential equation $x = a^y$, there is the equivalent logarithm equation $y = \log_a x$.

EXAMPLE: Because

$10^3 = 1000$

we have

$3 = \log_{10} 1000$ ■

EXAMPLE: Because

$3^{-1} = \frac{1}{3}$

we have

$\log_3 \frac{1}{3} = -1$ ■

EXAMPLE: Because

$25^{1/2} = 5$

we have

$\log_{25} 5 = \frac{1}{2}$ ■

For each $a > 1$, the graph of the equation $y = \log_a x$ has the same general shape as the graph of the equation $y = \log_2 x$. In Figure A–11 is the graph of $y = \log_e x$. This graph is the reflection of the graph of $y = e^x$ in the line $x = y$.

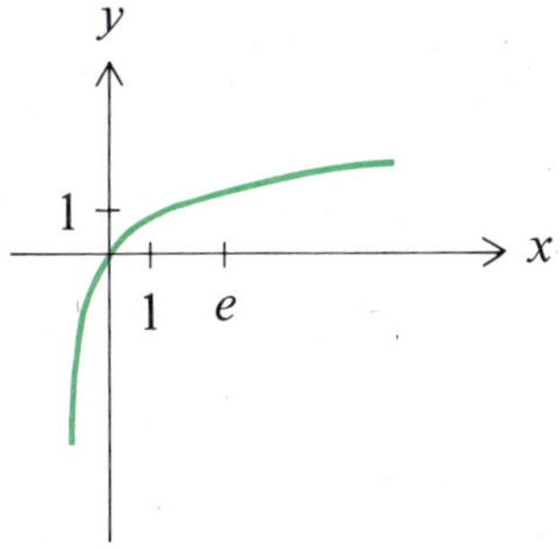

Figure A–11
$\{(x, y) \mid y = \log_e x\}$

If $0 < a < 1$, the graph has a different general shape. The graph of the function defined by $y = \log_{1/2} x$ is a typical example (see Figure A–12). This graph is the reflection of the graph of $y = (\frac{1}{2})^x$ in the line $y = x$.

No logarithmic function $\log_a$ can be defined for the base $a = 1$.

Because the function defined by $y = \log_a x$ is the inverse function to that defined by $y = a^x$, there are two important formulas relating logarithmic and exponential functions, as follows:

If $a > 0$ and $a \neq 1$, then

$$a^{\log_a x} = x$$

and

$$\log_a a^x = x$$

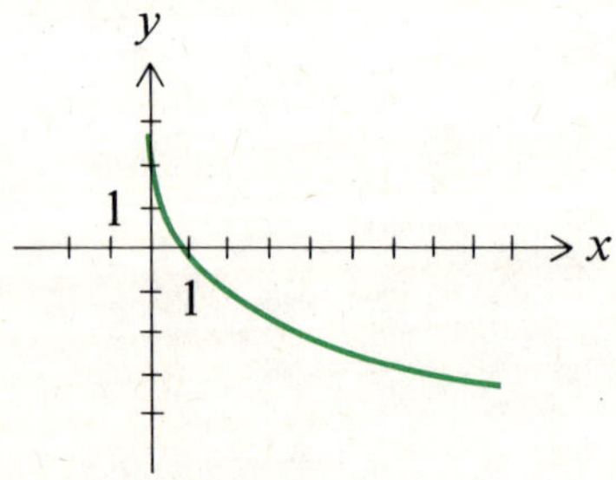

Figure A–12
$\{(x, y) \mid y = \log_{1/2} x\}$

EXAMPLE: $4^{\log_4 x} = x$ ■

EXAMPLE: $e^{\log_e 3} = 3$ ■

EXAMPLE: $\log_7 7^{12} = 12$ ■

EXAMPLE: $\log_e e^x = x$ ■

You are asked to justify the second formula in Exercise 33. Here is the pro of the first.

We let $y = \log_a x$. By the definition of the function $\log_a$, this equation equivalent to $a^y = x$. Substituting $\log_a x$ for y, we get $a^{\log_a x} = x$, as desired.

A-2 *Exercises*

Sketch the graph of the following logarithmic functions.

1. $y = \log_{10} x$

2. $y = \log_5 x$

3. Why is it impossible to define a logarithmic function with 1 as the base?

Express in exponential form.

4. $\log_{10} 100 = 2$

5. $\log_{36} 6 = \frac{1}{2}$

6. $\log_{10} 1 = 0$

7. $\log_e 1/e = -1$

8. $\log_{10} \frac{1}{100} = -2$

9. $\log_{10} 2 = 0.3010$ (approximately)

Express each exponential equation as a logarithmic statement.

10. $2^3 = 8$

11. $10^1 = 10$

12. $(\frac{1}{3})^{-2} = 9$

13. $10^3 = 1000$

14. $3^{1/2} = 1.7321$ (approximately)

15. $10^{-1} = \frac{1}{10}$

16. $5^4 = 625$

17. $10^{0.4771} = 3$ (approximately)

18. $216^{2/3} = 36$

19. $16^{-3/4} = \frac{1}{8}$

Evaluate each of the following.

20. $\log_2 32$

21. $\log_{10} 10$

22. $\log_2 \frac{1}{4}$

23. $10^{\log_{10} 2}$

24. $\log_{27} \frac{1}{9}$

25. $10^{\log_{10} 20}$

26. $\log_{1/5} 125$

27. $\log_{10} 3$ (see Exercise 17)

28. $\log_{16} 8$

29. $\log_{1/81} 27$

30. $\log_8 16$

31. $\log_\pi \pi^2$

32. $\log_e \sqrt[3]{e}$

33. Prove the formula $\log_a a^x = x$ by making the substitution $y = a^x$.

Use the graph of the function defined by $y = \log_2 x$ to solve the following inequalities.

34. $\log_2 x > 1$

35. $\log_2 x < 2$

36. $\log_2 x \leq 0$

37. $\log_2 x \geq -1$

38. Use the graph of the exponential function, base a, with $0 < a < 1$ (Figure A–6) to sketch the graph of the corresponding logarithmic function. What is the geometrical relationship between these two graphs?

39. Why is $\log_a 0$ not defined?

40. Why is $\log_a x$ not defined when x is negative?

-3 Properties of logarithmic functions

Each logarithmic function $\log_a$ is the inverse function to the corresponding exponential function defined by $y = a^x$. The laws of exponents give rise to several special properties of logarithmic functions. We now present these important properties of the function $\log_a$, where we work with one particular base a.

The first property states that the logarithm of a product of numbers may be computed as follows: Find the logarithm of each factor, and then add the logarithms. The result is the logarithm of the product. Stated formally, we have the following:

If x, y, and a are positive numbers with $a \neq 1$, then $\log_a xy = \log_a x + \log_a y$.

EXAMPLE: $\log_4 [(27)(0.364)] = \log_4 27 + \log_4 0.364$ ■

EXAMPLE: $\log_e [(42)(119{,}300)(0.53)] = \log_e 42 + \log_e 119{,}300 + \log_e 0.53$ ■

Here is how the above formula could be justified. We let $z = \log_a x$ and $w = \log_a y$. Then

$$a^z = x \quad \text{and} \quad a^w = y$$

Therefore,

$$a^z a^w = xy$$

and so, by the laws of exponents,

$$a^{z+w} = xy$$

But then

$$\log_a xy = z + w$$

Hence,

$$\log_a xy = \log_a x + \log_a y$$

The formula $\log_a xy = \log_a x + \log_a y$ is convenient for computations becau it is easier to add than to multiply. We can replace the multiplication problem finding xy by the addition problem of finding $\log_a x + \log_a y$. However, there is price which must be paid for this simplification. We will have to know how to fi the logarithm of a number so that we can find $\log_a x$ and $\log_a y$. Similarly, we ha to know how to find the number if we know its logarithm so that we will be able find xy after we have found $\log_a xy$. These aspects of computing with logarith are discussed in Section A–6.

We now turn to the second law of logarithms, which tells us how to find t logarithm of a number raised to a power. We first find the logarithm of that numb and then multiply by the exponent, which indicates the power. Stated formal we have the following:

If x and a are positive numbers with $a \neq 1$, and if c is any real number, th $\log_a x^c = c \log_a x$.

EXAMPLES: $\log_{10} 735^2 = 2 \log_{10} 735$
$\log_e 34^{0.01} = 0.01 \log_e 34$
$\log_2 (3.053)^{-7} = -7 \log_2 3.053$ ■

By this law, the complicated operation of raising x to a power in order to fi x^c is replaced by a simpler operation. We need only multiply $\log_a x$ by c, and th undo the process of taking logs.

The second law may be justified as follows. We let $y = \log_a x$, and therefc

$$x = a^y$$

Hence

$$x^c = (a^y)^c$$

but, by laws of exponents,

$$(a^y)^c = a^{cy}$$

Therefore,

$$x^c = a^{cy}$$

and so

$$cy = \log_a x^c$$

Therefore, because $y = \log_a x$,

$$c \log_a x = \log_a x^c$$

Logarithms can also be used to simplify computations involving division. To find the logarithm of a quotient, we take the logarithm of the numerator and subtract from it the logarithm of the denominator. Stated formally, the rule is as follows:

If x, y, and a are positive real numbers with $a \neq 1$, then $\log_a x/y = \log_a x - \log_a y$.

EXAMPLE: $\log_5 72/403 = \log_5 72 - \log_5 403$ ■

EXAMPLE: $\log_\pi 33.72/24.34 = \log_\pi 33.72 - \log_\pi 24.34$ ■

The third law of logarithms may be justified as follows, using the preceding two laws. Because $1/y = y^{-1}$, we have

$$\log_a \frac{x}{y} = \log_a xy^{-1}$$

By the first law of logarithms,

$$\log_a xy^{-1} = \log_a x + \log_a y^{-1}$$

By the second law of logarithms,

$$\log_a y^{-1} = -\log_a y$$

and so

$$\begin{aligned} \log_a \frac{x}{y} &= \log_a xy^{-1} \\ &= \log_a x + \log_a y^{-1} \\ &= \log_a x - \log_a y \end{aligned}$$

These three laws of logarithms should be memorized. We repeat them here for convenience. They are valid for any positive real numbers x, y, and a (with $a \neq 1$) and any real number c.

1. $\log_a xy = \log_a x + \log_a y$
2. $\log_a x^c = c \log_a x$
3. $\log_a \dfrac{x}{y} = \log_a x - \log_a y$

We can use any or all of the above laws to replace complicated operations by simpler ones.

EXAMPLE: We express $\log_{10}(35.2)^3(8309)$ using simpler arithmetic operations follows:

$$\log_{10}(35.2)^3(8309) = \log_{10}(35.2)^3 + \log_{10} 8309$$
$$= 3\log_{10} 35.2 + \log_{10} 8309 \blacksquare$$

EXAMPLE: We express

$$\log_e \frac{(.0027)^2}{(403.7)^4(38)^6}$$

using simpler arithmetic operations as follows:

$$\log_e \frac{(.0027)^2}{(403.7)^4(38)^6} = \log_e (.0027)^2 - \log_e (403.7)^4(38)^6$$
$$= 2\log_e .0027 - [\log_e (403.7)^4 + \log_e (38)^6]$$
$$= 2\log_e .0027 - 4\log_e 403.7 - 6\log_e 38 \blacksquare$$

We now study the connection between logarithmic functions with two differ bases, a and b. The following *change-of-base formula* expresses the relation betw them:

If a, b, and x are positive numbers with a and b both different from 1, then

$$\log_a x = \frac{\log_b x}{\log_b a}$$

EXAMPLE: $\log_{100} 2 = \dfrac{\log_{10} 2}{\log_{10} 100}$ ■

EXAMPLE: $\log_e x = \dfrac{\log_{10} x}{\log_{10} e}$ ■

The change-of-base formula can be used to evaluate logarithms to some bas if we know logarithms using some other base b. We are able to find an approxim value for $\log_{10} x$, for any positive number x, by using the table inside the back co (explained in Section A-4). From the change-of-base formula with $b = 10$, we ha

$$\log_a x = \frac{\log_{10} x}{\log_{10} a}$$

and so we can get an approximate value for $\log_a x$ no matter what base a we u

The change-of-base formula can be justified according to the following reas ing: If $y = \log_a x$, then we know that $a^y = x$. Let us apply the function $\log_b$ to e side of this equation. We then have

$$\log_b a^y = \log_b x$$

By the second law of logarithms,

$$\log_b a^y = y \log_b a$$

Therefore,

$$y \log_b a = \log_b x$$

where

$$y = \log_a x$$

Thus

$$(\log_a x)(\log_b a) = \log_b x$$

We divide by $\log_b a$ and obtain the desired result

$$\log_a x = \frac{\log_b x}{\log_b a}$$

In practice, the two logarithmic functions used most often are $\log_{10}$ and $\log_e$. Because 10 is the base of our system of numeration, $\log_{10}$ is especially convenient for arithmetic computation. The function $\log_{10}$ is usually called the *common logarithm* function; when we write log and do not specify any base, the base 10 is to be understood.

In more advanced mathematics, there are a number of theoretical advantages to the function $\log_e$. The function $\log_e$ is known as the *natural logarithm* function; in this book we use the abbreviation ln to stand for $\log_e$.

As we noted in an example above, for each positive number x,

$$\log_e x = \frac{\log_{10} x}{\log_{10} e}$$

or, using our abbreviations,

$$\ln x = \frac{\log x}{\log e}$$

Because an approximate value for $\log e$ is 0.4343, we have the following approximate formula for $\ln x$:

$$\ln x = \frac{\log x}{0.4343}$$

-3 *Exercises*

Use laws of logarithms to express using the simplest possible arithmetic operations. (Do not change the base.)

1. $\log (87 \times 0.541)$

2. $\ln (13.6 \times 8.2)$

3. $\log_a \dfrac{341}{72}$ **4.** $\log_b \dfrac{2.65}{139}$

5. $\ln 36^5$ **6.** $\log \sqrt{6.23}$

7. $\log_2 \sqrt[4]{11.5}$ **8.** $\log_3 17^{3/4}$

9. $\log_5 94.3^{-1/2}$ **10.** $\log_a 0.021^{-3}$

11. $\log_2 \dfrac{23^4 \times 17.2^{2/3}}{40.6^3}$ **12.** $\log \left(\dfrac{13.2}{11.6}\right)^4$

Express using the simplest possible algebraic operations. (Do not change the bas

13. $\log 2xyz$

14. $\log_2 \sqrt[3]{t}$

15. $\log_7 \dfrac{x^2 y}{z^3}$

16. $\log_a \dfrac{x^{-4} y^2}{z^5}$

17. $\ln \dfrac{\sqrt{x}}{\sqrt[3]{x}}$

18. $\ln \frac{4}{3}\pi r^3$

Let f be any logarithmic function. Use laws of logarithms to show that f has t following properties.

19. $f(xy) = f(x) + f(y)$

20. $f(x^c) = cf(x)$

21. $f(x/y) = f(x) - f(y)$

Let $A = 200(1.06)^7$, so that A is the amount an investment of \$200 will be wor after interest is compounded annually for 7 years at 6% per year.

22. Express $\log A$ in terms of $\log 200$ and $\log 1.06$.

23. Express $\ln A$ in terms of $\ln 200$ and $\ln 1.06$.

Express each of the following in terms of the logarithm function, base 10.

24. $\log_2 9$

25. $\ln 3$

26. $\log_5 7$

27. $\log_5 49$

28. $\ln \pi$

29. $\log_\pi e$

30. $\log_2 10$

31. Find a general formula for $\log_a 10$ in terms of logarithms, base 10.

32. Find a general relationship between $\log_a b$ and $\log_b a$.

4 The table of logarithms, base 10

As we pointed out in Section A–3, the three laws of logarithms permit us to use logarithms to simplify arithmetic computations. Because our decimal system of numeration is based on 10, the easiest logarithmic function to use for computational purposes is log ($= \log_{10}$).

To compute using logarithms, we need to have some method of finding log x if x is known. Similarly, we need to be able to find x whenever log x is known.

In Figure A–13 is a part of the table of logarithms, base 10, which appears inside the back cover of this book. Log x may well be an irrational number, and so the values given in the table are approximations. This table is called a *four-place logarithm table*, because it gives an approximate value of log x correct to four decimal places.

Because $10^0 = 1$ and $10^1 = 10$, we know that $\log 1 = 0$ and $\log 10 = 1$. If x lies between 1 and 10, its logarithm lies between 0 and 1. The logarithm of a number between 1 and 10 can be read off directly from the table, as shown in these examples.

x	0	1	2	3	4	5	6	7	8	9
1.0	.0000	.0043	.0086	.0128	.0170	.0212	.0253	.0294	.0334	.0374
1.1	.0414	.0453	.0492	.0531	.0569	.0607	.0645	.0682	.0719	.0755
1.2	.0792	.0828	.0864	.0899	.0934	.0969	.1004	.1038	.1072	.1106
1.3	.1139	.1173	.1206	.1239	.1271	.1303	.1335	.1367	.1399	.1430
1.4	.1461	.1492	.1523	.1553	.1584	.1614	.1644	.1673	.1703	.1732
1.5	.1761	.1790	.1818	.1847	.1875	.1903	.1931	.1959	.1987	.2014
1.6	.2041	.2068	.2095	.2122	.2148	.2175	.2201	.2227	.2253	.2279
1.7	.2304	.2330	.2355	.2380	.2405	.2430	.2455	.2480	.2504	.2529
1.8	.2553	.2577	.2601	.2625	.2648	.2672	.2695	.2718	.2742	.2765
1.9	.2788	.2810	.2833	.2856	.2878	.2900	.2923	.2945	.2967	.2989
2.0	.3010	.3032	.3054	.3075	.3096	.3118	.3139	.3160	.3181	.3201
2.1	.3222	.3243	.3263	.3284	.3304	.3324	.3345	.3365	.3385	.3404
2.2	.3424	.3444	.3464	.3483	.3502	.3522	.3541	.3560	.3579	.3598
2.3	.3617	.3636	.3655	.3674	.3692	.3711	.3729	.3747	.3766	.3784
2.4	.3802	.3820	.3838	.3856	.3874	.3892	.3909	.3927	.3945	.3962

Figure A–13

EXAMPLE: To find log 2.05, we first locate 2.0 in the left column of the table. Across from 2.0, directly under 5, we find the entry .3118 in the table. Thus $\log 2.05 = .3118$. ■

EXAMPLE: To find log 1.60, we first locate 1.6 in the left column of the table. Now, across from 1.6 and directly under 0, we find the entry .2041. Thus $\log 1.60 = .2041$ ■

EXAMPLE: To find log 1.48, we first locate 1.4 in the left column of the table. Across from 1.4 and directly under 8, we find the entry .1703. Thus $\log 1.48 = .1703$. ■

Strictly speaking, it is not accurate to say that log 1.48 = .1703, because .1 is only an approximate value (correct to four decimal places) for log 1.48. We s will use the "equals" symbol, although we should remember that in this cont "=" really means "approximately equal." For most applications, a four-pl logarithm table is sufficiently accurate. If more accuracy is needed, we could ge five-place (or larger) table from the library.

We have seen that the table can be used directly to find log x when x is a num between 1 and 10. To find logarithms of other positive numbers, we should fi express these numbers in *scientific notation*, as follows.

To write a number in scientific notation, we write it as a product of two factc The first of these factors must be a number which is at least one but less than t Thus it is a number which has just one digit to the left of the decimal point, a that digit is not zero. The second of the two factors must be expressed as a pov of 10.

EXAMPLE: Write 320 in scientific notation.

We write

$320 = 3.2 \times 10^2$

Note that $1 \leq 3.2 < 10$, and 10^2 is a power of 10. ■

EXAMPLE: Write 14,000,000,000,000 in scientific notation.

In scientific notation, this number is equal to 1.4×10^{13}. Note that $1 \leq 1.4 <$ and 10^{13} is a power of 10. ■

EXAMPLE: Write .0723 in scientific notation.

We obtain

$.0723 = 7.23 \times 10^{-2}$ ■

EXAMPLE: Write 5.3 in scientific notation.

We see that

$5.3 = 5.3 \times 10^0$

or simply

$5.3 = 5.3$ ■

EXAMPLE: Write .000000035 in scientific notation.

This number is equal to 3.5×10^{-8}. ■

Now we are ready to find logarithms of positive numbers which are not betwe 1 and 10. For example, let us find log 132. Here are the steps:

1. Write 132 in scientific notation. Thus we write $132 = 1.32 \times 10^2$.

2. Find log 1.32 in the table; log 1.32 = .1206. This number, .1206, is called the *mantissa* of log 132.
3. Add to the mantissa the number 2, which appeared as the exponent in the expression $132 = (1.32)(10^2)$. Thus, $\log 132 = .1206 + 2 = 2.1206$. This exponent 2 is called the *characteristic* of log 132.

Here is why this three-step procedure is valid. We know that

$$132 = (1.32)(10^2)$$

Therefore, by the first law of logarithms,

$$\log 132 = \log 1.32 + \log 10^2$$

However, we have seen that

$$\log 10^2 = 2$$

Thus,

$$\log 132 = \log 1.32 + 2 = .1206 + 2 = 2.1206$$

EXAMPLE: To find log 214,000, here are the steps:

1. In scientific notation, $214{,}000 = (2.14)(10^5)$.
2. The mantissa is log 2.14 = .3304.
3. The characteristic is 5, and therefore log 214,000 = 5.3304. ■

Some care must be exercised when the characteristic is negative, because *the mantissa is always positive.*

EXAMPLE: To find log .00193, here are the steps:

1. We write $.00193 = (1.93)(10^{-3})$ in scientific notation.
2. The mantissa is log 1.93 = .2856.
3. The characteristic is -3. Therefore, log .00193 = 0.2856 − 3. ■

Warning: In this last example, -3.2856 would not be correct, because -3.2856 indicates the sum of -3 and $-.2856$. The value which is desired is the sum of -3 and $+.2856$. Therefore, we simply write $0.2856 - 3$.

Often a characteristic of -3 is expressed as $7 - 10$; thus we might write

$$\log .00193 = 7.2856 - 10$$

EXAMPLE: To find log .0104, we follow these steps:

1. We write $.0104 = (1.04)(10^{-2})$.
2. The mantissa is log 1.04 = .0170.
3. The characteristic is -2. Therefore, log .0104 = 0.0170 − 2. ■

Alternatively, this answer might be written as $8.0170 - 10$.

Notice that when we find $\log x$, the mantissa is determined by the sequence digits which appear in x, but the characteristic is determined by the position of decimal point in x. Thus log 204, log 20.4, and log .00204 all have the same mantis but these three logarithms each have different characteristics (because of difference in position of the decimal point). Thus,

$$\log 204 = 2.3096$$

$$\log 20.4 = 1.3096$$

and

$$\log .00204 = 0.3096 - 3$$

The log table can also be used to find x when $\log x$ is known. The mantissa log x will determine the sequence of digits in x. The characteristic of $\log x$ is t used to locate the decimal point.

As an example, let us find x if $\log x = 3.3962$. Here are the steps:

1. We find the mantissa .3962 in the log table. The mantissa .3962 is found correspond to 2.49.
2. We express x in scientific notation, using the characteristic of $\log x$ as exponent. Thus $x = (2.49)(10^3)$, because 3 is the characteristic of log Therefore, $x = (2.49)(10^3) = 2490$.

The relationship

$$\log 2490 = 3.3962$$

is sometimes expressed by saying

$$2490 = \text{antilog } 3.3962$$

Thus to find antilog y means to find x such that $\log x = y$. Actually, antilog simply another notation for 10^y.

Here are additional examples in finding antilogarithms.

EXAMPLE: We find antilog $(0.2330 - 4)$ in the following steps:

1. We find the mantissa .2330 in the table. This mantissa corresponds to 1.
2. Therefore antilog $(0.2330 - 4) = (1.71)(10^{-4}) = .000171$. ■

EXAMPLE: To find antilog .1818, here are the steps:

1. We find the mantissa .1818. This mantissa corresponds to 1.52.
2. Because the characteristic is 0, antilog $.1818 = (1.52)(10^0) = (1.52)(1)$ 1.52. ■

EXAMPLE: We find antilog $(9.2900 - 10)$ as follows:

1. The mantissa .2900 is found to correspond to 1.95.

2. The characteristic is $9 - 10 = -1$. Therefore, antilog $(9.2900 - 10) = (1.95)(10^{-1}) = .195$. ■

Unfortunately, not all possible mantissas appear in the log table. We might well wonder how to evaluate antilog .1836, because .1836 is not in the table. This type of problem is solved by a method called *interpolation*, to be discussed in the next section.

A-4 *Exercises*

Write in scientific notation.

1. 362
2. 2.93
3. 0.037
4. 10
5. $\frac{1}{1000}$
6. $\frac{1}{2}$

Express in terms of log x.

7. $\log 10x$
8. $\log x/10$
9. $\log 100x$
10. $\log x/100$
11. $\log 10^n x$
12. $\log 10^{-n} x$

Find log x for the following values of x.

13. 2
14. 3
15. 4.75
16. 1.96
17. 39
18. 80
19. 0.41
20. 0.0368
21. 0.000051
22. 783,000
23. 8620
24. 49.4

Find antilog y for the following values of y.

25. 2.0000
26. 1.6021
27. 0.8451
28. 4.4232
29. $0.9763 - 1$
30. $0.0212 - 2$
31. $9.4425 - 10$
32. $8.3483 - 10$
33. 1.7597
34. $0.9015 - 3$
35. 6.3979
36. $17.0755 - 20$
37. Explain why log 1.79 and log 179 have the same mantissa.
38. How are antilog 2.8645 and antilog $(0.8645 - 1)$ related?
39. Explain in detail why $\log_{10}$ is the most useful logarithm function for computational purposes.

A-5 Interpolation

Interpolation is a process which increases the usefulness of the log table. Th method is used to find antilogarithms for mantissas which do not appear in th table. Interpolation is also used in finding logarithms, as in the following exampl

Suppose we wish to find log 2.563. The table tells us that log 2.56 = .4082 an also that log 2.57 = .4099. Because 2.563 lies between the number 2.56 = 2.560 an 2.57 = 2.570, we expect log 2.563 to lie between .4082 = log 2.560 and .4099 = log 2.570. Interpolation provides the means to estimate just where in this rang log 2.563 will lie.

For ease of computation, the data obtained from the table should be arrange as follows:

$$10\left[\begin{array}{l} 3\left[\begin{array}{l}\log 2.560 = .4082\\ \\ \log 2.563 = \ ?\end{array}\right] y \\ \quad \log 2.570 = .4099\end{array}\right] 17$$

The differences 3, 10, and 17 which appear in the above table are obtained by sub tracting the indicated numbers (ignoring the decimal points). Also, y indicates a unknown difference. We estimate y according to the fractional relationship

$$\frac{3}{10} = \frac{y}{17}$$

Solving for y, we find that $y = \frac{51}{10}$, and so $y = 5$ (correct to the nearest integer We use the value $y = 5$ for the unknown difference. Therefore, the unknown entry i the display is obtained by adding 5 in the last decimal place to .4082. This proces yields the result log 2.563 = .4087.

Now we should realize why it was possible to ignore the decimal points in th calculation. If we had used decimal points, we would have obtained the equatio

$$\frac{.003}{.010} = \frac{z}{.0017}$$

and found that the unknown difference z is $z = .0005$. However,

$$\frac{.003}{.010} = \frac{3}{10}$$

and

$$\frac{.0005}{.0017} = \frac{5}{17}$$

and so the (approximate) equality

$$\frac{.003}{.010} = \frac{.0005}{.0017}$$

is equivalent to the easier approximation

$$\frac{3}{10} = \frac{5}{17}$$

In applying the above interpolation method, we are actually assuming that the change in $\log x$ is approximately proportional to the change in x. Because the graph of the function $f(x) = \log x$ is not a straight line, this assumption introduces some error. However, the error is small enough to be neglected for the computations here.

Here are some additional examples of finding logarithms by interpolation.

EXAMPLE: To find log .03731, we assemble the following display from the table:

$$10\left[\begin{array}{l} 1\left[\begin{array}{l}\log .03730 = .5717 - 2\\ \cdot \\ \log .03731 = ?\end{array}\right] y \\ \quad \log .03740 = .5729 - 2\end{array}\right] 12$$

The unknown difference y is found by solving $1/10 = y/12$. Correct to the nearest integer, $y = 1$. Therefore, $\log .03731 = .5718 - 2$. ■

EXAMPLE: To find log 21.99, we proceed as follows:

$$10\left[\begin{array}{l} 9\left[\begin{array}{l}\log 21.90 = 1.3404\\ \\ \log 21.99 = ?\end{array}\right] y \\ \quad \log 22.00 = 1.3424\end{array}\right] 20$$

The unknown difference is found by solving $9/10 = y/20$ for y. We obtain $y = 18$, and so $\log 21.99 = 1.3422$. ■

Antilogarithms for mantissas not in the table are found by a similar process. For example, suppose we wish to find antilog 0.6324. The mantissa .6324 does not appear in the table. The closest values are .6314 (before .6324) and .6325 (after it). Thus we expect antilog 0.6324 to lie between 4.280 = antilog .6314 and 4.290 = antilog 0.6325. The computations should be arranged as follows:

$$11\left[\begin{array}{l} 10\left[\begin{array}{l}\text{antilog } 0.6314 = 4.280\\ \\ \text{antilog } 0.6324 = ?\end{array}\right] y \\ \quad \text{antilog } 0.6325 = 4.290\end{array}\right] 10$$

The proportion which must be solved for the unknown difference y is $10/11 = y/10$. Correct to the nearest integer, $y = 9$, and so antilog $0.6324 = 4.289$.

EXAMPLE: We now find antilog 2.9463. The mantissa, .9463, is not found in the table, and the work should be arranged as follows:

$$5\left[\begin{array}{l} 3\left[\begin{array}{l}\text{antilog } 2.9460 = 883.0\\ \\ \text{antilog } 2.9463 = \ ?\end{array}\right] y \\ \ \ \ \text{antilog } 2.9465 = 884.0\end{array}\right] 10$$

The unknown difference y is found by solving $3/5 = y/10$. We obtain $y = 6$, and so antilog $2.9463 = 883.6$. ■

EXAMPLE: To find antilog $(0.4142 - 2)$, where the mantissa .4142 is not found in the table, we arrange the work as follows:

$$17\left[\begin{array}{l} 9\left[\begin{array}{l}\text{antilog } (0.4133 - 2) = .02590\\ \\ \text{antilog } (0.4142 - 2) = \ ?\end{array}\right] y \\ \ \ \ \text{antilog } (0.4150 - 2) = .02600\end{array}\right] 10$$

The proportion to be solved is $9/17 = y/10$. To the nearest integer, $y = 5$. It follows that antilog $(0.4142 - 2) = .02595$. ■

The validity of the interpolation procedure for finding antilogs is again based on an assumption. We are assuming that the change in antilog x is approximately proportional to the change in x. For most applications, the error so introduced can be safely ignored. However, if more accuracy is needed, we can use a five-place table.

A-5 *Exercises*

Use interpolation to estimate the value of log x for the following values of x.

1. 4.136 **2.** 42.78
3. 0.8519 **4.** 384,100
5. 0.1347 **6.** 0.002614
7. $\pi(= 3.142)$ **8.** $e(= 2.718)$
9. $\sqrt{2}(= 1.414)$ **10.** $\sqrt{3}(= 1.732)$

Use interpolation to estimate antilog y for the following values of y.

11. 0.6584 **12.** 1.6470
13. 0.5903 **14.** 0.9516 − 1
15. 2.3193 **16.** 3.0650
17. 0.9914 − 2 **18.** 0.2362 − 3
19. 9.1051 − 10 **20.** 8.7641 − 10
21. 5.8374 **22.** 6.4529

23. **(a)** According to the table, log $2 = 0.3010$ and log $3 = 0.4771$. Therefore because $6 = 2 \times 3$, we should have log $6 = \log 2 + \log 3 = 0.3010 + 0.4771 = 0.7781$. However, according to the table, log $6 = 0.7782$. How do you explain this discrepancy?

(b) What do you think will be the effect of interpolation on the accuracy of computations?

24. (**a**) When we use interpolation to estimate log 2.587, we are using a straight line to approximate the graph of $\log x$ between the points (2.58, log 2.58) and (2.59, log 2.59). Do you think the value obtained by interpolation is smaller or larger than the true value of log 2.587?

(**b**) When we use interpolation to estimate antilog 0.2261, we are using a straight line to approximate the graph of $x = \text{antilog } y$ (or $x = 10^y$) between the points (0.2253, antilog 0.2253) and (0.2279, antilog 0.2279). How do you expect the interpolated value for antilog 0.2261 to compare with the true value?

-6 Computing with logarithms, base 10

Now that we know how to use the log tables, we are ready to use logarithms to simplify computational arithmetic. The procedures are based on the laws of logarithms, which we now recall from Section A–3:

1. $\log xy = \log x + \log y$
2. $\log x^c = c \log x$
3. $\log \dfrac{x}{y} = \log x - \log y$

As an example, here is how we would use logarithms to find (347)(.00435). Suppose we let

$$x = (347)(.00435)$$

By the first law of logarithms,

$$\log x = \log 347 + \log .00435$$

Thus we are going to find log 347 and log .00435 in the table. We then add these logarithms together, and so obtain $\log x$. Finally, we find x by using the log table again to find the antilog of $\log x$.

A great deal of confusion can be avoided by arranging the work in an orderly fashion. We might organize the computations as follows:

$$\begin{aligned} \log 347 &= 2.5403 \\ +\log .00435 &= 0.6385 - 3 \\ \hline \log x - 3.1788 - 3 &= 0.1788 \\ x &= \text{antilog } 0.1788 \\ x &= 1.509 \end{aligned}$$

Notice that interpolation was required to find $x = \text{antilog } 0.1788$, because the mantissa .1788 does not appear in the table.

We should always remember that computing with logarithms is an approximate procedure. The tables themselves are but approximations, and further errors may be introduced by interpolation. The exact value of (347)(.00435) is 1.50945, so the approximate value of 1.509 obtained using logarithms is sufficiently accurate f[or] most purposes.

The three operations which can be performed using logarithms are multiplic[a]tion, raising to a power, and division. *Logarithms cannot be used to do addition [or] subtraction.* The following examples illustrate the types of problems for whic[h] logarithms are often helpful.

EXAMPLE: We find x if

$$x = 2.834^5$$

as follows. By the second law of logarithms,

$$\log x = 5 \log 2.834$$

Therefore, the computation may be arranged as follows:

$$\begin{aligned} \log 2.834 &= 0.4524 \\ &\quad\ \times 5 \\ \hline 5 \log 2.834 &= 2.2620 \end{aligned}$$

Therefore,

$$x = \text{antilog } 2.2620$$

and so

$$x = 182.8 \ \blacksquare$$

EXAMPLE: Find x if

$$x = 371.2 \div 46.3$$

By the third law of logarithms,

$$\log x = \log 371.2 - \log 46.3$$

Therefore, the calculations may be arranged as follows:

$$\begin{aligned} \log 371.2 &= 2.5696 \\ -\log 46.3 &= -1.6656 \\ \hline \log x &= 0.9040 \\ x &= \text{antilog } 0.9040 \\ x &= 8.017 \ \blacksquare \end{aligned}$$

We note that logarithms can be used to extract roots, because $\sqrt[n]{a} = a^{1/n}$.

EXAMPLE: Let us find x if

$$x = \sqrt[4]{0.9312}$$

Because $x = (0.9312)^{1/4}$, it follows that

$\log x = \frac{1}{4} \log 0.9312$

We first find $\log 0.9312 = 0.9690 - 1$. Thus to find $\log x$, we must find $\frac{1}{4} \log 0.9312$, and so we must divide $0.9690 - 1$ by 4. To make the division easier, we shall first write $0.9690 - 1$ as $3.9690 - 4$. Now it is easy to divide by 4 to obtain

$\log x = \frac{1}{4}(3.9690 - 4) = 0.9923 - 1$

Therefore,

$$x = \text{antilog}\ (0.9923 - 1)$$

and so

$x = 0.9824$ ■

Logarithms are seen to best advantage in solving long problems involving several multiplications, divisions, or exponentiations (but no addition or subtraction!).

EXAMPLE: Find x if

$$x = \sqrt[3]{\frac{(329.3)(.074)}{8.421}}$$

Using all three laws of logarithms, we see that

$\log x = \frac{1}{3}(\log 329.3 + \log .074 - \log 8.421)$

The work may be arranged as follows:

$$\begin{aligned}
\log 329.3 &= 2.5176 \\
+\log .074 &= 0.8692 - 2 \\
&\overline{3.3868 - 2} \\
-\log 8.421 &= -0.9254 \\
&\overline{2.4614 - 2} = 0.4614 \\
\times \tfrac{1}{3} &= 0.1538
\end{aligned}$$

Thus,

$$\begin{aligned}
\log x &= 0.1538 \\
x &= \text{antilog}\ 0.1538 \\
x &= 1.425
\end{aligned}$$ ■

The formula for finding compound interest is perfectly suited to logarithmic computation. Suppose we invest P dollars at an annual interest rate of r. Then the amount A (that is, the worth of the investment after n years) is given by the formula

$A = P(1 + r)^n$

where r is expressed as a decimal. We are assuming here that the interest is compounded annually.

EXAMPLE: We shall compute how much an investment of \$500 is worth after years if interest is compounded annually at the rate of 6% per year.

The interest rate of 6% is expressed as the decimal 0.06. Therefore, the amount A given by the formula

$$A = P(1 + r)^n = 500(1.06)^8$$

Thus,

$$\log A = \log 500 + 8 \log 1.06$$

The computations are as follows:

$$\begin{aligned} \log 1.06 &= 0.0253 \\ &\quad \times 8 \\ \hline 8 \log 1.06 &= 0.2024 \\ +\log 500 &= 2.6990 \\ \hline \log A &= 2.9014 \\ A &= \text{antilog } 2.9014 \\ A &= 796.8, \text{ or approximately } \$796.80 \end{aligned}$$

A-6 *Exercises*

Use logarithms to compute the following.

1. 23.9×0.0376

2. $874 \div 295$

3. $(67.4)^3$

4. $(1.834)^8$

5. $\sqrt{37.42}$

6. $\sqrt[3]{0.01078}$

7. $\dfrac{475 \times 129}{0.0134 + 0.0625}$

8. $\dfrac{\sqrt[3]{304.1}}{842.1 - 628.3}$

9. $\dfrac{587 \times (1.62)^2}{(3.84)^3}$

10. $\sqrt{\dfrac{416 \times 0.262}{87.8}}$

11. How much will an investment of \$2000 be worth after 5 years if interest compounded annually at the rate of 4% per year?

12. When interest is compounded k times per year at an annual rate of r, th amount A of the original investment P after n years is given by the formula

$$A = P(1 + r/k)^{kn}$$

Use this formula to find out how much an investment of \$300 will be worth aft 10 years if interest is compounded semiannually at the rate of 5% per year.

13. Use the formula in Exercise 12 to determine how much an investment of \$5C will be worth after 8 years if interest is compounded quarterly at the rate of 6 per year. (According to the example worked in this section, if the interest is compounded annually, the investment will be worth \$796.80.)

Use the change-of-base formula (Section A–3) and the log table to obtain approximate values for the following.

14. $\log_3 7$
15. $\ln 10$ ($\ln = \log_e$, $e = 2.718$, approximately)
16. $\ln \pi$ ($\pi = 3.142$, approximately)
17. $\ln \pi^2$

-7 Tables of logarithms of trigonometric functions

When we solve a trigonometric problem by logarithmic methods, we may need to evaluate the logarithm of a trigonometric ratio. This is done most easily from a table.

A table of Logarithms of Trigonometric Ratios is found in Appendix B. This table gives four-place values of $\log \sin \theta$, $\log \cos \theta$, $\log \tan \theta$, and $\log \cot \theta$ for acute angles θ at intervals of 10 minutes. As with the table of trigonometric ratios, the headings at the top of the columns are used for $0 \le \theta \le 45°$, and the bottom labels apply when $45° \le \theta \le 90°$.

No values are given for $\log \sec \theta$ and $\log \csc \theta$. These are rarely used and unnecessary anyway, because $\sec \theta = 1/\cos \theta$ and $\csc \theta = 1/\sin \theta$. Therefore $\log \sec \theta = -\log \cos \theta$ and $\log \csc \theta = -\log \sin \theta$.

EXAMPLE: Find $\log \sin 23°\ 10'$.

From the table, across from $23°\ 10'$ in the left column, under log sin, we read off $\log \sin 23°\ 10' = 0.5948 - 1$. ■

EXAMPLE: Find $\log \sin 68°\ 20'$.

In the table, across from $68°\ 20'$ in the right column, above log sin, we read off $\log \sin 68°\ 20' = 0.9682 - 1$. ■

EXAMPLE: Find the acute angle θ such that $\log \cos \theta = 0.8125 - 1$.

We find $0.8125 - 1$ in the table, with log cos at the bottom of the column. Therefore θ is found in the right column, and so $\theta = 49°\ 30'$. ■

Interpolation (Section A–5) can be used, just as with the table of logarithms, as in the following examples.

EXAMPLE: Find $\log \sin 36°\ 43'$.

We assemble the following display:

$$10\left[\begin{array}{l} 3\left[\begin{array}{l} \log \sin 36°\ 40' = 0.7761 - 1 \\ \\ \log \sin 36°\ 43' = \ ? \end{array}\right] y \\ \quad \log \sin 36°\ 50' = 0.7778 - 1 \end{array}\right] 17$$

The unknown difference y is found by solving $3/10 = y/17$. We obtain $y = 5$, a
so $\log \sin 36^\circ\ 43' = 0.7766 - 1$. ■

EXAMPLE: Find $\log \tan 65^\circ\ 54'$.

We assemble the following display:

$$10\left[\begin{array}{l} 4\left[\begin{array}{l}\log \tan 65^\circ\ 50' = 0.3480 \\ \\ \log \tan 65^\circ\ 54' = \ ? \end{array}\right] y \\ \quad \log \tan 66^\circ \quad\ \ = 0.3514 \end{array}\right] 34$$

The unknown difference y is found by solving $4/10 = y/34$. We obtain $y =$
and so $\log \tan 65^\circ\ 54' = 0.3494$. ■

EXAMPLE: Find the acute angle θ such that $\log \sin \theta = 0.7164 - 1$.

We arrange information from the table as follows:

$$10\left[\begin{array}{l} y\left[\begin{array}{l}\log \sin 31^\circ\ 20' = 0.7160 - 1 \\ \\ \log \sin \theta \qquad\quad = 0.7164 - 1 \end{array}\right] 4 \\ \quad \log \sin 31^\circ\ 30' = 0.7181 - 1 \end{array}\right] 21$$

We have $y/10 = 4/21$, and so $y = 2$. Hence $\theta = 31^\circ\ 22'$. ■

EXAMPLE: Find the acute angle θ such that $\log \cot \theta = 0.9450 - 1$.

Our display from the table is as follows:

$$10\left[\begin{array}{l} y\left[\begin{array}{l}\log \cot 48^\circ\ 30' = 0.9468 - 1 \\ \\ \log \cot \theta \qquad\quad = 0.9450 - 1 \end{array}\right] 18 \\ \quad \log \cot 48^\circ\ 40' = 0.9443 - 1 \end{array}\right] 25$$

Thus $y/10 = 18/25$, and so $y = 7$. Hence $\theta = 48^\circ\ 37'$. ■

A-7 *Exercises*

Evaluate from the table.

1. $\log \sin 11^\circ$

2. $\log \cos 11^\circ$

3. $\log \tan 17^\circ\ 20'$

4. $\log \cot 17^\circ\ 20'$

5. $\log \sin 58^\circ\ 30'$

6. $\log \cos 58^\circ\ 30'$

7. $\log \sec 58^\circ\ 30'$

8. $\log \csc 58^\circ\ 30'$

Find the acute angle θ such that

9. $\log \sin \theta = 0.9816 - 2$

10. $\log \cot \theta = 0.4178 - 1$

11. $\log \cot \theta = 0.4083$

12. $\log \cos \theta = 0.8874 - 1$

Evaluate

13. log sin 42° 25′

14. log cos 42° 25′

15. log tan 67° 18′

16. log cot 67° 18′

17. log sin 73° 52′

18. log cos 73° 52′

Find the acute angle θ such that

19. $\log \sin \theta = 0.8620 - 1$

20. $\log \tan \theta = 0.2020$

21. $\log \cot \theta = 0.4241$

22. $\log \cos \theta = 0.9655 - 1$

23. $\log \cos \theta = 0.6331 - 1$

24. $\log \sin \theta = 0.7229 - 1$

-8 Trigonometric computations by logarithms

Often the computations needed to solve a triangle can be handled easily with logarithms.

Note how the table of Logarithms of Trigonometric Ratios saves us a step. In the following example, we find $\log \sin 24° = 0.6093 - 1$ directly from the table of Logarithms of Trigonometric Ratios. The longer, two-step process would be to first find $\sin 24° = 0.4067$ from the table of Trigonometric Ratios, then evaluate $\log 0.4067 = 0.6093 - 1$ from the table of Logarithms.

The solution of right triangles lends itself well to logarithmic computations, as we now illustrate.

EXAMPLE: Solve right $\triangle ABC$, given

$\angle A = 24°$ and $c = 1.16$.

Evidently $\angle B = 90° - \angle A = 90° - 24° = 66°$.
We have $a/c = \sin A$, and so $a = c \sin A = 1.16 \sin 24°$.

Thus,

$$\begin{aligned} \log a &= \log 1.16 + \log \sin 24° \\ &= 0.0645 + 0.6093 - 1 = 0.6738 - 1 \end{aligned}$$

Hence $a = 0.4719$.

Similarly, $b = c \cos A = 1.16 \cos 24°$, and so

$$\begin{aligned} \log b &= \log 1.16 + \log \cos 24° \\ &= 0.0645 + 0.9607 - 1 = 0.0252 \end{aligned}$$

and so $b = 1.060$. ■

EXAMPLE: In right $\triangle ABC$, we have $a = 67.2$ and $b = 38.4$. Find $\angle A$.

We have $\tan A = a/b = 67.2/38.4$. Therefore

$$\begin{aligned}\log \tan A &= \log 67.2 - \log 38.4 \\ &= 1.8274 - 1.5843 = 0.2431\end{aligned}$$

We use the table of Logarithms of Trigonometric Ratios directly to determine th $\angle A = 60° \ 16'$. ■

The law of sines is also well-suited to logarithmic computation.

EXAMPLE: Solve $\triangle ABC$, given $\angle A = 14° \ 20'$, $\angle B = 37° \ 30'$, and $c = 83$ feet.

First, $\angle C = 180° - 14° \ 20' - 37° \ 30' = 128° \ 10'$

Next, $(\sin A)/a = (\sin C)/c$, and so

$$a = \frac{c \sin A}{\sin C} = \frac{83 \sin 14° \ 20'}{\sin 128° \ 10'}$$

Therefore

$$\begin{aligned}\log a &= \log 83 + \log \sin 14° \ 20' - \log \sin 128° \ 10' \\ &= 1.9191 + (0.3937 - 1) - (0.8955 - 1) \\ &= 1.4173\end{aligned}$$

and $a = 26.14$.

Similarly

$$\begin{aligned}b &= \frac{c \sin B}{\sin C} \\ &= \frac{83 \sin 37° \ 30'}{\sin 128° \ 10'}\end{aligned}$$

and

$$\begin{aligned}\log b &= \log 83 + \log \sin 37° \ 30' - \log \sin 128° \ 10' \\ &= 1.9191 + (0.7844 - 1) - (0.8955 - 1) \\ &= 1.8080\end{aligned}$$

and

$b = 64.27$. ■

EXAMPLE: Solve $\triangle ABC$, given $\angle A = 63° \ 20'$, $a = 91.8$, and $c = 94.3$.

From the law of sines, we have

$$\sin C = \frac{c \sin A}{a}$$

$$= \frac{94.3 \sin 63° 20'}{91.8}$$

and so

$$\begin{aligned} \log \sin C &= \log 94.3 + \log \sin 63° 20' - \log 91.8 \\ &= 1.9745 + (0.9512 - 1) - 1.9628 \\ &= 0.9629 - 1 \end{aligned}$$

Thus $\angle C = 66° 40'$ or $\angle C = 180° - 66° 40' = 113° 20'$.

The given data (two sides and the angle opposite one of them) indicate the possibility of ambiguous solution. Each choice for $\angle C$ turns out to be valid, as follows. If $\angle C = 66° 40'$, we have

$$\angle B = 180° - 63° 20' - 66° 40' = 50°$$

and

$$b = \frac{a \sin B}{\sin A} = \frac{91.8 \sin 50°}{\sin 63° 20'}$$

Thus,

$$\begin{aligned} \log b &= \log 91.8 + \log \sin 50° - \log \sin 63° 20' \\ &= 1.9626 + (0.8843 - 1) - (0.9512 - 1) \\ &= 1.8957 \end{aligned}$$

and $b = 78.65$. Hence one solution is $\angle C = 66° 40'$, $\angle B = 50°$, and $b = 78.65$.

The other possibility is $\angle C = 113° 20'$. This gives

$$\angle B = 180° - 63° 20' - 113° 20' = 3° 20'$$

and

$$b = \frac{a \sin B}{\sin A} = \frac{91.8 \sin 3° 20'}{\sin 63° 20'}$$

Thus,

$$\begin{aligned} \log b &= \log 91.8 + \log \sin 3° 20' - \log \sin 63° 20' \\ &= 1.9626 + (0.7645 - 2) - (0.9512 - 1) \\ &= 0.7759 \end{aligned}$$

and $b = 5.969$. Hence the other solution is $\angle C = 113° 20'$, $\angle B = 3° 20'$, and $b = 5.969$. ■

-8 *Exercises*

Use logarithms to solve right triangle $\triangle ABC$ from the given data.

1. $\angle A = 46° 10'$, $a = 23.2$ meters

2. $\angle A = 83° 46'$, $b = 34.27$
3. $\angle B = 38° 27'$, $c = 16.46$ centimeters
4. $b = 196$, $c = 234$
5. $a = 172$ miles, $b = 146$ miles

Find the area of right triangle $\triangle ABC$ from the given data.
6. $\angle A = 41°$, $a = 34$ inches
7. $\angle B = 58° 20'$, $c = 85.3$
8. $a = 12.3$ meters, $b = 3.8$ meters

Solve triangle $\triangle ABC$ from the given data.
9. $\angle A = 53° 10'$, $\angle B = 38° 20'$, $c = 67.5$
10. $\angle B = 28° 42'$, $\angle C = 74° 17'$, $a = 18.5$ centimeters
11. $\angle A = 43° 40'$, $a = 57.2$, $b = 38.9$
12. $\angle B = 52° 14'$, $a = 23.48$, $b = 33.96$
13. $\angle C = 32° 40'$, $b = 11.5$ meters, $c = 8.1$ meters
14. $\angle C = 48° 29'$, $a = 1917$ feet, $c = 1875$ feet
15. $\angle A = 28° 20'$, $a = 29.4$, $c = 74.3$

A-9 Trigonometric formulas for logarithmic computation

As we saw in Chapter 5, when there is enough given data, the solution of triangle is possible by the law of sines and/or the law of cosines. Unfortunatel it is very awkward to use the law of cosines when the computations are to b done by logarithms or on a slide rule. The following formulas are often mo convenient. The proofs of these are somewhat complicated and are omitted.

Our first formula is usually known as the *first law of tangents*.
In any triangle $\triangle ABC$, we have

$$\frac{a-b}{a+b} = \frac{\tan \frac{1}{2}(A-B)}{\tan \frac{1}{2}(A+B)}$$

There are the alternate forms

$$\frac{a-c}{a+c} = \frac{\tan \frac{1}{2}(A-C)}{\tan \frac{1}{2}(A+C)}$$

and

$$\frac{b-c}{b+c} = \frac{\tan \frac{1}{2}(B-C)}{\tan \frac{1}{2}(B+C)}$$

The first law of tangents is most useful when we are given two sides and th angle included between them.

EXAMPLE: Solve triangle $\triangle ABC$, given $a = 43.1$, $b = 26.7$, and $\angle C = 68°$.

We use the first law of tangents, noting that

$$\tfrac{1}{2}(A + B) = \tfrac{1}{2}(180° - C) = \tfrac{1}{2}(180° - 68°) = 56°$$

Therefore

$$\tan \tfrac{1}{2}(A - B) = \frac{(a - b) \tan \frac{1}{2}(A + B)}{a + b}$$

$$= \frac{16.4 \tan 56°}{69.8}$$

Thus,

$$\begin{aligned}\log \tan \tfrac{1}{2}(A - B) &= \log 16.4 + \log \tan 56° - \log 69.8 \\ &= 1.2148 + 0.1710 - 1.8439 \\ &= 0.5419 - 1\end{aligned}$$

and so $\frac{1}{2}(A - B) = 19°\ 12'$.

Now, we solve the system

$$\begin{aligned}\tfrac{1}{2}(A + B) &= 56° \\ \tfrac{1}{2}(A - B) &= 19°\ 12'\end{aligned}$$

to obtain $\angle A = 75°\ 12'$ and $\angle B = 36°\ 48'$.

Finally, we compute c by the law of sines:

$$c = \frac{a \sin C}{\sin A}$$

$$= \frac{43.1 \sin 68°}{\sin 75°\ 12'}$$

and so

$$\begin{aligned}\log c &= \log 43.1 + \log \sin 68° - \log \sin 75°\ 12' \\ &= 1.6345 + (0.9672 - 1) - (0.9854 - 1) \\ &= 1.6163\end{aligned}$$

Hence $c = 41.33$. ■

For our next formula, it is convenient to define the *semiperimeter* s to be $s = \frac{1}{2}(a + b + c)$. We then have the *second law of tangents*:

In any triangle $\triangle ABC$, we have

$$\tan \frac{A}{2} = \sqrt{\frac{(s - b)(s - c)}{s(s - a)}}$$

The alternate forms are

$$\tan\frac{B}{2} = \sqrt{\frac{(s-a)(s-c)}{s(s-b)}}$$

and

$$\tan\frac{C}{2} = \sqrt{\frac{(s-a)(s-b)}{s(s-c)}}$$

The second law of tangents is most useful when we are given the three sic of the triangle.

EXAMPLE: Find $\angle B$, given $a = 14.6$, $b = 23.1$, and $c = 11.3$.

We have $s = \frac{1}{2}(14.6 + 23.1 + 11.3) = 24.5$, and so

$$\begin{aligned}\tan\frac{B}{2} &= \sqrt{\frac{(s-a)(s-c)}{s(s-b)}}\\ &= \sqrt{\frac{(24.5-14.6)(24.5-11.3)}{24.5(24.5-23.1)}}\\ &= \sqrt{\frac{(9.9)\cdot(13.2)}{(24.5)\cdot(1.4)}}\end{aligned}$$

Hence

$$\begin{aligned}\log\tan\tfrac{1}{2}B &= \tfrac{1}{2}(\log 9.9 + \log 13.2 - \log 24.5 - \log 1.4)\\ &= \tfrac{1}{2}(0.9956 + 1.1206 - 1.3892 - 0.1461)\\ &= 0.2905\end{aligned}$$

and so

$$\frac{B}{2} = 62^\circ\ 52'$$

and $\angle B = 125^\circ\ 44'$. ■

There is also a useful formula for the area of a triangle, known as *Heron Formula*. Like the second law of tangents, Heron's formula works best when w know the lengths of all three sides. Once again, s is the semiperimeter.

For any triangle $\triangle ABC$, the area K is given by

$$K = \sqrt{s(s-a)(s-b)(s-c)}$$

EXAMPLE: We calculate the area of the triangle of the previous example b Heron's formula.

We had $a = 14.6$, $b = 23.1$, $c = 11.3$, and $s = 24.5$.

Thus

$$K = \sqrt{(24.5) \cdot (9.9) \cdot (1.4) \cdot (13.2)}$$

and so

$$\begin{aligned}\log K &= \tfrac{1}{2}(\log 24.5 + \log 9.9 + \log 1.4 + \log 13.2)\\ &= \tfrac{1}{2}(1.3892 + 0.9956 + 0.1461 + 1.1206)\\ &= 1.8258\end{aligned}$$

Thus $K = 66.96$. ■

-9 Exercises

Solve $\triangle ABC$ using logarithms and the first law of tangents.

1. $\angle B = 48° \, 30'$, $a = 17.6$, $c = 22.2$
2. $\angle C = 73°$, $a = 11$ miles, $b = 17$ miles
3. $\angle A = 61° \, 21'$, $b = 14.52$ meters, $c = 33.84$ meters

Solve $\triangle ABC$ by the second law of tangents

4. $a = 56$, $b = 48$, $c = 62$
5. $a = 47.32$, $b = 32.21$, $c = 71.14$

Solve $\triangle ABC$ and find the area

6. $\angle C = 62° \, 50'$, $a = 76.9$ meters, $b = 83.4$ meters
7. $a = 129$ miles, $b = 97$ miles, $c = 167$ miles
8. $\angle B = 38° \, 2'$, $a = 45.73$, $c = 65.04$
9. $a = 85.3$ meters, $b = 64.0$ meters, $c = 65.2$ meters

eview exercises (*Appendix A*)

Explain, and illustrate with an example.

1. exponential function
2. logarithmic function
3. characteristic
4. mantissa
5. interpolation

Sketch the graph of the following functions.

6. $y = e^x$
7. $y = \ln x$

8. Find the inverse function f^{-1} if $f(x) = \log_{10} x$.

Express in terms of logarithms, base 10.

9. $\log_{100} 0.01$
10. $\ln \sqrt{5}$

Evaluate each of the following.

11. $\log_{16} 128$
12. log 1,000,000,000
13. log 361.8
14. log 0.000437
15. antilog 2.6160
16. antilog (8.3152 – 10)
17. antilog –1.3582
18. log sin 73° 28′
19. log cot 14° 22′

20. Find the acute angle θ such that $\log \cos \theta = 0.9631 - 1$
21. Find the acute angle θ such that $\log \tan \theta = 0.2100$

Compute the following using logarithms.

22. $\sqrt[5]{2}$
23. $(1.04)^{20}$
24. $\left(\frac{76.2}{49.1}\right)^2$
25. $\left(\frac{47.23 - 25.14}{13.37}\right)^{1/3}$
26. ln 2

27. A certain radioactive element decays in the following way. Suppose at ti $t = 0$ there is one unit of this element present. Then, at a later time t, the amou present will be e^{-3t} units, where t is measured in years.

(a) At what time t will exactly half the original amount remain (this val of t is called the *half-life* of the element)?

(b) What portion of the original amount will remain after 5 years?

28. It is estimated that the population of a certain fast-growing area is n increasing at the rate of 10% per year. This means that $P = P_0 e^{.1t}$, where P is t population at time t, P_0 is the current population, and t is time measured in yea If the present population is 100,000, and if growth continues at the current ra what will the population be in 20 years?

Use logarithms to solve $\triangle ABC$ and find its area.

29. $\angle A = 40° 10'$, $\angle B = 61° 30'$, $c = 54.3$ yards
30. $\angle A = 50° 20'$, $b = 35.8$ meters, $c = 47.0$ meters
31. $\angle A = 46°$, $a = 41$, $b = 49$
32. $a = 90.6$ centimeters, $b = 72.0$ centimeters, $c = 70.1$ centimeters.

B

Tables and formulas

the next few pages, we offer a recapitulation of the important formulas given in the text, as well as formulas from algebra. We also include the table of Trigonometric Ratios and the table of Logarithms of Trigonometric Ratios.

mbers

$N =$ the set of natural numbers $= \{1, 2, 3, \ldots\}$
$J =$ the set of integers $= \{0, \pm 1, \pm 2, \pm 3, \ldots\}$
$Q =$ the set of rational numbers $= \{a/b \mid a, b \in \text{J}, b \neq 0\}$
$R =$ the set of real numbers
$C =$ the set of complex numbers $= \{a + bi \mid a, b \in R\}$

ts

$A \cup B = \{x \mid x \in A \text{ or } x \in B\}$
$A \cap B = \{x \mid x \in A \text{ and } x \in B\}$

Constants

$$\pi = 3.1415926536 \qquad \text{(approximately)}$$
$$e = 2.7182818285 \qquad \text{(approximately)}$$
$$i = \sqrt{-1}$$

Algebra

$$\left.\begin{aligned} a + b &= b + a \\ ab &= ba \end{aligned}\right\}\text{commutative laws}$$

$$\left.\begin{aligned} a + (b + c) &= (a + b) + c \\ a(bc) &= (ab)c \end{aligned}\right\}\text{associative laws}$$

$$\left.\begin{aligned} a(b + c) &= ab + ac \\ (a + b)c &= ac + bc \end{aligned}\right\}\text{distributive laws}$$

$$a - b = a + (-b)$$
$$a - (b - c) = a - b + c$$
$$(-a)b = a(-b) = -ab$$
$$(-a)(-b) = ab$$

Fractions (*Reminder:* Division by zero is not defined.)

$$a/b = a(1/b)$$

$$\frac{a}{b} = \frac{c}{d} \quad \text{if and only if } ad = bc$$

$$\frac{ac}{bc} = \frac{a}{b}$$

$$\frac{-a}{-b} = \frac{a}{b}$$

$$\frac{-a}{b} = \frac{a}{-b} = -\frac{a}{b}$$

$$\frac{a}{b} \cdot \frac{c}{d} = \frac{ac}{bd}$$

$$\frac{a}{b} \div \frac{c}{d} = \frac{ad}{bc}$$

$$\frac{a}{c} + \frac{b}{c} = \frac{a + b}{c}$$

$$\frac{a}{b} + \frac{c}{d} = \frac{ad + bc}{db}$$

oducts and factoring

$$(x + y)^2 = x^2 + 2xy + y^2$$
$$(x - y)^2 = x^2 - 2xy + y^2$$
$$(x + y)(x - y) = x^2 - y^2$$
$$(x + y)(x^2 - xy + y^2) = x^3 + y^3$$
$$(x - y)(x^2 + xy + y^2) = x^3 - y^3$$

xponents and radicals (for positive numbers a and b)

$$a^0 = 1$$

$$a^{-1} = 1/a$$

$$a^{-n} = 1/a^n$$

$$a^{1/n} = \sqrt[n]{a}$$

$$a^{n/m} = (\sqrt[m]{a})^n = \sqrt[m]{a^n}$$

$$a^n a^m = a^{n+m}$$

$$(a^n)^m = a^{nm}$$

$$(ab)^n = a^n b^n$$

$$\frac{a^n}{a^m} = a^{n-m}$$

$$\left(\frac{a}{b}\right)^n = \frac{a^n}{b^n}$$

$$\sqrt{ab} = \sqrt{a}\sqrt{b}$$

$$\sqrt[n]{ab} = \sqrt[n]{a}\,\sqrt[n]{b}$$

bsolute value and distance

$$|a| = \sqrt{a^2} = \begin{cases} a \text{ if } a \geq 0 \\ -a \text{ if } a < 0 \end{cases}$$

$|a - b|$ = distance between a and b on the number line

$d = \sqrt{x^2 + y^2}$ = distance between (x, y) and the origin in the plane

$d = \sqrt{(x_2 - x_1)^2 + (y_2 - y_1)^2}$

$= \sqrt{(x_1 - x_2)^2 + (y_1 - y_2)^2}$

= distance between (x_1, y_1) and (x_2, y_2) in the plane

Quadratic expressions

The discriminant of the quadratic trinomial $ax^2 + bx + c$ is $b^2 - 4ac$. If a, b, an are integers, the trinomial can be factored into linear factors with integer coefficie if and only if the discriminant is a perfect square.

The solutions of the quadratic equation

$$ax^2 + bx + c = 0$$

are given by

$$x = \frac{-b \pm \sqrt{b^2 - 4ac}}{2a}$$

There are two real solutions if the discriminant is positive, one real solution if discriminant is zero, and two conjugate imaginary solutions if the discriminant negative. If a, b, and c are integers, the solutions are rational numbers if and only the discriminant is a perfect square.

Logarithms

If a is any positive number except 1, the function $\log_a$ defined by $y = \log_a x$ is t inverse function to the exponential function defined by $y = a^x$.

$$y = \log_a x \quad \text{if and only if } x = a^y$$

$$\log_a a^n = n$$

$$\log_a xy = \log_a x + \log_a y$$

$$\log_a x^c = c \log_a x$$

$$\log_a \frac{x}{y} = \log_a x - \log_a y$$

$$\log_a x = \frac{\log_b x}{\log_b a}$$

Right triangles

These formulas are valid for any right triangle $\triangle ABC$ with area K. The right ang is at $\angle C$.

Pythagorean theorem

$$c^2 = a^2 + b^2$$

igonometric ratios

$\sin A = a/c$
$\cos A = b/c$
$\tan A = a/b$
$\cot A = b/a$
$\sec A = c/b$
$\csc A = c/a$
$\sin B = \cos A = b/c$
$\cos B = \sin A = a/c$
$\tan B = \cot A = b/a$
$\cot B = \tan A = a/b$
$\sec B = \csc A = c/a$
$\csc B = \sec A = c/b$

rea

$$K = \frac{ab}{2} = \frac{bc}{2}\sin A = \frac{ac}{2}\sin B = \frac{c^2}{2}\sin A \cos A$$

$$= \frac{c^2}{2}\sin B \cos B = \frac{c^2}{2}\sin A \sin B = \frac{c^2}{2}\cos A \cos B$$

ircles

The graph of the equation

$x^2 + y^2 = r^2$

is a circle with radius r and center the origin.

Area of a circle of radius r

$A = \pi r^2$

Circumference of a circle of radius r

$C = 2\pi r$

ngle measurement

1 revolution $= 360° = 2\pi$ radians
1 radian $= (180/\pi)° = 1/2\pi$ revolutions
$1° = \pi/180$ radians $= 1/360$ revolution
$1° = 60'$
$1' = 60''$

Trigonometric ratios

The angle θ is in standard position. The point $P = (x, y)$ is on the terminal r
The distance from the origin to P is $r = \sqrt{x^2 + y^2}$

$\sin \theta = y/r$
$\cos \theta = x/r$
$\tan \theta = y/x,\ x \neq 0$
$\cot \theta = x/y,\ y \neq 0$
$\sec \theta = r/x,\ x \neq 0$
$\csc \theta = r/y,\ y \neq 0$

Signs of trigonometric ratios

Quadrant	I	II	III	IV
sin	+	+	−	−
cos	+	−	−	+
tan	+	−	+	−
cot	+	−	+	−
sec	+	−	−	+
csc	+	+	−	−

Reference angle

The reference angle θ^* is the acute angle between the x-axis and the termi
ray of the angle θ in standard position.

Quadrant	θ^*(radians)	θ^*(degrees)
I	θ	θ
II	$\pi - \theta$	$180° - \theta$
III	$\theta - \pi$	$\theta - 180°$
IV	$2\pi - \theta$	$360° - \theta$

The trigonometric point

The trigonometric point $T(\theta)$ is the point of intersection of the unit circle with th
terminal ray of the angle θ in standard position.

$T(\theta) = (x, y)$ if and only if $x = \cos \theta$ and $y = \sin \theta$

rigonometric identities

implification identities

$$\tan x = \frac{\sin x}{\cos x}$$

$$\cot x = \frac{\cos x}{\sin x}$$

$$\sec x = \frac{1}{\cos x}$$

$$\csc x = \frac{1}{\sin x}$$

ythagorean identities

$$\sin^2 x + \cos^2 x = 1$$
$$1 + \tan^2 x = \sec^2 x$$
$$1 + \cot^2 x = \csc^2 x$$

ddition/subtraction identities

$$\sin (x + y) = \sin x \cos y + \cos x \sin y$$
$$\sin (x - y) = \sin x \cos y - \cos x \sin y$$
$$\cos (x + y) = \cos x \cos y - \sin x \sin y$$
$$\cos (x - y) = \cos x \cos y + \sin x \sin y$$

$$\tan (x + y) = \frac{\tan x + \tan y}{1 - \tan x \tan y}$$

$$\tan (x - y) = \frac{\tan x - \tan y}{1 + \tan x \tan y}$$

eriodicity identities

$$\sin (x + 2\pi) = \sin x$$
$$\cos (x + 2\pi) = \cos x$$
$$\tan (x + \pi) = \tan x$$
$$\cot (x + \pi) = \cot x$$
$$\sec (x + 2\pi) = \sec x$$
$$\csc (x + 2\pi) = \csc x$$
$$T(\theta + 2\pi) = T(\theta)$$

Cofunction identities

$$\sin\left(\frac{\pi}{2} - x\right) = \cos x$$

$$\cos\left(\frac{\pi}{2} - x\right) = \sin x$$

$$\tan\left(\frac{\pi}{2} - x\right) = \cot x$$

$$\cot\left(\frac{\pi}{2} - x\right) = \tan x$$

$$\sec\left(\frac{\pi}{2} - x\right) = \csc x$$

$$\csc\left(\frac{\pi}{2} - x\right) = \sec x$$

Double-angle identities

$$\sin 2x = 2 \sin x \cos x$$

$$\cos 2x = \cos^2 x - \sin^2 x = 2 \cos^2 x - 1 = 1 - 2 \sin^2 x$$

$$\tan 2x = \frac{2 \tan x}{1 - \tan^2 x}$$

Half-angle identities

$$\sin \tfrac{1}{2}x = \pm \sqrt{\tfrac{1}{2}(1 - \cos x)}$$

$$\cos \tfrac{1}{2}x = \pm \sqrt{\tfrac{1}{2}(1 + \cos x)}$$

$$\tan \tfrac{1}{2}x = \pm \sqrt{(1 - \cos x)/(1 + \cos x)}$$

$$= \frac{1 - \cos x}{\sin x} = \frac{\sin x}{1 + \cos x}$$

Product-to-sum identities

$$\sin x \cos y = \tfrac{1}{2}[\sin (x + y) + \sin (x - y)]$$
$$\cos x \sin y = \tfrac{1}{2}[\sin (x + y) - \sin (x - y)]$$
$$\cos x \cos y = \tfrac{1}{2}[\cos (x + y) + \cos (x - y)]$$
$$\sin x \sin y = -\tfrac{1}{2}[\cos (x + y) - \cos (x - y)]$$

um-to-product identities

$$\sin x + \sin y = 2 \sin \tfrac{1}{2}(x + y) \cos \tfrac{1}{2}(x - y)$$
$$\sin x - \sin y = 2 \cos \tfrac{1}{2}(x + y) \sin \tfrac{1}{2}(x - y)$$
$$\cos x + \cos y = 2 \cos \tfrac{1}{2}(x + y) \cos \tfrac{1}{2}(x - y)$$
$$\cos x - \cos y = -2 \sin \tfrac{1}{2}(x + y) \sin \tfrac{1}{2}(x - y)$$

liscellaneous identities

$$\sin(-x) = \sin(2\pi - x) = -\sin x$$
$$\cos(-x) = \cos(2\pi - x) = \cos x$$
$$\tan(-x) = \tan(2\pi - x) = -\tan x$$
$$\sin(\pi - x) = \sin x$$
$$\cos(\pi - x) = -\cos x$$
$$\tan(\pi - x) = -\tan x$$
$$\sin(x - \pi) = -\sin x$$
$$\cos(x - \pi) = -\cos x$$
$$\tan(x - \pi) = \tan x$$
$$\sin^2 x = \tfrac{1}{2}(1 - \cos 2x)$$
$$\cos^2 x = \tfrac{1}{2}(1 + \cos 2x)$$

riangles

These formulas are valid for any triangle $\triangle ABC$. The area is denoted by K, and $s = \frac{1}{2}(a + b + c)$.

Sum of angles

$$\angle A + \angle B + \angle C = 180^\circ$$

Length of sides

$$a + b > c$$
$$a + c > b$$
$$b + c > a$$

aw of cosines

$$a^2 = b^2 + c^2 - 2bc \cos A$$
$$b^2 = a^2 + c^2 - 2ac \cos B$$
$$c^2 = a^2 + b^2 - 2ab \cos C$$

aw of sines

$$\frac{\sin A}{a} = \frac{\sin B}{b} = \frac{\sin C}{c}$$

$$\frac{a}{\sin A} = \frac{b}{\sin B} = \frac{c}{\sin C}$$

First law of tangents

$$\frac{a-b}{a+b} = \frac{\tan \frac{1}{2}(A-B)}{\tan \frac{1}{2}(A+B)}$$

$$\frac{a-c}{a+c} = \frac{\tan \frac{1}{2}(A-C)}{\tan \frac{1}{2}(A+C)}$$

$$\frac{b-c}{b+c} = \frac{\tan \frac{1}{2}(B-C)}{\tan \frac{1}{2}(B+C)}$$

Second law of tangents

$$\tan \tfrac{1}{2}A = \sqrt{\frac{(s-b)(s-c)}{s(s-a)}}$$

$$\tan \tfrac{1}{2}B = \sqrt{\frac{(s-a)(s-c)}{s(s-b)}}$$

$$\tan \tfrac{1}{2}C = \sqrt{\frac{(s-a)(s-b)}{s(s-c)}}$$

Area

$$\begin{aligned} K &= \tfrac{1}{2}(\text{base}) \cdot (\text{altitude}) \\ &= \tfrac{1}{2}bc \sin A = \tfrac{1}{2}ab \sin C = \tfrac{1}{2}ac \sin B \\ &= \tfrac{1}{2}c^2 \frac{\sin A \sin B}{\sin C} = \tfrac{1}{2}b^2 \frac{\sin A \sin C}{\sin B} \\ &= \tfrac{1}{2}a^2 \frac{\sin B \sin C}{\sin A} \\ &= \sqrt{s(s-a)(s-b)(s-c)} \end{aligned}$$

Sinusoidal functions

An equation of form $y = a \sin b(x - c)$ defines a sinusoidal function. The graph is a sine wave with amplitude $|a|$, period $2\pi/b$, and phase shift c.

A function defined by an equation of form $y = a \cos b(x - c)$ is also a sinusoidal function. The graph is a sine wave with amplitude $|a|$, period $2\pi/b$, and phase shift $c - \pi/2b$.

Complex numbers

Rectangular form

$$(a + bi) + (c + di) = (a + c) + (b + d)i$$
$$(a + bi)(c + di) = (ac - bd) + (ad + bc)i$$
$$i^2 = -1$$
$$i^{-1} = -i$$
$$i = \sqrt{-1}$$
$$\sqrt{-a} = i\sqrt{a} \qquad \text{if } a \text{ is a positive real number}$$
$$\frac{1}{a+bi} = \frac{a}{a^2+b^2} + \frac{b}{a^2+b^2}i \qquad \text{if } a + bi \neq 0$$
$$|a + bi| = \sqrt{a^2 + b^2}$$
$$= \text{the distance from the origin to } a + bi \text{ in the complex plane}$$
$$(a + bi)(a - bi) = |a + bi|^2 = a^2 + b^2$$

Trigonometric form

$$\text{cis } \theta = \cos \theta + i \sin \theta$$
$$(r_1 \text{ cis } \theta_1)(r_2 \text{ cis } \theta_2) = r_1 r_2 \text{ cis } (\theta_1 + \theta_2)$$
$$(r_1 \text{ cis } \theta_1) \div (r_2 \text{ cis } \theta_2) = \frac{r_1}{r_2} \text{ cis } (\theta_1 - \theta_2)$$
$$(r \text{ cis } \theta)^n = r^n \text{ cis } n\theta, \qquad n \in J$$

If $n \in N$ and $r > 0$, the nth roots of r cis θ are:

$$\sqrt[n]{r} \text{ cis } \frac{\theta}{n}$$
$$\sqrt[n]{r} \text{ cis } \frac{\theta + 2\pi}{n}$$
$$\vdots$$
$$\sqrt[n]{r} \text{ cis } \frac{\theta + (n-1)2\pi}{n}$$

Conversion between rectangular and trigonometric forms

The rectangular form for r cis θ is $a + bi$, where:

$$a = r \cos \theta$$
$$b = r \sin \theta$$

The trigonometric form for $a + bi$ is $r \text{ cis } \theta$, where:

$$r = |a + bi| = \sqrt{a^2 + b^2}$$

$$\sin \theta = \frac{b}{r}$$

$$\cos \theta = \frac{a}{r}$$

$$\tan \theta = \frac{b}{a}$$

Inverse trigonometric functions

$y = \text{Arcsin } x$ if and only if $x = \sin y$ and $-\frac{\pi}{2} \le y \le \frac{\pi}{2}$

$y = \text{Arccos } x$ if and only if $x = \cos y$ and $0 \le y \le \pi$

$y = \text{Arctan } x$ if and only if $x = \tan y$ and $-\frac{\pi}{2} < y < \frac{\pi}{2}$

$y = \text{Arccot } x$ if and only if $x = \cot y$ and $0 < y < \pi$

$y = \text{Arcsec } x$ if and only if $x = \sec y$ and $0 \le y \le \pi$

$y = \text{Arccsc } x$ if and only if $x = \csc y$ and $-\frac{\pi}{2} \le y \le \frac{\pi}{2}$

rigonometric Ratios

Degrees	Radians	sin	cos	tan	cot	sec	csc		
0° 00′	.0000	.0000	1.0000	.0000	—	1.000	–	1.5708	90° 00′
10	.0029	.0029	1.0000	.0029	343.8	1.000	343.8	1.5679	50
20	.0058	.0058	1.0000	.0058	171.9	1.000	171.9	1.5650	40
30	.0087	.0087	1.0000	.0087	114.6	1.000	114.6	1.5621	30
40	.0116	.0116	.9999	.0116	85.94	1.000	85.95	1.5592	20
50	.0145	.0145	.9999	.0145	68.75	1.000	68.76	1.5563	10
1° 00′	.0175	.0175	.9998	.0175	57.29	1.000	57.30	1.5533	89° 00′
10	.0204	.0204	.9998	.0204	49.10	1.000	49.11	1.5504	50
20	.0233	.0233	.9997	.0233	42.96	1.000	42.98	1.5475	40
30	.0262	.0262	.9997	.0262	38.19	1.000	38.20	1.5446	30
40	.0291	.0291	.9996	.0291	34.37	1.000	34.38	1.5417	20
50	.0320	.0320	.9995	.0320	31.24	1.001	31.26	1.5388	10
2° 00′	.0349	.0349	.9994	.0349	28.64	1.001	28.65	1.5359	88° 00′
10	.0378	.0378	.9993	.0378	26.43	1.001	26.45	1.5330	50
20	.0407	.0407	.9992	.0407	24.54	1.001	24.56	1.5301	40
30	.0436	.0436	.9990	.0437	22.90	1.001	22.93	1.5272	30
40	.0465	.0465	.9989	.0466	21.47	1.001	21.49	1.5243	20
50	.0495	.0494	.9988	.0495	20.21	1.001	20.23	1.5213	10
3° 00′	.0524	.0523	.9986	.0524	19.08	1.001	19.11	1.5184	87° 00′
10	.0553	.0552	.9985	.0553	18.07	1.002	18.10	1.5155	50
20	.0582	.0581	.9983	.0582	17.17	1.002	17.20	1.5126	40
30	.0611	.0610	.9981	.0612	16.35	1.002	16.38	1.5097	30
40	.0640	.0640	.9980	.0641	15.60	1.002	15.64	1.5068	20
50	.0669	.0669	.9978	.0670	14.92	1.002	14.96	1.5039	10
4° 00′	.0698	.0698	.9976	.0699	14.30	1.002	14.34	1.5010	86° 00′
10	.0727	.0727	.9974	.0729	13.73	1.003	13.76	1.4981	50
20	.0756	.0756	.9971	.0758	13.20	1.003	13.23	1.4952	40
30	.0785	.0785	.9969	.0787	12.71	1.003	12.75	1.4923	30
40	.0814	.0814	.9967	.0816	12.25	1.003	12.29	1.4893	20
50	.0844	.0843	.9964	.0846	11.83	1.004	11.87	1.4864	10
5° 00′	.0873	.0872	.9962	.0875	11.43	1.004	11.47	1.4835	85° 00′
10	.0902	.0901	.9959	.0904	11.06	1.004	11.10	1.4806	50
20	.0931	.0929	.9957	.0934	10.71	1.004	10.76	1.4777	40
30	.0960	.0958	.9954	.0963	10.39	1.005	10.43	1.4748	30
40	.0989	.0987	.9951	.0992	10.08	1.005	10.13	1.4719	20
50	.1018	.1016	.9948	.1022	9.788	1.005	9.839	1.4690	84° 10′
		cos	sin	cot	tan	csc	sec	Radians	Degrees

Trigonometric Ratios (*Continued*)

Degrees	Radians	sin	cos	tan	cot	sec	csc		
6° 00′	.1047	.1045	.9945	.1051	9.514	1.006	9.567	1.4661	84° 00′
10	.1076	.1074	.9942	.1080	9.255	1.006	9.309	1.4632	50
20	.1105	.1103	.9939	.1110	9.010	1.006	9.065	1.4603	40
30	.1134	.1132	.9936	.1139	8.777	1.006	8.834	1.4573	30
40	.1164	.1161	.9932	.1169	8.556	1.007	8.614	1.4544	20
50	.1193	.1190	.9929	.1198	8.345	1.007	8.405	1.4515	10
7° 00′	.1222	.1219	.9925	.1228	8.144	1.008	8.206	1.4486	83° 00′
10	.1251	.1248	.9922	.1257	7.953	1.008	8.016	1.4457	50
20	.1280	.1276	.9918	.1287	7.770	1.008	7.834	1.4428	40
30	.1309	.1305	.9914	.1317	7.596	1.009	7.661	1.4399	30
40	.1338	.1334	.9911	.1346	7.429	1.009	7.496	1.4370	20
50	.1367	.1363	.9907	.1376	7.269	1.009	7.337	1.4341	10
8° 00′	.1396	.1392	.9903	.1405	7.115	1.010	7.185	1.4312	82° 00′
10	.1425	.1421	.9899	.1435	6.968	1.010	7.040	1.4283	50
20	.1454	.1449	.9894	.1465	6.827	1.011	6.900	1.4254	40
30	.1484	.1478	.9890	.1495	6.691	1.011	6.765	1.4224	30
40	.1513	.1507	.9886	.1524	6.561	1.012	6.636	1.4195	20
50	.1542	.1536	.9881	.1554	6.435	1.012	6.512	1.4166	10
9° 00′	.1571	.1564	.9877	.1584	6.314	1.012	6.392	1.4137	81° 00′
10	.1600	.1593	.9872	.1614	6.197	1.013	6.277	1.4108	50
20	.1629	.1622	.9868	.1644	6.084	1.013	6.166	1.4079	40
30	.1658	.1650	.9863	.1673	5.976	1.014	6.059	1.4050	30
40	.1687	.1679	.9858	.1703	5.871	1.014	5.955	1.4021	20
50	.1716	.1708	.9853	.1733	5.769	1.015	5.855	1.3992	10
10° 00′	.1745	.1736	.9848	.1763	5.671	1.015	5.759	1.3963	80° 00′
10	.1774	.1765	.9843	.1793	5.576	1.016	5.665	1.3934	50
20	.1804	.1794	.9838	.1823	5.485	1.016	5.575	1.3904	40
30	.1833	.1822	.9833	.1853	5.396	1.107	5.487	1.3875	30
40	.1862	.1851	.9827	.1883	5.309	1.018	5.403	1.3846	20
50	.1891	.1880	.9822	.1914	5.226	1.018	5.320	1.3817	10
11° 00′	.1920	.1908	.9816	.1944	5.145	1.019	5.241	1.3788	79° 00′
10	.1949	.1937	.9811	.1974	5.066	1.019	5.164	1.3759	50
20	.1978	.1965	.9805	.2004	4.989	1.020	5.089	1.3730	40
30	.2007	.1994	.9799	.2035	4.915	1.020	5.016	1.3701	30
40	.2036	.2022	.9793	.2065	4.843	1.021	4.945	1.3672	20
50	.2065	.2051	.9787	.2095	4.773	1.022	4.876	1.3643	78° 10′
		cos	sin	cot	tan	csc	sec	Radians	Degrees

igonometric Ratios (*Continued*)

Degrees	Radians	sin	cos	tan	cot	sec	csc		
12° 00′	.2094	.2079	.9781	.2126	4.705	1.022	4.810	1.3614	78° 00′
10	.2123	.2108	.9775	.2156	4.638	1.023	4.745	1.3584	50
20	.2153	.2136	.9769	.2186	4.574	1.024	4.682	1.3555	40
30	.2182	.2164	.9763	.2217	4.511	1.024	4.620	1.3526	30
40	.2211	.2193	.9757	.2247	4.449	1.025	4.560	1.3497	20
50	.2240	.2221	.9750	.2278	4.390	1.026	4.502	1.3468	10
13° 00′	.2269	.2250	.9744	.2309	4.331	1.026	4.445	1.3439	77° 00′
10	.2298	.2278	.9737	.2339	4.275	1.027	4.390	1.3410	50
20	.2327	.2306	.9730	.2370	4.219	1.028	4.336	1.3381	40
30	.2356	.2334	.9724	.2401	4.165	1.028	4.284	1.3352	30
40	.2385	.2363	.9717	.2432	4.113	1.029	4.232	1.3323	20
50	.2414	.2391	.9710	.2462	4.061	1.030	4.182	1.3294	10
14° 00′	.2443	.2419	.9703	.2493	4.011	1.031	4.134	1.3265	76° 00′
10	.2473	.2447	.9696	.2524	3.962	1.031	4.086	1.3235	50
20	.2502	.2476	.9689	.2555	3.914	1.032	4.039	1.3206	40
30	.2531	.2504	.9681	.2586	3.867	1.033	3.994	1.3177	30
40	.2560	.2532	.9674	.2617	3.821	1.034	3.950	1.3148	20
50	.2589	.2560	.9667	.2648	3.776	1.034	3.906	1.3119	10
15° 00′	.2618	.2588	.9659	.2679	3.732	1.035	3.864	1.3090	75° 00′
10	.2647	.2616	.9652	.2711	3.689	1.036	3.822	1.3061	50
20	.2676	.2644	.9644	.2742	3.647	1.037	3.782	1.3032	40
30	.2705	.2672	.9636	.2773	3.606	1.038	3.742	1.3003	30
40	.2734	.2700	.9628	.2805	3.566	1.039	3.703	1.2974	20
50	.2763	.2728	.9621	.2836	3.526	1.039	3.665	1.2945	10
16° 00′	.2793	.2756	.9613	.2867	3.487	1.040	3.628	1.2915	74° 00′
10	.2822	.2784	.9605	.2899	3.450	1.041	3.592	1.2886	50
20	.2851	.2812	.9596	.2931	3.412	1.042	3.556	1.2857	40
30	.2880	.2840	.9588	.2962	3.376	1.043	3.521	1.2828	30
40	.2909	.2868	.9580	.2994	3.340	1.044	3.487	1.2799	20
50	.2938	.2896	.9572	.3026	3.305	1.045	3.453	1.2770	10
17° 00′	.2967	.2924	.9563	.3057	3.271	1.046	3.420	1.2741	73° 00′
10	.2996	.2952	.9555	.3089	3.237	1.047	3.388	1.2712	50
20	.3025	.2979	.9546	.3121	3.204	1.048	3.356	1.2683	40
30	.3054	.3007	.9537	.3153	3.172	1.049	3.326	1.2654	30
40	.3083	.3035	.9528	.3185	3.140	1.049	3.295	1.2625	20
50	.3113	.3062	.9520	.3217	3.108	1.050	3.265	1.2595	72° 10′
		cos	sin	cot	tan	csc	sec	Radians	Degrees

Trigonometric Ratios *(Continued)*

Degrees	Radians	sin	cos	tan	cot	sec	csc		
18° 00′	.3142	.3090	.9511	.3249	3.078	1.051	3.236	1.2566	72° 00′
10	.3171	.3118	.9502	.3281	3.047	1.052	3.207	1.2537	50
20	.3200	.3145	.9492	.3314	3.018	1.053	3.179	1.2508	40
30	.3229	.3173	.9483	.3346	2.989	1.054	3.152	1.2479	30
40	.3258	.3201	.9474	.3378	2.960	1.056	3.124	1.2450	20
50	.3287	.3228	.9465	.3411	2.932	1.057	3.098	1.2421	10
19° 00′	.3316	.3256	.9455	.3443	2.904	1.058	3.072	1.2392	71° 00′
10	.3345	.3283	.9446	.3476	2.877	1.059	3.046	1.2363	50
20	.3374	.3311	.9436	.3508	2.850	1.060	3.021	1.2334	40
30	.3403	.3338	.9426	.3541	2.824	1.061	2.996	1.2305	30
40	.3432	.3365	.9417	.3574	2.798	1.062	2.971	1.2275	20
50	.3462	.3393	.9407	.3607	2.773	1.063	2.947	1.2246	10
20° 00′	.3491	.3420	.9397	.3640	2.747	1.064	2.924	1.2217	70° 00′
10	.3520	.3448	.9387	.3673	2.723	1.065	2.901	1.2188	50
20	.3549	.3475	.9377	.3706	2.699	1.066	2.878	1.2159	40
30	.3578	.3502	.9367	.3739	2.675	1.068	2.855	1.2130	30
40	.3607	.3529	.9356	.3772	2.651	1.069	2.833	1.2101	20
50	.3636	.3557	.9346	.3805	2.628	1.070	2.812	1.2072	10
21° 00′	.3665	.3584	.9336	.3839	2.605	1.071	2.790	1.2043	69° 00′
10	.3694	.3611	.9325	.3872	2.583	1.072	2.769	1.2014	50
20	.3723	.3638	.9315	.3906	2.560	1.074	2.749	1.1985	40
30	.3752	.3665	.9304	.3939	2.539	1.075	2.729	1.1956	30
40	.3782	.3692	.9293	.3973	2.517	1.076	2.709	1.1926	20
50	.3811	.3719	.9283	.4006	2.496	1.077	2.689	1.1897	10
22° 00′	.3840	.3746	.9272	.4040	2.475	1.079	2.669	1.1868	68° 00′
10	.3869	.3773	.9261	.4074	2.455	1.080	2.650	1.1839	50
20	.3898	.3800	.9250	.4108	2.434	1.081	2.632	1.1810	40
30	.3927	.3827	.9239	.4142	2.414	1.082	2.613	1.1781	30
40	.3956	.3854	.9228	.4176	2.394	1.084	2.595	1.1752	20
50	.3985	.3881	.9216	.4210	2.375	1.085	2.577	1.1723	10
23° 00′	.4014	.3907	.9205	.4245	2.356	1.086	2.559	1.1694	67° 00′
10	.4043	.3934	.9194	.4279	2.337	1.088	2.542	1.1665	50
20	.4072	.3961	.9182	.4314	2.318	1.089	2.525	1.1636	40
30	.4102	.3987	.9171	.4348	2.300	1.090	2.508	1.1606	30
40	.4131	.4014	.9159	.4383	2.282	1.092	2.491	1.1577	20
50	.4160	.4041	.9147	.4417	2.264	1.093	2.475	1.1548	66° 10′
		cos	sin	cot	tan	csc	sec	Radians	Degrees

igonometric Ratios (*Continued*)

Degrees	Radians	sin	cos	tan	cot	sec	csc		
24° 00′	.4189	.4067	.9135	.4452	2.246	1.095	2.459	1.1519	66° 00′
10	.4218	.4094	.9124	.4487	2.229	1.096	2.443	1.1490	50
20	.4247	.4120	.9112	.4522	2.211	1.097	2.427	1.1461	40
30	.4276	.4147	.9100	.4557	2.194	1.099	2.411	1.1432	30
40	.4305	.4173	.9088	.4592	2.177	1.100	2.396	1.1403	20
50	.4334	.4200	.9075	.4628	2.161	1.102	2.381	1.1374	10
25° 00′	.4363	.4226	.9063	.4663	2.145	1.103	2.366	1.1345	65° 00′
10	.4392	.4253	.9051	.4699	2.128	1.105	2.352	1.1316	50
20	.4422	.4279	.9038	.4734	2.112	1.106	2.337	1.1286	40
30	.4451	.4305	.9026	.4770	2.097	1.108	2.323	1.1257	30
40	.4480	.4331	.9013	.4806	2.081	1.109	2.309	1.1228	20
50	.4509	.4358	.9001	.4841	2.066	1.111	2.295	1.1199	10
26° 00′	.4538	.4384	.8988	.4877	2.050	1.113	2.281	1.1170	64° 00′
10	.4567	.4410	.8975	.4913	2.035	1.114	2.268	1.1141	50
20	.4596	.4436	.8962	.4950	2.020	1.116	2.254	1.1112	40
30	.4625	.4462	.8949	.4986	2.006	1.117	2.241	1.1083	30
40	.4654	.4488	.8936	.5022	1.991	1.119	2.228	1.1054	20
50	.4683	.4514	.8923	.5059	1.977	1.121	2.215	1.1025	10
27° 00′	.4712	.4540	.8910	.5095	1.963	1.122	2.203	1.0996	63° 00′
10	.4741	.4566	.8897	.5132	1.949	1.124	2.190	1.0966	50
20	.4771	.4592	.8884	.5169	1.935	1.126	2.178	1.0937	40
30	.4800	.4617	.8870	.5206	1.921	1.127	2.166	1.0908	30
40	.4829	.4643	.8857	.5243	1.907	1.129	2.154	1.0879	20
50	.4858	.4669	.8843	.5280	1.894	1.131	2.142	1.0850	10
28° 00′	.4887	.4695	.8829	.5317	1.881	1.133	2.130	1.0821	62° 00′
10	.4916	.4720	.8816	.5354	1.868	1.134	2.118	1.0792	50
20	.4945	.4746	.8802	.5392	1.855	1.136	2.107	1.0763	40
30	.4974	.4772	.8788	.5430	1.842	1.138	2.096	1.0734	30
40	.5003	.4797	.8774	.5467	1.829	1.140	2.085	1.0705	20
50	.5032	.4823	.8760	.5505	1.816	1.142	2.074	1.0676	10
29° 00′	.5061	.4848	.8746	.5543	1.804	1.143	2.063	1.0647	61° 00′
10	.5091	.4874	.8732	.5581	1.792	1.145	2.052	1.0617	50
20	.5120	.4899	.8718	.5619	1.780	1.147	2.041	1.0588	40
30	.5149	.4924	.8704	.5658	1.767	1.149	2.031	1.0559	30
40	.5178	.4950	.8689	.5696	1.756	1.151	2.020	1.0530	20
50	.5207	.4975	.8675	.5735	1.744	1.153	2.010	1.0501	60° 10′
		cos	sin	cot	tan	csc	sec	Radians	Degrees

Trigonometric Ratios (*Continued*)

Degrees	Radians	sin	cos	tan	cot	sec	csc		
30° 00′	.5236	.5000	.8660	.5774	1.732	1.155	2.000	1.0472	60° 00′
10	.5265	.5025	.8646	.5812	1.720	1.157	1.990	1.0443	50
20	.5294	.5050	.8631	.5851	1.709	1.159	1.980	1.0414	40
30	.5323	.5075	.8616	.5890	1.698	1.161	1.970	1.0385	30
40	.5352	.5100	.8601	.5930	1.686	1.163	1.961	1.0356	20
50	.5381	.5125	.8587	.5969	1.675	1.165	1.951	1.0327	10
31° 00′	.5411	.5150	.8572	.6009	1.664	1.167	1.942	1.0297	59° 00′
10	.5440	.5175	.8557	.6048	1.653	1.169	1.932	1.0268	50
20	.5469	.5200	.8542	.6088	1.643	1.171	1.923	1.0239	40
30	.5498	.5225	.8526	.6128	1.632	1.173	1.914	1.0210	30
40	.5527	.5250	.8511	.6168	1.621	1.175	1.905	1.0181	20
50	.5556	.5275	.8496	.6208	1.611	1.177	1.896	1.0152	10
32° 00′	.5585	.5299	.8480	.6249	1.600	1.179	1.887	1.0123	58° 00′
10	.5614	.5324	.8465	.6289	1.590	1.181	1.878	1.0094	50
20	.5643	.5348	.8450	.6330	1.580	1.184	1.870	1.0065	40
30	.5672	.5375	.8434	.6371	1.570	1.186	1.861	1.0036	30
40	.5701	.5398	.8418	.6412	1.560	1.188	1.853	1.0007	20
50	.5730	.5422	.8403	.6453	1.550	1.190	1.844	.9977	10
33° 00′	.5760	.5446	.8387	.6494	1.540	1.192	1.836	.9948	57° 00′
10	.5789	.5471	.8371	.6536	1.530	1.195	1.828	.9919	50
20	.5818	.5495	.8355	.6577	1.520	1.197	1.820	.9890	40
30	.5847	.5519	.8339	.6619	1.511	1.199	1.812	.9861	30
40	.5876	.5544	.8323	.6661	1.501	1.202	1.804	.9832	20
50	.5905	.5568	.8307	.6703	1.492	1.204	1.796	.9803	10
34° 00′	.5934	.5592	.8290	.6745	1.483	1.206	1.788	.9774	56° 00′
10	.5963	.5616	.8274	.6787	1.473	1.209	1.781	.9745	50
20	.5992	.5640	.8258	.6830	1.464	1.211	1.773	.9716	40
30	.6021	.5664	.8241	.6873	1.455	1.213	1.766	.9687	30
40	.6050	.5688	.8225	.6916	1.446	1.216	1.758	.9657	20
50	.6080	.5712	.8208	.6959	1.437	1.218	1.751	.9628	10
35° 00′	.6109	.5736	.8192	.7002	1.428	1.221	1.743	.9599	55° 00′
10	.6138	.5760	.8175	.7046	1.419	1.223	1.736	.9570	50
20	.6167	.5783	.8158	.7089	1.411	1.226	1.729	.9541	40
30	.6196	.5807	.8141	.7133	1.402	1.228	1.722	.9512	30
40	.6225	.5831	.8124	.7177	1.393	1.231	1.715	.9483	20
50	.6254	.5854	.8107	.7221	1.385	1.233	1.708	.9454	54° 10′
		cos	sin	cot	tan	csc	sec	Radians	Degrees

igonometric Ratios (*Continued*)

Degrees	Radians	sin	cos	tan	cot	sec	csc		
36° 00′	.6283	.5878	.8090	.7265	1.376	1.236	1.701	.9425	54° 00′
10	.6312	.5901	.8073	.7310	1.368	1.239	1.695	.9396	50
20	.6341	.5925	.8056	.7355	1.360	1.241	1.688	.9367	40
30	.6370	.5948	.8039	.7400	1.351	1.244	1.681	.9338	30
40	.6400	.5972	.8021	.7445	1.343	1.247	1.675	.9308	20
50	.6429	.5995	.8004	.7490	1.335	1.249	1.668	.9279	10
37° 00′	.6458	.6018	.7986	.7536	1.327	1.252	1.662	.9250	53° 00′
10	.6487	.6041	.7969	.7581	1.319	1.255	1.655	.9221	50
20	.6516	.6065	.7951	.7627	1.311	1.258	1.649	.9192	40
30	.6545	.6088	.7934	.7673	1.303	1.260	1.643	.9163	30
40	.6574	.6111	.7916	.7720	1.295	1.263	1.636	.9134	20
50	.6603	.6134	.7898	.7766	1.288	1.266	1.630	.9105	10
38° 00′	.6632	.6157	.7880	.7813	1.280	1.269	1.624	.9076	52° 00′
10	.6661	.6180	.7862	.7860	1.272	1.272	1.618	.9047	50
20	.6690	.6202	.7844	.7907	1.265	1.275	1.612	.9018	40
30	.6720	.6225	.7826	.7954	1.257	1.278	1.606	.8988	30
40	.6749	.6248	.7808	.8002	1.250	1.281	1.601	.8959	20
50	.6778	.6271	.7790	.8050	1.242	1.284	1.595	.8930	10
39° 00′	.6807	.6293	.7771	.8098	1.235	1.287	1.589	.8901	51° 00′
10	.6836	.6316	.7753	.8146	1.228	1.290	1.583	.8872	50
20	.6865	.6338	.7735	.8195	1.220	1.293	1.578	.8843	40
30	.6894	.6361	.7716	.8243	1.213	1.296	1.572	.8814	30
40	.6923	.6383	.7698	.8292	1.206	1.299	1.567	.8785	20
50	.6952	.6406	.7679	.8342	1.199	1.302	1.561	.8756	10
40° 00′	.6981	.6428	.7660	.8391	1.192	1.305	1.556	.8727	50° 00′
10	.7010	.6450	.7642	.8441	1.185	1.309	1.550	.8698	50
20	.7039	.6472	.7623	.8491	1.178	1.312	1.545	.8668	40
30	.7069	.6494	.7604	.8541	1.171	1.315	1.540	.8639	30
40	.7098	.6517	.7585	.8591	1.164	1.318	1.535	.8610	20
50	.7127	.6539	.7566	.8642	1.157	1.322	1.529	.8581	10
41° 00′	.7156	.6561	.7547	.8693	1.150	1.325	1.524	.8552	49° 00′
10	.7185	.6583	.7528	.8744	1.144	1.328	1.519	.8523	50
20	.7214	.6604	.7509	.8796	1.137	1.332	1.514	.8494	40
30	.7243	.6626	.7490	.8847	1.130	1.335	1.509	.8465	30
40	.7272	.6648	.7470	.8899	1.124	1.339	1.504	.8436	20
50	.7301	.6670	.7451	.8952	1.117	1.342	1.499	.8407	48° 10′
		cos	sin	cot	tan	csc	sec	Radians	Degrees

Trigonometric Ratios (*Continued*)

Degrees	Radians	sin	cos	tan	cot	sec	csc		
42° 00′	.7330	.6691	.7431	.9004	1.111	1.346	1.494	.8378	48° 00′
10	.7359	.6713	.7412	.9057	1.104	1.349	1.490	.8348	50
20	.7389	.6734	.7392	.9110	1.098	1.353	1.485	.8319	40
30	.7418	.6756	.7373	.9163	1.091	1.356	1.480	.8290	30
40	.7447	.6777	.7353	.9217	1.085	1.360	1.476	.8261	20
50	.7476	.6799	.7333	.9271	1.079	1.364	1.471	.8232	10
43° 00′	.7505	.6820	.7314	.9325	1.072	1.367	1.466	.8203	47° 00′
10	.7534	.6841	.7294	.9380	1.066	1.371	1.462	.8174	50
20	.7563	.6862	.7274	.9435	1.060	1.375	1.457	.8145	40
30	.7592	.6884	.7254	.9490	1.054	1.379	1.453	.8116	30
40	.7621	.6905	.7234	.9545	1.048	1.382	1.448	.8087	20
50	.7650	.6926	.7214	.9601	1.042	1.386	1.444	.8058	10
44° 00′	.7679	.6947	.7193	.9657	1.036	1.390	1.440	.8029	46° 00′
10	.7709	.6967	.7173	.9713	1.030	1.394	1.435	.7999	50
20	.7738	.6988	.7153	.9770	1.024	1.398	1.431	.7970	40
30	.7767	.7009	.7133	.9827	1.018	1.402	1.427	.7941	30
40	.7796	.7030	.7112	.9884	1.012	1.406	1.423	.7912	20
50	.7825	.7050	.7092	.9942	1.006	1.410	1.418	.7883	10
45° 00′	.7854	.7071	.7071	1.000	1.000	1.414	1.414	.7854	45° 00′
		cos	sin	cot	tan	csc	sec	Radians	Degrees

garithms of Trigonometric Ratios

egrees	$\log_{10}$ sin	$\log_{10}$ tan	$\log_{10}$ cot	$\log_{10}$ cos	
° 00′					90° 00′
10′	.4637—3	.4637—3	2.5363	.0000	50′
20′	.7648—3	.7648—3	2.2352	.0000	40′
30′	.9408—3	.9409—3	2.0591	.0000	30′
40′	.0658—2	.0658—2	1.9342	.0000	20′
50′	.1627—2	.1627—2	1.8373	.0000	10′
° 00′	.2419—2	.2419—2	1.7581	.9999—1	89° 00′
10′	.3088—2	.3089—2	1.6911	.9999—1	50′
20′	.3668—2	.3669—2	1.6331	.9999—1	40′
30′	.4179—2	.4181—2	1.5819	.9999—1	30′
40′	.4637—2	.4638—2	1.5362	.9998—1	20′
50′	.5050—2	.5053—2	1.4947	.9998—1	10′
° 00′	.5428—2	.5431—2	1.4569	.9997—1	88° 00′
10′	.5776—2	.5779—2	1.4221	.9997—1	50′
20′	.6097—2	.6101—2	1.3899	.9996—1	40′
30′	.6397—2	.6401—2	1.3599	.9996—1	30′
40′	.6677—2	.6682—2	1.3318	.9995—1	20′
50′	.6940—2	.6945—2	1.3055	.9995—1	10′
° 00′	.7188—2	.7194—2	1.2806	.9994—1	87° 00′
10′	.7423—2	.7429—2	1.2571	.9993—1	50′
20′	.7645—2	.7652—2	1.2348	.9993—1	40′
30′	.7857—2	.7865—2	1.2135	.9992—1	30′
40′	.8059—2	.8067—2	1.1933	.9991—1	20′
50′	.8251—2	.8261—2	1.1739	.9990—1	10′
° 00′	.8436—2	.8446—2	1.1554	.9989—1	86° 00′
10′	.8613—2	.8624—2	1.1376	.9989—1	50′
20′	.8783—2	.8795—2	1.1205	.9988—1	40′
30′	.8946—2	.8960—2	1.1040	.9987—1	30′
40′	.9104—2	.9118—2	1.0882	.9986—1	20′
50′	.9256—2	.9272—2	1.0728	.9985—1	10′
° 00′	.9403—2	.9420—2	1.0580	.9983—1	85° 00′
10′	.9545—2	.9563—2	1.0437	.9982—1	50′
20′	.9682—2	.9701—2	1.0299	.9981—1	40′
30′	.9816—2	.9836—2	1.0164	.9980—1	30′
40′	.9945—2	.9966—2	1.0034	.9979—1	20′
50′	.0070—1	.0093—1	.9907	.9977—1	10′
° 00′	.0192—1	.0216—1	.9784	.9976—1	84° 00′
10′	.0311—1	.0336—1	.9664	.9975—1	50′
20′	.0426—1	.0453—1	.9547	.9973—1	40′
30′	.0539—1	.0567—1	.9433	.9972—1	30′
40′	.0648—1	.0678—1	.9322	.9971—1	20′
50′	.0755—1	.0786—1	.9214	.9969—1	83° 10′
	$\log_{10}$ cos	$\log_{10}$ cot	$\log_{10}$ tan	$\log_{10}$ sin	Degrees

Logarithms of Trigonometric Ratios *(Continued)*

Degrees	$\log_{10}$ sin	$\log_{10}$ tan	$\log_{10}$ cot	$\log_{10}$ cos	
7° 00′	.0859—1	.0891—1	.9109	.9968—1	83°
10′	.0961—1	.0995—1	.9005	.9966—1	
20′	.1060—1	.1096—1	.8904	.9964—1	
30′	.1157—1	.1194—1	.8806	.9963—1	
40′	.1252—1	.1291—1	.8709	.9961—1	
50′	.1345—1	.1385—1	.8615	.9959—1	
8° 00′	.1436—1	.1478—1	.8522	.9958—1	82°
10′	.1525—1	.1569—1	.8431	.9956—1	
20′	.1612—1	.1658—1	.8342	.9954—1	
30′	.1697—1	.1745—1	.8255	.9952—1	
40′	.1781—1	.1831—1	.8169	.9950—1	
50′	.1863—1	.1915—1	.8085	.9948—1	
9° 00′	.1943—1	.1997—1	.8003	.9946—1	81°
10′	.2022—1	.2078—1	.7922	.9944—1	
20′	.2100—1	.2158—1	.7842	.9942—1	
30′	.2176—1	.2236—1	.7764	.9940—1	
40′	.2251—1	.2313—1	.7687	.9938—1	
50′	.2324—1	.2389—1	.7611	.9936—1	
10° 00′	.2397—1	.2463—1	.7537	.9934—1	80°
10′	.2468—1	.2536—1	.7464	.9931—1	
20′	.2538—1	.2609—1	.7391	.9929—1	
30′	.2606—1	.2680—1	.7320	.9927—1	
40′	.2674—1	.2750—1	.7250	.9924—1	
50′	.2740—1	.2819—1	.7181	.9922—1	
11° 00′	.2806—1	.2887—1	.7113	.9919—1	79°
10′	.2870—1	.2953—1	.7047	.9917—1	
20′	.2934—1	.3020—1	.6980	.9914—1	
30′	.2997—1	.3085—1	.6915	.9912—1	
40′	.3058—1	.3149—1	.6851	.9909—1	
50′	.3119—1	.3212—1	.6788	.9907—1	
12° 00′	.3179—1	.3275—1	.6725	.9904—1	78°
10′	.3238—1	.3336—1	.6664	.9901—1	
20′	.3296—1	.3397—1	.6603	.9899—1	
30′	.3353—1	.3458—1	.6542	.9896—1	
40′	.3410—1	.3517—1	.6483	.9893—1	
50′	.3466—1	.3576—1	.6424	.9890—1	
13° 00′	.3521—1	.3634—1	.6366	.9887—1	77°
10′	.3575—1	.3691—1	.6309	.9884—1	
20′	.3629—1	.3748—1	.6252	.9881—1	
30′	.3682—1	.3804—1	.6196	.9878—1	
40′	.3734—1	.3859—1	.6141	.9875—1	
50′	.3786—1	.3914—1	.6086	.9872—1	76°
	$\log_{10}$ cos	$\log_{10}$ cot	$\log_{10}$ tan	$\log_{10}$ sin	Degrees

garithms of Trigonometric Ratios *(Continued)*

grees	$\log_{10}$ sin	$\log_{10}$ tan	$\log_{10}$ cot	$\log_{10}$ cos	
° 00′	.3837—1	.3968—1	.6032	.9869—1	76° 00′
10′	.3887—1	.4021—1	.5979	.9866—1	50′
20′	.3937—1	.4074—1	.5926	.9863—1	40′
30′	.3986—1	.4127—1	.5873	.9859—1	30′
40′	.4035—1	.4178—1	.5822	.9856—1	20′
50′	.4083—1	.4230—1	.5770	.9853—1	10′
° 00′	.4130—1	.4281—1	.5719	.9849—1	75° 00′
10′	.4177—1	.4331—1	.5669	.9846—1	50′
20′	.4223—1	.4381—1	.5619	.9843—1	40′
30′	.4269—1	.4430—1	.5570	.9839—1	30′
40′	.4314—1	.4479—1	.5521	.9836—1	20′
50′	.4359—1	.4527—1	.5473	.9832—1	10′
° 00′	.4403—1	.4575—1	.5425	.9828—1	74° 00′
10′	.4447—1	.4622—1	.5378	.9825—1	50′
20′	.4491—1	.4669—1	.5331	.9821—1	40′
30′	.4533—1	.4716—1	.5284	.9817—1	30′
40′	.4576—1	.4762—1	.5238	.9814—1	20′
50′	.4618—1	.4808—1	.5192	.9810—1	10′
° 00′	.4659—1	.4853—1	.5147	.9806—1	73° 00′
10′	.4700—1	.4898—1	.5102	.9802—1	50′
20′	.4741—1	.4943—1	.5057	.9798—1	40′
30′	.4781—1	.4987—1	.5013	.9794—1	30′
40′	.4821—1	.5031—1	.4969	.9790—1	20′
50′	.4861—1	.5075—1	.4925	.9786—1	10′
° 00′	.4900—1	.5118—1	.4882	.9782—1	72° 00′
10′	.4939—1	.5161—1	.4839	.9778—1	50′
20′	.4977—1	.5203—1	.4797	.9774—1	40′
30′	.5015—1	.5245—1	.4755	.9770—1	30′
40′	.5052—1	.5287—1	.4713	.9765—1	20′
50′	.5090—1	.5329—1	.4671	.9761—1	10′
° 00′	.5126—1	.5370—1	.4630	.9757—1	71° 00′
10′	.5163—1	.5411—1	.4589	.9752—1	50′
20′	.5199—1	.5451—1	.4549	.9748—1	40′
30′	.5235—1	.5491—1	.4509	.9743—1	30′
40′	.5270—1	.5531—1	.4469	.9739—1	20′
50′	.5306—1	.5571—1	.4429	.9734—1	10′
° 00′	.5341—1	.5611—1	.4389	.9730—1	70° 00′
10′	.5375—1	.5650—1	.4350	.9725—1	50′
20′	.5409—1	.5689—1	.4311	.9721—1	40′
30′	.5443—1	.5727—1	.4273	.9716—1	30′
40′	.5477—1	.5766—1	.4234	.9711—1	20′
50′	.5510—1	.5804—1	.4196	.9706—1	69° 10′
	$\log_{10}$ cos	$\log_{10}$ cot	$\log_{10}$ tan	$\log_{10}$ sin	Degrees

Logarithms of Trigonometric Ratios (*Continued*)

Degrees	$\log_{10}$ sin	$\log_{10}$ tan	$\log_{10}$ cot	$\log_{10}$ cos	
21° 00′	.5543—1	.5842—1	.4158	.9702—1	69°
10′	.5576—1	.5879—1	.4121	.9697—1	
20′	.5609—1	.5917—1	.4083	.9692—1	
30′	.5641—1	.5954—1	.4046	.9687—1	
40′	.5673—1	.5991—1	.4009	.9682—1	
50′	.5704—1	.6028—1	.3972	.9677—1	
22° 00′	.5736—1	.6064—1	.3936	.9672—1	68°
10′	.5767—1	.6100—1	.3900	.9667—1	
20′	.5798—1	.6136—1	.3864	.9661—1	
30′	.5828—1	.6172—1	.3828	.9656—1	
40′	.5859—1	.6208—1	.3792	.9651—1	
50′	.5889—1	.6243—1	.3757	.9646—1	
23° 00′	.5919—1	.6279—1	.3721	.9640—1	67°
10′	.5948—1	.6314—1	.3686	.9635—1	
20′	.5978—1	.6348—1	.3652	.9629—1	
30′	.6007—1	.6383—1	.3617	.9624—1	
40′	.6036—1	.6417—1	.3583	.9618—1	
50′	.6065—1	.6452—1	.3548	.9613—1	
24° 00′	.6093—1	.6486—1	.3514	.9607—1	66°
10′	.6121—1	.6520—1	.3480	.9602—1	
20′	.6149—1	.6553—1	.3447	.9596—1	
30′	.6177—1	.6587—1	.3413	.9590—1	
40′	.6205—1	.6620—1	.3380	.9584—1	
50′	.6232—1	.6654—1	.3346	.9579—1	
25° 00′	.6259—1	.6687—1	.3313	.9573—1	65°
10′	.6286—1	.6720—1	.3280	.9567—1	
20′	.6313—1	.6752—1	.3248	.9561—1	
30′	.6340—1	.6785—1	.3215	.9555—1	
40′	.6366—1	.6817—1	.3183	.9549—1	
50′	.6392—1	.6850—1	.3150	.9543—1	
26° 00′	.6418—1	.6882—1	.3118	.9537—1	64°
10′	.6444—1	.6914—1	.3086	.9530—1	
20′	.6470—1	.6946—1	.3054	.9524—1	
30′	.6495—1	.6977—1	.3023	.9518—1	
40′	.6521—1	.7009—1	.2991	.9512—1	
50′	.6546—1	.7040—1	.2960	.9505—1	
27° 00′	.6570—1	.7072—1	.2928	.9499—1	63°
10′	.6595—1	.7103—1	.2897	.9492—1	
20′	.6620—1	.7134—1	.2866	.9486—1	
30′	.6644—1	.7165—1	.2835	.9479—1	
40′	.6668—1	.7196—1	.2804	.9473—1	
50′	.6692—1	.7226—1	.2774	.9466—1	62°
	$\log_{10}$ cos	$\log_{10}$ cot	$\log_{10}$ tan	$\log_{10}$ sin	Deg

garithms of Trigonometric Ratios (*Continued*)

grees	$\log_{10}$ sin	$\log_{10}$ tan	$\log_{10}$ cot	$\log_{10}$ cos	
° 00′	.6716—1	.7257—1	.2743	.9459—1	62° 00′
10′	.6740—1	.7287—1	.2713	.9453—1	50′
20′	.6763—1	.7317—1	.2683	.9446—1	40′
30′	.6787—1	.7348—1	.2652	.9439—1	30′
40′	.6810—1	.7378—1	.2622	.9432—1	20′
50′	.6833—1	.7408—1	.2592	.9425—1	10′
° 00′	.6856—1	.7438—1	.2562	.9418—1	61° 00′
10′	.6878—1	.7467—1	.2533	.9411—1	50′
20′	.6901—1	.7497—1	.2503	.9404—1	40′
30′	.6923—1	.7526—1	.2474	.9397—1	30′
40′	.6946—1	.7556—1	.2444	.9390—1	20′
50′	.6968—1	.7585—1	.2415	.9383—1	10′
° 00′	.6990—1	.7614—1	.2386	.9375—1	60° 00′
10′	.7012—1	.7644—1	.2356	.9368—1	50′
20′	.7033—1	.7673—1	.2327	.9361—1	40′
30′	.7055—1	.7701—1	.2299	.9353—1	30′
40′	.7076—1	.7730—1	.2270	.9346—1	20′
50′	.7097—1	.7759—1	.2241	.9338—1	10′
° 00′	.7118—1	.7788—1	.2212	.9331—1	59° 00′
10′	.7139—1	.7816—1	.2184	.9323—1	50′
20′	.7160—1	.7845—1	.2155	.9315—1	40′
30′	.7181—1	.7873—1	.2127	.9308—1	30′
40′	.7201—1	.7902—1	.2098	.9300—1	20′
50′	.7222—1	.7930—1	.2070	.9292—1	10′
° 00′	.7242—1	.7958—1	.2042	.9284—1	58° 00′
10′	.7262—1	.7986—1	.2014	.9276—1	50′
20′	.7282—1	.8014—1	.1986	.9268—1	40′
30′	.7302—1	.8042—1	.1958	.9260—1	30′
40′	.7322—1	.8070—1	.1930	.9252—1	20′
50′	.7342—1	.8097—1	.1903	.9244—1	10′
° 00′	.7361—1	.8125—1	.1875	.9236—1	57° 00′
10′	.7380—1	.8153—1	.1847	.9228—1	50′
20′	.7400—1	.8180—1	.1820	.9219—1	40′
30′	.7419—1	.8208—1	.1792	.9211—1	30′
40′	.7438—1	.8235—1	.1765	.9203—1	20′
50′	.7457—1	.8263—1	.1737	.9194—1	10′
° 00′	.7476—1	.8290—1	.1710	.9186—1	56° 00′
10′	.7494—1	.8317—1	.1683	.9177—1	50′
20′	.7513—1	.8344—1	.1656	.9169—1	40′
30′	.7531—1	.8371—1	.1629	.9160—1	30′
40′	.7550—1	.8398—1	.1602	.9151—1	20′
50′	.7568—1	.8425—1	.1575	.9142—1	55° 10′
	$\log_{10}$ cos	$\log_{10}$ cot	$\log_{10}$ tan	$\log_{10}$ sin	Degrees

Logarithms of Trigonometric Ratios (*Continued*)

Degrees	$\log_{10}$ sin	$\log_{10}$ tan	$\log_{10}$ cot	$\log_{10}$ cos	
35° 00′	.7586—1	.8452—1	.1548	.9134—1	55° 0
10′	.7604—1	.8479—1	.1521	.9125—1	5
20′	.7622—1	.8506—1	.1494	.9116—1	4
30′	.7640—1	.8533—1	.1467	.9107—1	3
40′	.7657—1	.8559—1	.1441	.9098—1	2
50′	.7675—1	.8586—1	.1414	.9089—1	1
36° 00′	.7692—1	.8613—1	.1387	.9080—1	54° 0
10′	.7710—1	.8639—1	.1361	.9070—1	5
20′	.7727—1	.8666—1	.1334	.9061—1	4
30′	.7744—1	.8692—1	.1308	.9052—1	3
40′	.7761—1	.8718—1	.1282	.9042—1	2
50′	.7778—1	.8745—1	.1255	.9033—1	1
37° 00′	.7795—1	.8771—1	.1229	.9023—1	53° 0
10′	.7811—1	.8797—1	.1203	.9014—1	5
20′	.7828—1	.8824—1	.1176	.9004—1	4
30′	.7844—1	.8850—1	.1150	.8995—1	3
40′	.7861—1	.8876—1	.1124	.8985—1	2
50′	.7877—1	.8902—1	.1098	.8975—1	1
38° 00′	.7893—1	.8928—1	.1072	.8965—1	52° 0
10′	.7910—1	.8954—1	.1046	.8955—1	5
20′	.7926—1	.8980—1	.1020	.8945—1	4
30′	.7941—1	.9006—1	.0994	.8935—1	3
40′	.7957—1	.9032—1	.0968	.8925—1	2
50′	.7973—1	.9058—1	.0942	.8915—1	1
39° 00′	.7989—1	.9084—1	.0916	.8905—1	51° 0
10′	.8004—1	.9110—1	.0890	.8895—1	5
20′	.8020—1	.9135—1	.0865	.8884—1	4
30′	.8035—1	.9161—1	.0839	.8874—1	3
40′	.8050—1	.9187—1	.0813	.8864—1	2
50′	.8066—1	.9212—1	.0788	.8853—1	1
40° 00′	.8081—1	.9238—1	.0762	.8843—1	50° 0
10′	.8096—1	.9264—1	.0736	.8832—1	5
20′	.8111—1	.9289—1	.0711	.8821—1	4
30′	.8125—1	.9315—1	.0685	.8810—1	3
40′	.8140—1	.9341—1	.0659	.8800—1	2
50′	.8155—1	.9366—1	.0634	.8789—1	1
41° 00′	.8169—1	.9392—1	.0608	.8778—1	49° 0
10′	.8184—1	.9417—1	.0583	.8767—1	5
20′	.8198—1	.9443—1	.0557	.8756—1	4
30′	.8213—1	.9468—1	.0532	.8745—1	3
40′	.8227—1	.9494—1	.0506	.8733—1	2
50′	.8241—1	.9519—1	.0481	.8722—1	48° 1
	$\log_{10}$ cos	$\log_{10}$ cot	$\log_{10}$ tan	$\log_{10}$ sin	Degree

garithms of Trigonometric Ratios *(Continued)*

egrees	$\log_{10}\sin$	$\log_{10}\tan$	$\log_{10}\cot$	$\log_{10}\cos$	
2° 00′	.8255—1	.9544—1	.0456	.8711—1	48° 00′
10′	.8269—1	.9570—1	.0430	.8699—1	50′
20′	.8283—1	.9595—1	.0405	.8688—1	40′
30′	.8297—1	.9621—1	.0379	.8676—1	30′
40′	.8311—1	.9646—1	.0354	.8665—1	20′
50′	.8324—1	.9671—1	.0329	.8653—1	10′
3° 00′	.8338—1	.9697—1	.0303	.8641—1	47° 00′
10′	.8351—1	.9722—1	.0278	.8629—1	50′
20′	.8365—1	.9747—1	.0253	.8618—1	40′
30′	.8378—1	.9772—1	.0228	.8606—1	30′
40′	.8391—1	.9798—1	.0202	.8594—1	20′
50′	.8405—1	.9823—1	.0177	.8582—1	10′
4° 00′	.8418—1	.9848—1	.0152	.8569—1	46° 00′
10′	.8431—1	.9874—1	.0126	.8557—1	50′
20′	.8444—1	.9899—1	.0101	.8545—1	40′
30′	.8457—1	.9924—1	.0076	.8532—1	30′
40′	.8469—1	.9949—1	.0051	.8520—1	20′
50′	.8482—1	.9975—1	.0025	.8507—1	10′
5° 00′	.8495—1	.0000	.0000	.8495—1	45° 00′
	$\log_{10}\cos$	$\log_{10}\cot$	$\log_{10}\tan$	$\log_{10}\sin$	Degrees

C

Answers to selected odd-numbered exercises

Section 1-1

7. y, x

9. false

11. false

13. false

15. false

17. true

19. true

21. $1 \ngeq 4$

23. $7\frac{1}{2} \nleq -3$

25. $-3 \geq -5$

27. $2 \nleq y$

29. $x \geq y + \frac{1}{2}$

31. true

33. false

35. true

37. no

Section 1-2

1. false

3. true

5. true

7. false

9. true

11. false

13. false

15. false

33. $\{-1, 1, 2\}$
35. $\{x \mid 1 < x \leq 3\}$
37. $\{x \mid -1 < x\}$
39. $\{2\}$
41. no
43. no
45. yes

Section 1-3

1. 3
3. $2\frac{1}{4}$
5. 1
7. $2\frac{1}{2}$
9. $\{0\}$
11. $\{x \mid -1 < x < 1\}$
13. $\{0\}$
15. distance between 8 and 3
17. distance between 4 and x
19. distance between z and 0
21. $|-2+3|$
23. $|3+y|$
25. they are equal
27. they are equal
31. $\{x \mid |x| \geq a\}$

Section 1-4

7. $a = (7, 0)$; $b = (1, 3)$; $c = (4, -2)$; $d = (0, 6)$; $e = (5, 4\frac{1}{2})$; $f = (-4\frac{1}{2}, 3\frac{1}{2})$; $g = (-6, 3)$; $h = (-4\frac{1}{2}, 1)$; $i = (-3, 0)$; $j = (-2\frac{1}{2}, -3\frac{1}{2})$
13. $y = 0$
15. $x > 0$
17. $x = 2$
19. $x < 0$, $y > 0$
21. $x = -3$
23. $-5 < x < -2$

Section 1-5

1. yes
3. yes
5. Each vertical line meets the graph in at most one point.
7. yes
9. no
11. 4
13. $t + 2$
15. $4x + 2$
17. Exercise 11, 2; Exercise 13, $3t - 4$; Exercise 15, $12x - 4$
19. Exercise 11, 5; Exercise 13, $|t| + 3$; Exercise 15, $|4x| + 3$
21. domain $= R$, range $= \{y \mid 3 \leq y \leq 5\}$
23. domain $= R$, range $= R$
25. domain $= \{x \mid -4 \leq x \leq 4\}$, range $= \{y \mid -4 \leq y \leq 4\}$
39. Exercise 27, domain $= \{3, 4\}$, range $= \{1, 2, 3, 4\}$;
Exercise 29, domain $= \{1, 2, 3, 4\}$, range $= \{1, 3, 5, 7\}$;
Exercise 31, domain $= R$, range $= R$;
Exercise 33, domain $= R$, range $= R$;

Exercise 35, domain $= R$, range $= \{3\}$;
Exercise 37, domain $= R$, range $= \{y \mid y < 0\}$

Review exercises (Chapter 1)

21. false
23. false
25. false
27. no
29. yes
31. $\{y \mid y \geq 0\}$
33. 3
35. 7
37. $|5(x + y) - 2|$

Section 2-1

1. $\frac{1}{2}$
3. $\sqrt{3}$
5. $2/\sqrt{3}$
7. no
9. yes, $\angle C$
11. yes, $\angle C$
13. $3/\sqrt{10}$
15. 3
17. $\sqrt{10}$
19. $1/\sqrt{10}$
21. $\sqrt{10}/3$
23. Exercise 13, $\frac{1}{4}\sqrt{15}$; Exercise 15, $\sqrt{15}$; Exercise 17, 4; Exercise 19, $\frac{1}{4}$; Exercise 21, $4/\sqrt{15}$
25. $\sqrt{5}$
27. $\sqrt{5}/3$
29. 1
31. $5/\sqrt{26}$
33. 1
35. $\frac{1}{2}\sqrt{5}$
37. 1

Section 2-2

11. $1/\cos A$
13. $(1 + \cos A)/\sin A$
15. $1/\sin^2 A$
17. $(\cos 32°)/\sin 32°$
19. $1/(\sin 42° \cos 42°)$
21. $\sqrt{1 - \sin^2 11°}$
23. $1/\sin 25°$
25. $\sqrt{40}/7$
27. $\sqrt{40}/3$
29. $7/3$
31. $\sqrt{33}/7$
33. $\sqrt{33}/4$
35. $7/4$
37. $1/\sqrt{1 + \tan^2 A}$
39. $\sqrt{1 + \tan^2 A}$

Section 2-3

1. 0.1392
3. 1.589
5. 0.4540
11. 0.6561
13. 0.8693
19. 0.3420
21. 0.3640

Section 2-4

1. 41°
3. 15.09
5. 24.09
7. 38°
9. 77
11. 3865 yards
13. 31°
15. 121.8 meters
17. 102.6 feet
19. $\angle A = 90° - \angle B$; $a = b \cot B$; $c = b \csc B$; $K = \frac{1}{2}b^2 \cot B$

Section 2-5

1. 17.3 feet
3. 170 feet
5. 92 feet
7. 4944 yards
9. 213 feet
11. 115 feet
13. 58°, 9.4 miles per hour

Review exercises (Chapter 2)

1. $2/\sqrt{53} = 2\sqrt{53}/53$
3. $\sqrt{53}/7$
5. 7/2
7. 1/2
9. $1/\sqrt{2} = \sqrt{2}/2$
11. sec A
13. 0.5446
15. 9.936
17. 24°

Section 3-1

1. 5
3. $\sqrt{34}$
5. $\sqrt{13}$
9. 5
11. 3

13. $\sqrt{5}$
15. 13
17. $x^2 + y^2 = 4$
19. $x^2 + y^2 = 64$
21. $x^2 + y^2 = 15$
29. Exercise 23, no; Exercise 25, no; Exercise 27, no
31. **(a)** $\sqrt{(x - x_0)^2 + (y - y_0)^2}$
(b) $\sqrt{(x - x_0)^2 + (y - y_0)^2} = r$
(c) $(x - x_0)^2 + (y - y_0)^2 = r^2$

Section 3-2

13. 180°
15. 225°
17. 18°
19. $3\pi/2$
21. $\pi/36$
23. $7\pi/6$
31. 2π
33. 5π
35. $\pi/5$
37. $\pi/6$

Section 3-3

1. positive
3. negative
5. positive
7. positive
9. positive
11. $\sin\theta = -4/5$, $\cos\theta = 3/5$, $\tan\theta = -4/3$, $\cot\theta = -3/4$, $\sec\theta = 5/3$, $\csc\theta$ $-5/4$
13. $\sin\theta = 2/\sqrt{5} = (2\sqrt{5})/5$, $\cos\theta = 1/\sqrt{5} = \sqrt{5}/5$, $\tan\theta = 2$, $\cot\theta = 1/2$, $\sec\theta$ $\sqrt{5}$, $\csc\theta = \sqrt{5}/2$
15. $\sin 90° = 1$, $\cos 90° = 0$, $\tan 90°$ is undefined, $\cot 90° = 0$, $\sec 90°$ is undefine $\csc 90° = 1$
17. -1
19. tan 280°
21. $\cos 9\pi/5$
23. $\csc 11\pi/9$
25. $\sqrt{2}/2$
27. $-\sqrt{2}/2$
29. 0

Section 3-4

1. -1
3. 1
5. $\sqrt{3}/2$
7. $\sin 5\pi/4$

9. sec $7\pi/6$
17. defined
19. undefined
21. none
23. $\sin\theta = 0.84$, $\cos\theta = 0.54$
25. $\sin\theta = -0.14$, $\cos\theta = -0.99$
27. The x-coordinates are the same. The y-coordinate of $T(-\theta)$ is the negative of the y-coordinate of $T(\theta)$.
29. $\sin(-\theta) = -\sin\theta$

Section 3-5

1. $20°$
3. $57°$
5. $\pi/4$
7. $51°$
9. $2\pi/5$
11. $\sin 37°$
13. $-\tan\pi/8$
15. $\csc 60°$
17. $-1/2$
19. $-\sqrt{2}/2$
21. -1
23. -0.9925
25. -0.9848
27. -1.305
29. $-y = -\sin\theta$*
31. $-y/(-x) = \tan\theta$*
33. $1/(-x) = -\sec\theta$*

Section 3-6

1. $\{\pi/2 + 2k\pi,\ k \in J\}$
3. $\{3\pi/2 + 2k\pi,\ k \in J\}$
5. $\{\pi/2 + k\pi,\ k \in J\}$
7. **(b)** $\{\pi/4 + k\pi,\ k \in J\}$
11. $\sin(-x) = -\sin x$

Review exercises (Chapter 3)

9. $2\sqrt{13}$
11. $x^2 + y^2 = 36$
13. $\sin\theta = 7/\sqrt{53}$, $\cos\theta = -2/\sqrt{53}$, $\tan\theta = -7/2$, $\cot\theta = -2/7$, $\sec\theta = -\sqrt{53}/2$, $\csc\theta = \sqrt{53}/7$
15. They differ by a multiple of 360°.

Section 4-1

1. $4/5$
3. $-12/5$
5. $\sqrt{3}/2$
7. $\sqrt{34}/3$
9. $\sqrt{1 - \sin^2\theta}$
11. $\sqrt{\sec^2\theta - 1}$
13. $-\sqrt{\csc^2\theta - 1}$
15. $1 - 2\cos^2\theta + \cos^4\theta$

Section 4-2

1. $\cos x \cos y + \sin x \sin y$
3. $-\cos \theta$
5. $-\sin x$
7. $\frac{1}{2}\sqrt{3} \cos \theta + \frac{1}{2} \sin \theta$
9. $\frac{1}{2}\sqrt{2} \cos \theta - \frac{1}{2}\sqrt{2} \sin \theta$
13. $\frac{1}{4}(\sqrt{6} + \sqrt{2})$
15. $\frac{1}{4}(\sqrt{6} - \sqrt{2})$
17. $\frac{1}{4}(\sqrt{6} + \sqrt{2})$
19. $\frac{1}{4}(\sqrt{2} - \sqrt{6})$
21. $\frac{1}{4}(\sqrt{6} - \sqrt{2})$
23. $-\frac{1}{2}\sqrt{2}$
25. $\frac{1}{4}(\sqrt{2} - \sqrt{6})$
27. $\frac{1}{4}(\sqrt{2} + \sqrt{6})$

Section 4-3

1. $-\cos \theta$
3. $\frac{1}{2}\sqrt{3} \cos x - \frac{1}{2} \sin x$
5. $\frac{1}{2} \sin \theta + \frac{1}{2}\sqrt{3} \cos \theta$
7. $\cos x$
11. $\cos^3 \theta - 3 \cos \theta \sin^2 \theta$
21. $\frac{1}{4}(\sqrt{3} - 1)$
23. $-\frac{1}{4}(\sqrt{2} + 1)$
25. 0

Section 4-4

15. $\frac{1}{2}\sqrt{2 - \sqrt{2}}$
17. $\sqrt{3 - 2\sqrt{2}}$
19. $\frac{1}{2}\sqrt{2 + \sqrt{3}}$
21. $\frac{1}{2}\sqrt{2 + \sqrt{2 + \sqrt{3}}}$

Section 4-6

1. $\{\pi/4, 5\pi/4\}$
3. $\{\pi/2, 7\pi/6, 3\pi/2, 11\pi/6\}$
5. $\{\pi/8, 5\pi/8, 9\pi/8, 13\pi/8\}$
7. $\{7\pi/6, 11\pi/6\}$
9. $\{\pi/2, 3\pi/2\}$
11. $\theta = 210°, 330°$
13. $\theta = 72°, 252°$
15. $\theta = 162°, 342°$
17. $\theta = 45°, 72°, 225°, 252°$
19. $\theta = 0°$
21. $\theta = 0°, 90°, 180°, 270°$

Review exercises (Chapter 4)

1. 1
3. $\csc^2 x$
5. $\cos x$
7. $\cos 2x$
9. $\cos 2x$
11. $\cos x \cos y - \sin x \sin y$
13. $\frac{3}{5} \sin A - \frac{4}{5} \cos A$
15. $(3 \tan A + 4)/(3 - 4 \tan A)$
17. $-24/7$
19. 2
27. $\theta = 42°, 138°$

Section 5-1

1. 9.6
3. 10.8 meters
5. 4.3
7. 89 feet

9. $41°$
11. $83°$
13. $39°$
17. $(\cos\alpha - \cos\beta)^2 + (\sin\alpha - \sin\beta)^2$

Section 5-2

1. 1.90
3. 13.1 feet
5. 28.4 meters
7. $48°$
9. no solution
11. $43°$
15. 462 square feet
17. 1640 square meters
21. 69.3 square feet
23. 245 square inches

Section 5-3

1. $\angle C = 100°$, $a = 90.9$, $b = 22.2$
3. $\angle C = 29°$, $a = 99.9$ feet, $c = 48.9$ feet, $K = 1950$ square feet
5. $\angle A = 49°$, $\angle B = 59°$, $\angle C = 72°$, $K = 34$
7. $\angle A = 110°$, $\angle B = 18°$, $\angle C = 51°$
9. $a = 41$ feet, $\angle B = 98°$, $\angle C = 19°$, $K = 307$ square feet
11. $c = 125$, $\angle A = 33°$, $\angle B = 39°$, $K = 3050$
13. $\angle B = 145°$, $\angle C = 15°$, $b = 50$ miles
15. $\angle A = 35°$, $\angle B = 113°$, $b = 19$ feet or $\angle A = 145°$, $\angle B = 3°$, $b = 1$ foot

Section 5-4

1. **(a)** 10.8 miles; **(b)** $60°$
3. 3
5. **(a)** 242 miles; **(b)** $246°$
7. $64°$
9. **(a)** 66 feet 9 inches; **(b)** 63 feet 9 inches
11. $96°$ or $264°$

Review exercises (Chapter 5)

1. $c^2 = a^2 + b^2 - 2ab\cos C$
3. $K = \frac{1}{2}ab\sin C = \frac{1}{2}ac\sin B = \frac{1}{2}bc\sin A$
5. $\angle A = 72°$, $b = 526$ meters, $c = 613$ meters
7. $\angle B = 93°$, $\angle C = 55°$, $b = 122$ meters or $\angle B = 23°$, $\angle C = 125°$, $b = 48$ meters
9. one triangle, provided $\angle B + \angle C < 180°$
11. one triangle, provided $a + b > c$, $a + c > b$, and $b + c > a$
13. 8.4 nautical miles

Section 6-1

1. $\{k\pi,\ k \in J\}$
3. $\{\pi/2 + k\pi,\ k \in J\}$
5. $\{2k\pi,\ k \in J\}$

7. $\{\pi/2 + 2k\pi,\ k \in J\}$
9. $\{3\pi/4 + k\pi,\ k \in J\}$
11. $\{\pi/4,\ 3\pi/4,\ 5\pi/4,\ 7\pi/4\}$
13. $\{0.67, 2.47\}$

Section 6-2

1. amplitude $= 2$, period $= 2\pi$, phase shift $= 0$
3. amplitude $= 2$, period $= 2\pi$, phase shift $= 0$
5. amplitude $= 1$, period $= 2\pi/5$, phase shift $= 0$
7. amplitude $= 1$, period $= 2\pi$, phase shift $= \pi/3$
9. amplitude $= 3$, period $= 4\pi$, phase shift $= 0$
11. amplitude $= 1$, period $= 2\pi/5$, phase shift $= 0$
13. amplitude $= \frac{1}{2}$, period $= 4\pi$, phase shift $= \pi$
15. amplitude $= \pi$, period $= 2$, phase shift $= 1$
17. amplitude $= 3$, period $= 4\pi$, phase shift $= -2\pi/3$

Section 6-4

1. **(a)** 4π radians per second; **(b)** $\frac{1}{2}$ second
3. $13\frac{1}{27}$ radians per second
5. $y = 5 \sin (\pi/6)(t - 3)$
7. **(a)** $p = \sin 880\pi t$; **(b)** 1320 cycles per second
11. **(a)** 984 feet; **(b)** 10 feet

Section 7-1

1. complex, imaginary
3. complex, imaginary, pure imagina
5. complex, imaginary
7. complex, imaginary
9. $2, 5, 2 - 5i$
11. $0, 6, -6i$
13. $0, 0, 0$
15. $\sqrt{2}, 0, \sqrt{2}$
17. $-2 - 4i$
19. $4 + i$
21. $1 + 13i$
23. $6 - 24i$
25. $(\sqrt{6} + \sqrt{2}) + (\sqrt{3} - 2)i$
27. $\frac{2}{5} + \frac{1}{5}i$
29. $2 - i$
31. $-\frac{19}{113} + \frac{54}{113}i$
33. $\dfrac{8}{21} + \dfrac{2\sqrt{5}}{21}i$
35. $3i$
37. $i\sqrt{2}$
39. -2
41. -4
43. $-\sqrt{11}$
45. $i^2 = -1;\ i^3 = -i;\ i^4 = 1,\ i^5 = i$
47. **(a)** 17;
(c) $(a + bi)(a - bi) = a^2 + b^2$
49. -1

Section 7-2

15. 2
17. 13

19. 5
21. $2\sqrt{2}$

Section 7-3

1. $-\sqrt{3} + i$
3. $\frac{1}{4}\sqrt{3} - \frac{1}{4}i$
5. $-\frac{1}{2}\sqrt{2} - \frac{1}{2}i\sqrt{6}$
7. $1.6570 + 0.7033i$
9. $r = 18;\ \theta = 0 + 2\pi k,\ k \in J$
11. $r = \sqrt{5};\ \theta = \pi + 2\pi k,\ k \in J$
13. $r = \sqrt{34};\ \theta = 301° + k \cdot 360°,\ k \in J$
15. $3 \text{ cis } \pi$
17. $3\sqrt{2} \text{ cis } 5\pi/4$
19. $5 \text{ cis } (-53°)$
21. $5\sqrt{2} \text{ cis } (-82°)$
23. $6 \text{ cis } 90°$
25. $3 \text{ cis } 90°$
27. $r \text{ cis } (-\theta)$
29. $r \text{ cis } (\pi - \theta)$

Section 7-4

1. $20 \text{ cis } 105°$
3. $3\sqrt{5} \text{ cis } 3\pi/8$
5. $14 - 14i$
7. $-8i$
9. -4
11. $2 - 2i\sqrt{3}$
13. $r^2 \text{ cis } 2\theta$
15. $r^n \text{ cis } n\theta$
17. $a^2 + b^2$
19. $\frac{11}{4} \text{ cis } (-100°)$
21. $\frac{2}{5} \text{ cis } (-\pi/8)$

Section 7-5

1. $9 \text{ cis } 48°$
3. -16
5. $-2 + 2i$
7. $-8i$
9. $4 \text{ cis } 9°,\ 4 \text{ cis } 189°$
11. $2 \text{ cis } 53°,\ 2 \text{ cis } 173°,\ 2 \text{ cis } 293°$
13. $\text{cis } 0,\ \text{cis } 2\pi/5,\ \text{cis } 4\pi/5,\ \text{cis } 6\pi/5,\ \text{cis } 8\pi/5$
15. $-2\sqrt{2} + 2i\sqrt{2},\ 2\sqrt{2} - 2i\sqrt{2}$
17. $3,\ -\frac{3}{2} + (\frac{3}{2}\sqrt{3})i,\ -\frac{3}{2} - (\frac{3}{2}\sqrt{3})i$
19. $1,\ \frac{1}{2} + \left(\frac{\sqrt{3}}{2}\right)i,\ -\frac{1}{2} + \left(\frac{\sqrt{3}}{2}\right)i,\ -1,\ -\frac{1}{2} - \left(\frac{\sqrt{3}}{2}\right)i,\ \frac{1}{2} - \left(\frac{\sqrt{3}}{2}\right)i$

25. yes

Review exercises (Chapter 7)

13. $\frac{1}{2} - \left(\frac{\sqrt{3}}{2}\right) i$

15. $-\frac{18}{41} - \frac{43}{41}i$

17. $10 \text{ cis } (-62°)$

19. -1

21. $2, -1 + (\sqrt{3})i, -1 - (\sqrt{3})i$

Section 8-1

1. one-to-one, f^{-1} exists with domain $= \{-2, 0, 2, 3, 5, 7\}$ and range $\{-3, -2, -1, 0, 1, 2\}$

3. not one-to-one, f^{-1} does not exist

5. not one-to-one, f^{-1} does not exist

7. This graph is not the graph of a function.

9. $f^{-1}(b) = a$

11. -16

13. 2

15. 0

17. 12

19. 1

21. $\frac{1}{4}$

23. $f^{-1}(x) = \frac{1}{3}(x + 8)$ with domain $= R$ and range $= R$

25. $f^{-1}(x) = \sqrt[3]{x}$ with domain $= R$ and range $= R$

27. $f^{-1}(x) = \sqrt[4]{x}$ with domain $= \{x \mid x \geq 0\}$ and range $= \{x \mid x \geq 0\}$

29. $f^{-1}(x) = \dfrac{x^3}{8}$ with domain $= R$ and range $= R$

Section 8-2

1. $\pi/6$

3. π

5. $\pi/4$

7. $135°$

9. $21°$

11. $111°$

13. $\sqrt{5}/3$

15. $\sqrt{5}/3$

17. $\sqrt{5}/3$

21. $\sqrt{1 - x^2}$

23. $\pi/2$ (or $90°$)

25. $\pi/2$

27. $\{x \mid 0 \leq x \leq \pi\}$

Section 8-3

1. $\pi/3$ (or $60°$)

3. $5\pi/6$ (or $150°$)

5. $14°$

7. 0 (or $0°$)

9. $-\pi/4$ (or $-45°$)

17. $\pi/2$ (or $90°$)

19. $y = 0$ and $y = \pi$

21. $y = 0$

Section 8-4

13. $y = -8 \cos (x + \pi/6)$
15. $y = -13 \cos (x - 23°)$

Review exercises (Chapter 8)

3. $y = f^{-1}(x) = \frac{1}{4}x - \frac{1}{4}$
5. $y = f^{-1}(x) = \sqrt[5]{1 + x}$
7. $\pi/6$
9. $\pi/4$
11. domain $= \{x \mid -1 \leq x \leq 1\}$, range $= \{y \mid -\pi/2 \leq y \leq \pi/2\}$
13. domain $= R$, range $= \{y \mid -\pi/2 \leq y \leq \pi/2\}$

Section A-1

5. Exercise 1, domain $= R$, range $=$ {positive real numbers}; Exercise 3, domain $=$ R, range $=$ {positive real numbers}
7. $>$
9. $=$
11. domain $= R$, range $=$ {positive real numbers}
17. **(a)** 3200; **(b)** less than 17 weeks

Section A-2

5. $36^{1/2} = 6$
7. $e^{-1} = 1/e$
9. $10^{0.3010} = 2$, approximately
11. $\log_{10} 10 = 1$
13. $\log_{10} 1000 = 3$
15. $\log_{10} \frac{1}{10} = -1$
17. $\log_{10} 3 = 0.4771$, approximately
19. $\log_{16} \frac{1}{8} = -\frac{3}{4}$
21. 1
23. 2
25. 20
27. 0.4771, approximately
29. $-\frac{3}{4}$
31. 2
35. $\{x \mid 0 < x < 4\}$
37. $\{x \mid x \geq \frac{1}{2}\}$

Section A-3

1. $\log 87 + \log 0.541$
3. $\log_a 341 - \log_a 72$
5. $5 \ln 36$
7. $\frac{1}{4} \log_2 11.5$
9. $-\frac{1}{7} \log_5 94.3$
11. $4 \log_2 23 + \frac{2}{3} \log_2 17.2 - 3 \log_2 40.6$
13. $\log 2 + \log x + \log y + \log z$
15. $2 \log_7 x + \log_7 y - 3 \log_7 z$
17. $\frac{1}{6} \ln x$
23. $\ln A = \ln 200 + 7 \ln 1.06$
25. $\dfrac{\log 3}{\log e}$

27. $\dfrac{\log 49}{\log 5}$

29. $\dfrac{\log e}{\log \pi}$

31. $\dfrac{1}{\log a}$

Section A-4

1. 3.62×10^2
3. 3.7×10^{-2}
5. 1×10^{-3}
7. $1 + \log x$
9. $2 + \log x$
11. $n + \log x$
13. 0.3010
15. 0.6767
17. 1.5911
19. 0.6128 − 1
21. 0.7076 − 5
23. 3.9355
25. 100
27. 7
29. 0.947
31. 0.277
33. 57.5
35. 2,500,000

Section A-5

1. 0.6166
3. 0.9304 − 1
5. 0.1293 − 1
7. 0.4972
9. 0.1504
11. 4.554
13. 3.893
15. 208.6
17. 0.09804
19. 0.1274
21. 687,700

Section A-6

1. 0.8986
3. 306,300
5. 6.118
7. 807,400
9. 27.21
11. $2432
13. $807.10
15. 2.303
17. 2.29

Section A-7

1. 0.2806 − 1
3. 0.4943 − 1
5. 0.9308 − 1
7. 0.2819
9. 5° 30′
11. 21° 20′
13. 0.8290 − 1
15. 0.3785
17. 0.9826 − 1
19. 46° 42′
21. 20° 38′
23. 64° 33′

Section A-8

1. $\angle B = 43° 50'$, $b = 22.3$ meters, $c = 32.2$ meters
3. $\angle A = 51° 33'$, $a = 12.89$ centimeters, $b = 10.24$ centimeters
5. $\angle A = 49° 40'$, $\angle B = 40° 20'$, $c = 226$ miles
7. 1630
9. $\angle C = 88° 30'$, $a = 54.0$, $b = 41.9$
11. $\angle B = 28°$, $\angle C = 108° 20'$, $c = 78.7$
13. $\angle A = 17° 20'$, $\angle B = 130°$, $a = 4.47$ meters or $\angle A = 97° 20'$, $\angle B = 50°$, $a =$ 14.9 meters
15. no solution

Section A-9

1. $\angle A = 51° 20'$, $\angle C = 80° 10'$, $b = 16.9$
3. $\angle B = 25° 21'$, $\angle C = 93° 18'$, $a = 29.76$ meters
5. $\angle A = 32° 40'$, $\angle B = 21° 32'$, $\angle C = 125° 48'$
7. $\angle A = 50° 24'$, $\angle B = 35° 24'$, $\angle C = 94° 12'$, $K = 6240$ square miles
9. $\angle A = 82° 40'$, $\angle B = 48° 10'$, $\angle C = 49° 20'$, $K = 2070$ square meters

Review exercises (Appendix A)

9. (log 0.01)/(log 100)
11. 7/4
13. 2.5585
15. 413
17. 0.04383
19. 0.5915
21. 58° 21′
23. 2.188
25. 1.182
27. **(a)** $t = (\ln 2)/3 = 0.231$, approximately; **(b)** e^{-15} (approximately 0.0000003)
29. $\angle C = 78° 20'$, $a = 35.8$ yards, $b = 48.7$ yards, $K = 853$ square yards
31. $\angle B = 59°$, $\angle C = 75°$, $c = 55$, $K = 970$

Index

76 77 78 79 9 8 7 6 5 4 3 2

Logarithms, Base 10

x	0	1	2	3	4	5	6	7	8	9
1.0	.0000	.0043	.0086	.0128	.0170	.0212	.0253	.0294	.0334	.0374
1.1	.0414	.0453	.0492	.0531	.0569	.0607	.0645	.0682	.0719	.0755
1.2	.0792	.0828	.0864	.0899	.0934	.0969	.1004	.1038	.1072	.1106
1.3	.1139	.1173	.1206	.1239	.1271	.1303	.1335	.1367	.1399	.1430
1.4	.1461	.1492	.1523	.1553	.1584	.1614	.1644	.1673	.1703	.1732
1.5	.1761	.1790	.1818	.1847	.1875	.1903	.1931	.1959	.1987	.2014
1.6	.2041	.2068	.2095	.2122	.2148	.2175	.2201	.2227	.2253	.2279
1.7	.2304	.2330	.2355	.2380	.2405	.2430	.2455	.2480	.2504	.2529
1.8	.2553	.2577	.2601	.2625	.2648	.2672	.2695	.2718	.2742	.2765
1.9	.2788	.2810	.2833	.2856	.2878	.2900	.2923	.2945	.2967	.2989
2.0	.3010	.3032	.3054	.3075	.3096	.3118	.3139	.3160	.3181	.3201
2.1	.3222	.3243	.3263	.3284	.3304	.3324	.3345	.3365	.3385	.3404
2.2	.3424	.3444	.3464	.3483	.3502	.3522	.3541	.3560	.3579	.3598
2.3	.3617	.3636	.3655	.3674	.3692	.3711	.3729	.3747	.3766	.3784
2.4	.3802	.3820	.3838	.3856	.3874	.3892	.3909	.3927	.3945	.3962
2.5	.3979	.3997	.4014	.4031	.4048	.4065	.4082	.4099	.4116	.4133
2.6	.4150	.4166	.4183	.4200	.4216	.4232	.4249	.4265	.4281	.4298
2.7	.4314	.4330	.4346	.4362	.4378	.4393	.4409	.4425	.4440	.4456
2.8	.4472	.4487	.4502	.4518	.4533	.4548	.4564	.4579	.4594	.4609
2.9	.4624	.4639	.4654	.4669	.4683	.4698	.4713	.4728	.4742	.4757
3.0	.4771	.4786	.4800	.4814	.4829	.4843	.4857	.4871	.4886	.4900
3.1	.4914	.4928	.4942	.4955	.4969	.4983	.4997	.5011	.5024	.5038
3.2	.5051	.5065	.5079	.5092	.5105	.5119	.5132	.5145	.5159	.5172
3.3	.5185	.5198	.5211	.5224	.5237	.5250	.5263	.5276	.5289	.5302
3.4	.5315	.5328	.5340	.5353	.5366	.5378	.5391	.5403	.5416	.5428
3.5	.5441	.5453	.5465	.5478	.5490	.5502	.5514	.5527	.5539	.5551
3.6	.5563	.5575	.5587	.5599	.5611	.5623	.5635	.5647	.5658	.5670
3.7	.5682	.5694	.5705	.5717	.5729	.5740	.5752	.5763	.5775	.5786
3.8	.5798	.5809	.5821	.5832	.5843	.5855	.5866	.5877	.5888	.5899
3.9	.5911	.5922	.5933	.5944	.5955	.5966	.5977	.5988	.5999	.6010
4.0	.6021	.6031	.6042	.6053	.6064	.6075	.6085	.6096	.6107	.6117
4.1	.6128	.6138	.6149	.6160	.6170	.6180	.6191	.6201	.6212	.6222
4.2	.6232	.6243	.6253	.6263	.6274	.6284	.6294	.6304	.6314	.6325
4.3	.6335	.6345	.6355	.6365	.6375	.6385	.6395	.6405	.6415	.6425
4.4	.6435	.6444	.6454	.6464	.6474	.6484	.6493	.6503	.6513	.6522
4.5	.6532	.6542	.6551	.6561	.6571	.6580	.6590	.6599	.6609	.6618
4.6	.6628	.6637	.6646	.6656	.6665	.6675	.6684	.6693	.6702	.6712
4.7	.6721	.6730	.6739	.6749	.6758	.6767	.6776	.6785	.6794	.6803
4.8	.6812	.6821	.6830	.6839	.6848	.6857	.6866	.6875	.6884	.6893
4.9	.6902	.6911	.6920	.6928	.6937	.6946	.6955	.6964	.6972	.6981
5.0	.6990	.6998	.7007	.7016	.7024	.7033	.7042	.7050	.7059	.7067
5.1	.7076	.7084	.7093	.7101	.7110	.7118	.7126	.7135	.7143	.7152
5.2	.7160	.7168	.7177	.7185	.7193	.7202	.7210	.7218	.7226	.7235
5.3	.7243	.7251	.7259	.7267	.7275	.7284	.7292	.7300	.7308	.7316
5.4	.7324	.7332	.7340	.7348	.7356	.7364	.7372	.7380	.7388	.7396